Advanced Video Communications over Wireless Networks

Advanced Video Communications over Wireless Networks

Edited by

Ce Zhu and Yuenan Li

CRC Press
Taylor & Francis Group
Boca Raton London New York

CRC Press is an imprint of the
Taylor & Francis Group, an **informa** business

CRC Press
Taylor & Francis Group
6000 Broken Sound Parkway NW, Suite 300
Boca Raton, FL 33487-2742

First issued in paperback 2017

© 2013 by Taylor & Francis Group, LLC
CRC Press is an imprint of Taylor & Francis Group, an Informa business

No claim to original U.S. Government works

Version Date: 20121115

ISBN 13: 978-1-138-07290-9 (pbk)
ISBN 13: 978-1-4398-7998-6 (hbk)

Contents

Preface

Video communication has evolved from a simple tool for visual communication to a key enabler for various video applications. A number of exciting video applications have been successfully deployed in recent years, with the goal of providing users with more flexible, personalized, and content-rich viewing experience. Accompanied with the ubiquitous video applications, we have also experienced a paradigm shift from passive, wired, and centralized video content access to interactive, wireless, and distributed content access. Undoubtedly, wireless video communications have paved the way for advanced applications. However, given the distributed, resource-constraint, and heterogeneous nature of wireless networks, the support of quality video communications over wireless networks is still challenging. Video coding is one of the indispensable components in various wireless video applications, whereas the wireless network condition always imposes more stringent requirements on coding technologies. To cope with the limited transmission bandwidth and to offer adaptivity to the harsh wireless channels, rate control, packet scheduling, as well as error control mechanisms are usually incorporated in the design of codecs to enable efficient and reliable video communications. At the same time, due to energy constraint in wireless systems, video coding algorithms should operate with the lowest possible power consumption. Therefore, video coding over wireless networks is inherently a complex optimization problem with a set of constraints. In addition, the high heterogeneity and user mobility associated with wireless networks are also key issues to be tackled for a seamless delivery of quality-of-experience supported video streams.

To sum up, wireless video communications encompass a broad range of challenges and opportunities that provide the catalyst for technical innovations. To disseminate the most recent advances in this challenging yet exciting field, we bring forth this book as a compilation of high-quality chapters. This book is intended to be an up-to-date reference book on wireless video communications, providing the fundamentals, recent technical achievements, challenges, and some emerging trends. We hope that the book will be accessible to various audiences, ranging from those in academia and industry to senior undergraduates and postgraduates. To achieve this goal, we have solicited chapters from a number of researchers who are experts in diverse aspects of wireless video communications. We received a good response and, finally, after peer review and revision, 15 chapters were selected. These chapters cover a wide spectrum of topics, including the underlying theoretical fundamentals associated with wireless video communications, transmission schemes tailored to mobile and wireless networks, quality metrics, architectures of practical systems, as well as some novel directions. In what follows, we present a summary of each chapter.

In Chapter 1, "Network-Aware Error-Resilient Video Coding," a network-aware Intra coding refresh method is presented. This method increases the error robustness of H.264/AVC bitstreams, considering the network packet loss rate and the encoding bit rate, by efficiently taking into account the rate-distortion impact of Intra coding decisions while guaranteeing that errors do not propagate.

Chapter 2, "Distributed Video Coding: Principles and Challenges," is a tutorial on distributed video coding (DVC). In contrast to conventional video compression schemes featuring an encoder that is significantly more complex than the decoder, in DVC the complexity distribution is the reverse. This chapter provides an overview of the basic principles, state of the art, current problems, and trends in DVC.

Chapter 3, "Computer Vision Aided Video Coding," studies video coding from the perspective of computer vision. Motivated by the fact that the human visual system (HVS) is the ultimate receiver of the majority of compressed videos and that there is a scope to remove unimportant information through HVS, the chapter proposes a computer vision–aided video coding technique by exploiting the spatial and temporal redundancies with visually unimportant information.

In Chapter 4, "Macroblock Classification Method for Computation Control Video Coding and Other Video Applications Involving Motions," a new macroblock (MB) classification method is proposed, which classifies MBs into different classes according to their temporal and spatial motion and texture information. Furthermore, the implementations of the proposed MB classification method into complexity-scalable video coding as well as other video applications are also discussed in detail in the chapter.

Chapter 5, "Transmission Rate Adaptation in Multimedia WLAN: A Dynamic Games Approach," considers the scheduling, rate adaptation, and buffer management in a multiuser wireless local area network (WLAN), where each user transmits scalable video payload. Based on opportunistic scheduling, users access the available medium (channel) in a decentralized manner. The rate adaptation problem of the WLAN multimedia networks is then formulated as a general-sum switching control dynamic Markovian game.

In Chapter 6, "Energy and Bandwidth Optimization in Mobile Video Streaming Systems," the authors consider the problem of multicasting multiple variable bit rate video streams from a wireless base station to many mobile receivers over a common wireless channel. This chapter presents a sequence of increasingly sophisticated streaming protocols for optimizing energy usage and utilization of the wireless bandwidth.

Chapter 7, "Resource Allocation for Scalable Videos over Cognitive Radio Networks," investigates the challenging problem of video communication over cognitive radio (CR) networks. It first addresses the problem of scalable video over infrastructure-based CR networks and then considers the problem of scalable video over multihop CR networks.

Chapter 8, "Cooperative Video Provisioning in Mobile Wireless Environments," focuses on the challenging scenario of cooperative video provisioning in mobile wireless environments. On one hand, it provides a general overview about the state-of-the-art literature on collaborative mobile networking. On the other hand, it provides technical details and reports about the RAMP middleware case study, practically showing that node cooperation can properly achieve streaming adaptation.

Chapter 9, "Multilayer Iterative FEC Decoding for Video Transmission over Wireless Networks," develops a novel multilayer iterative decoding scheme using deterministic bits to lower the decoding threshold of low-density parity-check (LDPC) codes. These deterministic bits serve as known information in the LDPC decoding process to reduce redundancy during data transmission. Unlike the existing work, the proposed scheme addresses controllable deterministic bits, such as MPEG null packets, rather than widely investigated protocol headers.

Chapter 10, "Network-Adaptive Rate and Error Controls for WiFi Video Streaming," investigates the fundamental issues for network-adaptive mobile video streaming over WiFi networks. Specifically, it highlights the practical aspects of network-adaptive rate and error control schemes to overcome the dynamic variations of underlying WiFi networks.

Chapter 11, "State of the Art and Challenges for 3D Video Delivery over Mobile Broadband Networks," examines the technologies underlying the delivery of 3D video content to wireless subscribers over mobile broadband networks. The incorporated study covers key issues, such as the effective delivery of 3D video content in a system that has limited resources in comparison to wired networks, network design issues, as well as scalability and backward compatibility concepts.

In Chapter 12, "A New Hierarchical 16-QAM-Based UEP Scheme for 3-D Video with Depth Image–Based Rendering," an unequal error protection (UEP) scheme based on hierarchical quadrature amplitude modulation (HQAM) for 3-D video transmission is proposed. The proposed scheme exploits the unique characteristics of the color plus depth map stereoscopic video where the color sequence has a significant impact on the reconstructed video quality.

Chapter 13, "2D-to-3D Video Conversion: Techniques and Applications in 3D Video Communications," provides an overview of the main techniques for 2D-to-3D conversion, which includes different depth cues and state-of-the-art schemes. In the 3D video communications context, 2D-to-3D conversion has been used to improve the coding efficiency and the error resiliency and concealment for the 2D video plus depth format.

Chapter 14, "Combined CODEC and Network Parameters for an Enhanced Quality of Experience in Video Streaming," presents the research involved in bridging the gap between the worlds of video compression/encoding and network traffic engineering by (i) using enriched video trace formats in scheduling and traffic control, (ii) using prioritized and error-resilience features in H.264, and (iii) optimizing the combination of the network performance indices with codec-specific distortion parameters for an increased quality of the received video.

In Chapter 15, "Video QoS Analysis over Wi-Fi Networks," the authors present a detailed end-to-end QoS analysis for video applications over wireless networks, both infrastructure and ad hoc networks. Several networking scenarios are carefully configured with variations in network sizes, applications, codecs, and routing protocols to extensively analyze network performance.

MATLAB® is a registered trademark of The MathWorks, Inc. For product information, please contact:

The MathWorks, Inc.
3 Apple Hill Drive
Natick, MA, 01760-2098 USA
Tel: 508-647-7000
Fax: 508-647-7001
E-mail: info@mathworks.com
Web: www.mathworks.com

Acknowledgments

We sincerely appreciate all the authors for their dedication that made this book possible and our effort rewarding. We also owe special thanks to all the reviewers for devoting their time to help us evaluate the submissions. Their expertise and insights have greatly strengthened the quality of this book. Finally, we extend our deep gratitude to the staff of CRC Press, in particular Leong Li-Ming, Joselyn Banks-Kyle, and Michele Smith, whose continued assistance and guidance have been vital to the production of this book.

Contributors

Omar Abdul-Hameed
Faculty of Engineering and Physical
 Sciences
I-Lab: Multimedia Communications
 Research
Department of Electronic
 Engineering
Centre for Vision, Speech and Signal
 Processing
University of Surrey
Surrey, United Kingdom

Khalid Mohamed Alajel
Faculty of Engineering and
 Surveying
University of Southern Queensland
Toowoomba, Queensland, Australia

Raad Alturki
Department of Computer Science
Al Imam Mohammad Ibn Saud
 Islamic University
Riyadh, Saudi Arabia

Paolo Bellavista
Department of Electronics,
 Computer Science, and Systems
University of Bologna
Bologna, Italy

Antonio Corradi
Department of Electronics,
 Computer Science, and Systems
University of Bologna
Bologna, Italy

Jan De Cock
Multimedia Lab
Department of Electronics and
 Information Systems
Ghent University—IBBT
Ghent, Belgium

Erhan Ekmekcioglu
Faculty of Engineering and Physical
 Sciences
I-Lab: Multimedia Communications
 Research
Department of Electronic
 Engineering
Centre for Vision, Speech and Signal
 Processing
University of Surrey
Surrey, United Kingdom

Dalia Fayek
School of Engineering
University of Guelph
Guelph, Ontario, Canada

Gilles Gagnon
Branch of Broadcast Technologies
 Research
Communications Research Centre
 Canada
Ottawa, Ontario, Canada

Carlo Giannelli
Department of Electronics,
 Computer Science, and Systems
University of Bologna
Bologna, Italy

Mohamed Hefeeda
School of Computing Science
Simon Fraser University
Surrey, British Columbia, Canada

Cheng-Hsin Hsu
Department of Computer Science
National Tsing Hua University
Hsin Chu, Taiwan, Republic of
 China

Donglin Hu
Department of Electrical and
    Computer Engineering
Auburn University
Auburn, Alabama

Jane Wei Huang
Electrical Computer Engineering
    Department
University of British Columbia
Vancouver, British Columbia,
    Canada

Araz Jahaniaval
School of Engineering
University of Guelph
Guelph, Ontario, Canada

Dong Jiang
Institute of Microelectronics
Chinese Academy of Sciences
Haidian, Beijing, People's Republic
    of China

JongWon Kim
School of Information and
    Communications
Gwangju Institute of Science and
    Technology (GIST)
Gwangju, South Korea

Ahmet Kondoz
Faculty of Engineering and Physical
    Sciences
I-Lab: Multimedia Communications
    Research
Department of Electronic
    Engineering
Centre for Vision, Speech and Signal
    Processing
University of Surrey
Surrey, United Kingdom

Vikram Krishnamurthy
Electrical Computer Engineering
    Department
University of British Columbia
Vancouver, British Columbia,
    Canada

Peter Lambert
Multimedia Lab
Department of Electronics and
    Information Systems
Ghent University—IBBT
Ghent, Belgium

Chunyu Lin
Multimedia Lab
Department of Electronics and
    Information Systems
Interdisciplinary Institute for
    Broadband Technology
Ghent University
Ghent, Belgium

and

Institute of Information Science
Beijing Jiaotong University
Haidian, Beijing, People's Republic
    of China

Weisi Lin
School of Computer Engineering
Nanyang Technological University
Singapore, Singapore

Weiyao Lin
Department of Electronic
    Engineering
Shanghai Jiao Tong University
Xuhui, Shanghai, People's Republic
    of China

Hassan Mansour
Electrical Computer Engineering
    Department
University of British Columbia
Vancouver, British Columbia,
    Canada

Shiwen Mao
Department of Electrical and
    Computer Engineering
Auburn University
Auburn, Alabama

Rashid Mehmood
School of Computing and
    Engineering
University of Huddersfield
Huddersfield, West Yorkshire,
    United Kingdom

Paulo Nunes
Instituto de Telecomunicações
and
Instituto Universitário de Lisboa
Lisboa, Portugal

Sang-Hoon Park
Communications R&D Center
Samsung Thales Co., Ltd.
Seongnam-Si, South Korea

Manoranjan Paul
School of Computing and
    Mathematics
Charles Sturt University
Bathurst, New South Wales,
    Australia

Joseph Peters
School of Computing Science
Simon Fraser University
Surrey, British Columbia, Canada

Bo Rong
Branch of Broadcast Technologies
    Research
Communications Research Centre
    Canada
Ottawa, Ontario, Canada

Jürgen Slowack
Multimedia Lab
Department of Electronics and
    Information Systems
Ghent University—IBBT
Ghent, Belgium

Luís Ducla Soares
Instituto de Telecomunicações
and
Instituto Universitário de Lisboa
    (ISCTE-IUL)
Lisboa, Portugal

Rik Van de Walle
Multimedia Lab
Department of Electronics and
    Information Systems
Ghent University—IBBT
Ghent, Belgium

Yiyan Wu
Branch of Broadcast Technologies
    Research
Communications Research Centre
    Canada
Ottawa, Ontario, Canada

Wei Xiang
Faculty of Engineering and
    Surveying
University of Southern Queensland
Toowoomba, Queensland, Australia

Chongyang Zhang
Department of Electronic
    Engineering
Shanghai Jiao Tong University
Xuhui, Shanghai, People's Republic
    of China

Bing Zhou
School of Information Engineering
Zhengzhou University
Zhengzhou, Henan, People's
    Republic of China

1

Network-Aware Error-Resilient Video Coding

Luís Ducla Soares and Paulo Nunes

CONTENTS

1.1 Introduction

With the growing demand for universal accessibility to video content, more and more different networks are being used to deploy video services. However, in order to make these video services efficiently available with an acceptable quality in error-prone environments, such as mobile networks, appropriate error resilience techniques are necessary. Since these error-prone environments can typically have very different characteristics, which can also vary over time, it is important that the considered error resilience techniques are network-aware and can adapt to the varying characteristics of the used networks.

In order to extend the useful lifetime of a video coding standard, standardization bodies usually specify the minimum set of tools that are essential for guaranteeing interoperability between devices or applications of different manufacturers. With this strategy, the standard may evolve continuously through the development and improvement of its nonnormative parts. Error resilience is an example of a video coding tool that is not completely specified in a normative way, in any of the currently available and emerging video coding standards. The reason for this is that it is simply not necessary for interoperability and, therefore, it is one of the main degrees of freedom to improve the performance of standard-based systems, even after the standard has been finalized. Nevertheless, recognizing the paramount importance of this type of tool, standardization initiatives always include a minimum set of error-resilient hooks (e.g., in the form of bitstream syntax elements) in order to facilitate the development of effective error resilience techniques, as needed for the particular application envisaged.

Error-resilience techniques are usually seen as playing a role at the decoder side of the communication chain. However, by using preventive error resilience techniques at the encoder side, which involve the intelligent design of the encoder, it is also possible to make the task of the decoder much easier in terms of dealing with errors. In fact, the performance of the decoder can greatly vary depending on the amount of error resilience help provided in the bitstream generated by the encoder. This way, at the encoder, the challenge is to develop techniques that make video bitstreams more resilient to errors, in order to allow the decoder to better recover in case errors occur; these techniques may be called preventive error resilience techniques. At the decoder, the challenge is to develop techniques that make it possible for the decoder to take all the available received data (correct and, eventually, corrupted) and decode it with the best possible video quality, thus minimizing the negative subjective impact of the errors on the video quality offered to the user; these techniques may be called corrective error resilience techniques.

Video communication systems, in order to be globally more error-resilient to channel errors, typically include both preventive and corrective error-resilient techniques. An important class of preventive techniques is error-resilient source coding, which consists of providing redundancy at the source coding level in order to prevent error propagation and consequently reduce the distortion caused by data corruption/loss. Error-resilient source coding techniques include data partitioning, resynchronization and reversible variable length codes [1,2], redundant coding schemes, such as sending the same information predicted from different references [3], scalable video coding [4–6], or multiple description coding [7,8]. Besides source coding redundancy, channel coding redundancy can also be used, where a good example is the case of forward error correction [9]. In terms of corrective error-resilient techniques, error concealment techniques correspond to one of the most important classes, but other important techniques also exist, such as error detection and error localization techniques [10]. Error concealment techniques consist essentially of postprocessing methods aiming at recovering missing or corrupted data from neighboring data (either spatially or temporally) [11], but for these techniques to be truly effective, an error detection technique should be first used to detect if an error has indeed occurred, followed by an error localization technique to determine where the error occurred and which parts of the video content were affected [10]. For a good review of the many different preventive and corrective error-resilient video coding techniques that have been proposed in the literature, the reader can refer to Refs. [12,13].

This chapter addresses the problem of error-resilient encoding, in particular of how to efficiently improve the resilience of compressed video bitstreams, while adaptively considering the network characteristics in terms of information loss.

Video coding systems that rely on predictive (inter) coding to remove temporal redundancy, such as those based on the H.264/AVC standard [14], are strongly affected by transmission errors/information loss due to the error propagation caused by the prediction mechanisms. Therefore, typical approaches to make bitstreams generated by the encoder more error-resilient rely on the adaptation of the video coding mode decisions, at various levels (e.g., picture, slice, or macroblock level), to the underlying network characteristics, trying to establish an adequate trade-off between predictive and non-predictive encoding modes. This is done because nonpredictive modes are less efficient in terms of compression but can provide higher error resilience. In this context, controlling the amount of nonpredictive versus predictive encoded data is an efficient and highly scalable error resilience tool.

The intracoding refresh schemes available in the literature [2,15–22] are a typical example of efficient error resilience techniques to improve the video quality over error-prone environments without requiring changes to the bitstream syntax, thus allowing to continuously improve the performance of standard video codecs without compromising interoperability. However, a permanently open issue related to these techniques is how to achieve the best trade-off between error resilience and coding efficiency.

Since these schemes work by selectively coding in intra mode different parts of the video content at different time instants, they are able to avoid long-term propagation of transmission or storage errors that could make the decoded quality decay very rapidly. This way, these intracoding refresh schemes are able to significantly improve the error resilience of the coded bitstreams and increase the overall subjective impact of the decoded video. While some schemes do not require any specific knowledge of what is being done at the decoder in terms of error concealment [16–18], other approaches try to estimate the distortion experienced at the decoder given a certain probability of data corruption/loss and the concealment techniques adopted [2,22].

The problem with most video coding mode decision approaches, including typical intracoding refresh schemes, is that they can significantly decrease the coding efficiency if they make their decisions without taking into account the rate-distortion (RD) cost of such decisions. This problem can be dealt with by combining the error-resilient coding mode decisions with the video encoder rate control module [23], where the usual coding mode decisions are taken [24,25]. This way, coding-efficient error robustness can be achieved. In the specific case of intracoding refresh schemes, a clever solution for this combination, is to compare the RD cost of coding macroblocks (MBs) in intra and inter modes; if the cost of intracoding is only slightly larger than the cost of intercoding, then the coding mode could be changed to intra, providing error robustness almost for free. This strategy is able to reduce error propagation and, thus, to increase error robustness when transmission errors occur, at a very limited RD cost increase and without the huge complexity of estimating the expected distortion experienced at the decoder.

Nevertheless, in order for these error-resilient video coding mode decision schemes to be really useful in an adaptive way, the current error characteristics of the underlying network being used for transmission should be taken into account. For example, in the case of intracoding refresh schemes, this will allow the bit rate resources allocated to intracoding refresh to be adequately adapted to the error characteristics of the network [26]. After all, networks with small amounts of channel errors only need small amounts of intracoding refresh and vice versa. Thus, efficient bit rate allocation in an error-resilient way has to depend on the feedback received from the network about its current error characteristics, which define the error robustness needed.

Therefore, network awareness makes it possible to dynamically vary the amount of error resilience resources to better suit the current state of the network and, therefore, further improve the decoded video quality without reducing the error robustness [26,27]. This problem is nowadays more relevant than ever, since more and more audiovisual content is accessed over error-prone networks, such as mobile networks, and these networks can have extremely varying error characteristics (over time).

As an illustrative insightful example, this chapter presents a fully automatic network-aware MB intracoding refresh technique for error-resilient H.264/AVC video coding, which also dynamically adjusts the amount of cyclically intra refreshed MBs according to the network conditions, guaranteeing that endless error propagation is avoided.

The rest of the chapter is organized as follows. Section 1.2 describes the general video coding framework that was used for implementing the considered error-resilient network-aware MB intracoding refresh scheme. Section 1.3 introduces the concept of efficient intracoding refresh, which will later be needed in Section 1.4, where the considered network-aware intracoding refresh scheme itself is described. Section 1.5 presents some relevant performance results for the considered scheme in typical mobile network conditions and, finally, Section 1.6 concludes the chapter.

1.2 Video Coding Framework

The network-aware error-resilient scheme described in this chapter relies on the rate control scheme proposed by Li et al. [24,28], as well as on the RD optimization (RDO) framework and the random intra refresh technique included in the H.264/AVC reference software [25]. Since the main contributions and novelty of network-aware error-resilient scheme described in this chapter regard the latter two techniques, it is useful to first briefly review the RDO and the random intra refresh techniques included in the H.264/AVC reference software in order for the reader to better understand the described solutions.

1.2.1 Rate-Distortion Optimization

The H.264/AVC video coding standard owes its major performance gains, relatively to previous standards, essentially to the many different intra and inter MB coding modes supported by the video coding syntax. Although not all modes are allowed in every H.264/AVC profile [14], even for the simplest profiles, such as the Baseline Profile, the encoder has a plethora of possibilities to encode each MB, which makes it difficult to accomplish optimal MB coding mode decisions with low (encoding) complexity. Besides the MB coding mode decision, for motion-compensated inter coded MBs, finding the optimal motion vectors and MB partitions is also not a straightforward task. In this context, RDO becomes a powerful tool, allowing the encoder to optimally select the best MB coding modes and motion vectors (if applicable) [28,29].

In the H.264/AVC reference software [25], the best MB mode decision is accomplished through the RDO technique, where the best MB mode is selected by minimizing the following Lagrangian cost function:

$$J_{MODE} = D(MODE, QP) + \lambda_{MODE} \times R(MODE, QP) \qquad (1.1)$$

where
  $MODE$ is one of the allowable MB coding modes (e.g., SKIP, INTER 16×16, INTER 16×8, INTER 8×16, INTER 8×8, INTRA 4×4, INTRA 16×16)
  QP is the quantization parameter
  $D(MODE, QP)$ and $R(MODE, QP)$ are, respectively, the distortion (between the original and the reconstructed MB) and the number of bits that will be achieved by applying the corresponding $MODE$ and QP

In Ref. [28], it is recommended that, for intra (*I*) and inter predicted (*P*) slices, λ_{MODE} be computed as follows:

$$\lambda_{MODE} = 0.85 \times 2^{(QP-12)/3} \tag{1.2}$$

Motion estimation can also be accomplished through the same framework. In this case, the best motion vector and reference frame can be selected by minimizing the following Lagrangian cost function:

$$J_{MOTION} = D\big(mv(REF)\big) + \lambda_{MOTION} \times R\big(mv(REF)\big) \tag{1.3}$$

where
  $mv(REF)$ is the motion vector for the frame reference *REF*

  $D\big(mv(REF)\big)$ is the residual error measure, such as the sum of absolute differences (SAD) between the original and the reference

  $R\big(mv(REF)\big)$ is the number of bits necessary to encode the corresponding motion vector (i.e., the motion vector difference between the selected motion vector and its prediction) and to signal the selected reference frame

In a similar way, Ref. [28] also recommends that, for P-slices, λ_{MOTION} be computed as

$$\lambda_{MOTION} = \sqrt{\lambda_{MODE}} \tag{1.4}$$

when the SAD measure is used.

Since the quantization parameter is required for computing the Lagrangian multipliers λ_{MODE} and λ_{MOTION}, as well as for computing the number of bits to encode the residue for a given MB, a rate control mechanism must be used that can efficiently compute for each MB (or set of MBs, such as a slice) an adequate quantization parameter in order to maximize the decoded video quality for a given bit rate budget. In this case, the method proposed by Li et al. [24,28] has been used since it is the one implemented in the H.264/AVC reference software [25].

1.2.2 Random Intra Refresh

As mentioned earlier, the H.264/AVC reference software [25] includes a (nonnormative) technique for intra refreshing MBs. Although this technique is called random intra refresh (RIR), it is not really a purely random refresh technique. This technique is basically a cyclic intra refresh (CIR) technique for which the refresh order is not simply the raster scan order. The refresh order is randomly defined once before encoding, but afterward intra refresh proceeds cyclically, following the determined order, with *n* MBs for each time instant.

95	94	24	60	97	22	90	21	61	77	43
9	65	28	41	50	11	23	18	82	26	73
87	36	71	15	29	52	70	10	64	44	45
63	8	66	88	35	32	33	39	0	80	72
78	25	93	67	68	85	48	19	89	1	7
53	75	49	46	54	42	13	4	81	30	40
57	3	79	84	86	55	98	56	58	16	47
91	69	92	59	62	5	51	83	38	76	27
34	96	17	37	31	6	20	14	12	2	74

FIGURE 1.1
Example of random intra refresh order for QCIF spatial resolution. (From Nunes, P. et al., Error resilient macroblock rate control for H.264/AVC video coding, *Proceedings of the IEEE International Conference on Image Processing*, San Diego, CA, p. 2133, October 2008. With permission. © 2008 IEEE.)

An example of a randomly determined intra refresh order, for QCIF spatial resolution, may be seen in Figure 1.1.

Since the RIR technique used in the H.264/AVC reference software and also considered here is basically a CIR technique, in the remainder of this chapter, the acronyms RIR and CIR will be used interchangeably.

One of the main advantages of this technique is that, being cyclic, it guarantees that all MBs will be refreshed, at least, once in each cycle, thus guaranteeing that there are no MBs where errors can propagate indefinitely. However, this technique also has disadvantages, one of which is the fact that all MBs are refreshed exactly the same number of times. This basically means that it is not possible to refresh more often MBs that are more likely to be lost or are harder to conceal at the decoder if an error does occur.

Another important aspect of this technique is that MBs are refreshed according to the predetermined order, without taking into account the eventual RD cost of intra refreshing a given MB, as opposed to letting the rate control module decide which encoding mode is best in terms of RD cost. This is exactly where there is room for improvement: Intra refresh should be performed by taking into account the RD cost of a given MB.

1.3 Efficient Intracoding Refresh

When deciding the best MB coding mode, notably between inter- and intracoding modes, the RDO framework, as briefly described in Section 1.2.1, simply selects the mode that has lower RD cost, given by Equation 1.1. This RDO framework, as implemented in the H.264/AVC reference software, does not take into account other dimensions, besides rate and distortion optimization, such as the robustness of the bitstream in error-prone environments. Therefore, some MBs are simply inter coded because their best inter mode RD cost is slightly lower than the best intra mode RD cost. For these cases, selecting the intra mode, although not optimal in a strict RD sense, can prove to be a much better decision when the bitstream becomes corrupted by errors (e.g., due to packet losses in packet networks), and the intra coded MBs can be used to stop error propagation due to the (temporal) predictive coding modes. Moreover, if additional error robustness is introduced through an intra refresh technique, for example, as the one described in Section 1.2.2, some MBs can be highly penalized in a RD sense, since they can be blindly forced to be encoded in an intra mode, without taking into account the RD cost of that decision.

1.3.1 Error-Resilient RDO-Driven Intra Refresh

The main idea of a network-aware error-resilient scheme is to perform RDO in a resilient manner, using the relative RD cost of the best intra mode and the best inter mode for each MB. Therefore, whenever coding a given MB in intra mode does not cost significantly more than the best intercoding mode, the given MB is gracefully forced to be encoded in its best intra mode.

This error-resilient RDO provides an efficient intra refresh scheme, thus guaranteeing that the generated bitstream will be more robust to channel errors, without having to spend a lot of bits on intra coded MBs, which typically reduces the decoded video quality when there are no errors in the channel. This scheme can be described through the MB-level mode decision architecture depicted in Figure 1.2.

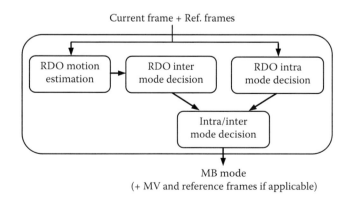

FIGURE 1.2
Architecture of the error-resilient MB intra/inter mode decision scheme. (From Nunes, P. et al., Error resilient macroblock rate control for H.264/AVC video coding, *Proceedings of the IEEE International Conference on Image Processing,* San Diego, CA, p. 2134, October 2008. With permission. © 2008 IEEE.)

1.3.1.1 RDO Intra and Inter Mode Decision

Before deciding the best mode to encode a given MB, the best inter mode RD cost, J_{INTER}, is computed from the set of all possible inter modes, and the best intra mode RD cost, J_{INTRA}, is computed from the set of all possible intra modes through RDO, i.e., Equations 1.1 and 1.3, where

$$J_{INTER} = \min_{MODE \in S_{INTER}} (J_{MODE}) \qquad (1.5)$$

and

$$J_{INTRA} = \min_{MODE \in S_{INTRA}} (J_{MODE}) \qquad (1.6)$$

where
  S_{INTER} is the set of allowed inter modes (i.e., SKIP, INTER 16×16, INTER 16×8, INTER 8×16, INTER 8×8, INTER 8×4, INTER 4×8, and INTER 4×4)
  S_{INTRA} is the set of allowed intra modes (i.e., INTRA 4×4, INTRA 16×16, INTRA PCM, or INTRA 8×8)

The best intra and inter modes are the ones with the lowest intra and inter RD costs, respectively.

1.3.1.2 Error-Resilient Intra/Inter Mode Decision

To control the amount of MBs that will be gracefully forced to be encoded in intra mode, a control parameter, α_{RD} (which basically specifies the tolerable RD cost increase for replacing an inter by an intra MB) is used in such a way that

$$\text{if } \left(J_{INTRA}/J_{INTER} \leq \alpha_{RD} \right) \text{ Intra mode is selected} \qquad (1.7)$$

$$\text{if } \left(J_{INTRA}/J_{INTER} > \alpha_{RD} \right) \text{ Inter mode is selected} \qquad (1.8)$$

Notice that, for $\alpha_{RD}=1$, no particular mode is favored in an RD sense, while for $\alpha_{RD}>1$, the intra modes are favored relatively to the inter modes (see Figure 1.3). Therefore, the amount of gracefully forced intra encoded MBs can be controlled by the α_{RD} parameter. The MBs that end up being forced to intra mode are the MBs for which the RD costs of intra and inter modes are similar, which typically correspond to MBs that have high inter RD cost and, therefore, would be difficult to conceal at the decoder if lost.

1.3.2 Random Intra Refresh

Notice that the previous scheme does not guarantee that all MBs are periodically refreshed, which, if not properly handled, could lead to an

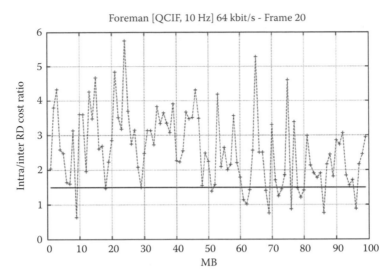

FIGURE 1.3
MBs with an intra/inter RD cost ratio below the line will be gracefully forced to intra mode. (From Nunes, P. et al., Error resilient macroblock rate control for H.264/AVC video coding, *Proceedings of the IEEE International Conference on Image Processing*, San Diego, CA, p. 2134, October 2008. With permission. © 2008 IEEE.)

endless propagation of errors along time for some MBs in the video sequence. To handle this issue, an RIR can also be concurrently applied, but with a lower number of refreshed MBs per frame when compared with solely applying the RIR technique, in order not to compromise dramatically the RD efficiency.

1.4 Network-Aware Error-Resilient Video Coding Method

The main limitation of the MB coding mode decision method described in Section 1.3 is that the control parameter, α_{RD}, is not dynamically adapted to the actual network error conditions. However, when feedback about the network error conditions is available, it would be possible to use this information to adjust the α_{RD} control parameter in order to maximize the decoded video quality while dynamically providing adequate error resilience.

1.4.1 Intra/Inter Mode Decision with Constant α_{RD}

When a constant α_{RD} value is used without considering the current network error conditions in terms of packet loss rate (PLR), the benefits of the technique described in Section 1.3 (and proposed in Ref. [23]) are not fully exploited. This is clear from Figure 1.4, where the *Foreman* sequence has been encoded

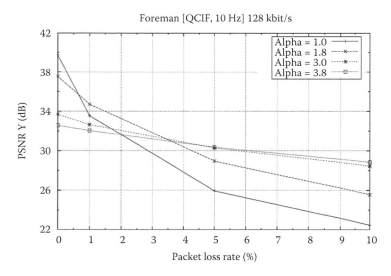

FIGURE 1.4
PSNR versus PLR for a constant α_{RD} parameter for the *Foreman* sequence. (From Soares, L.D. et al., Efficient network-aware macroblock mode decision for error resilient H.264/AVC video coding, *Proceedings of the SPIE Conference on Applications of Digital Image Processing*, vol. 7073, San Diego, CA, August 2008.)

with the Baseline Profile of H.264/AVC with different α_{RD} values, including $\alpha_{RD} = 1$. In Figure 1.4, as well as in the remainder of Section 1.4, CIR is not used in order to avoid biasing the behavior associated with the α_{RD} parameter. Notice, however, that the use of CIR is typically recommended, as mentioned in Section 1.2.2. As can be seen, in these conditions, the optimal α_{RD} (i.e., the one that leads to the highest PSNR) is highly dependent on the network PLR.

As expected, when there are no errors (*PLR* = 0%), the highest decoding quality is achieved when no intra MBs are forced (i.e., $\alpha_{RD} = 1.0$). However, for this α_{RD} value, the decoded video quality decays very rapidly as the PLR increases. On the other hand, if only a small amount of intra MBs are forced (i.e., $\alpha_{RD} = 1.8$), the decoded video quality is slightly improved for the higher PLR values, when compared to the case with no forced intra MBs, but will be slightly penalized for error-free transmission. This effect is even more evident as the α_{RD} value increases, which corresponds to the situation where more and more intra MBs are gracefully forced, depending on the α_{RD} value. For example, for $\alpha_{RD} = 3.8$ and for a PLR of 10%, the decoded video quality is highly improved relatively to the situation with no forced intra MBs (i.e., 6.36 dB), because the error propagation is significantly reduced. However, for lower PLRs, the decoded video quality is penalized due to the excessive use of intracoding (i.e., 7.19 dB for *PLR* = 0% and 1.50 dB for *PLR* = 1%), still for $\alpha_{RD} = 3.8$.

Therefore, from what has been presented earlier, it is possible to conclude that the optimal amount of intra coded MBs is highly dependent on the error characteristics of the underlying network and, thus, the error resilience

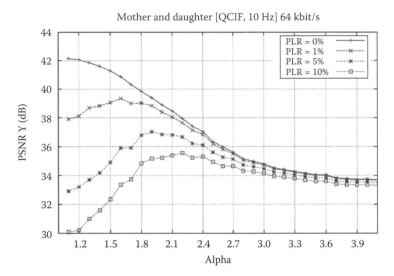

FIGURE 1.5
PSNR versus α_{RD} (alpha in the x-axis label) parameter for various PLRs for the *Mother and Daughter* sequence. (From Soares, L.D. et al., Efficient network-aware macroblock mode decision for error resilient H.264/AVC video coding, *Proceedings of the SPIE Conference on Applications of Digital Image Processing*, vol. 7073, San Diego, CA, August 2008.)

control parameter α_{RD} should be dynamically adjusted to the channel error conditions to maximize the decoded quality.

In order to illustrate the influence of the α_{RD} parameter on the decoded PSNR, Figure 1.5 shows the decoded video quality, in terms of PSNR, versus the α_{RD} parameter for several PLRs for the *Mother and Daughter* sequence (QCIF, 10 Hz) encoded at 64 kbit/s. Clearly, for each PLR condition, there is an α_{RD} value that maximizes the decoded video quality. For example, for a PLR of 10%, the maximum PSNR value is achieved for $\alpha_{RD}=2.2$. To further illustrate the importance of a proper selection of the α_{RD} parameter and how it can significantly improve the overall decoded video quality under severe error conditions, it should be noted that, for a PLR of 10%, the PSNR difference between having $\alpha_{RD}=2.2$ and $\alpha_{RD}=1.1$ is 5.47 dB.

1.4.2 Intra/Inter Mode Decision with Network-Aware α_{RD} Selection

A possible approach to address the problem of adapting the α_{RD} parameter to the channel error conditions is to use the information in the receiver reports (RR) of the real-time transport protocol (RTP) control protocol (RTCP) [30] to provide the encoder with the actual error characteristics of the underlying network. This makes it possible to adaptively and efficiently select the amount of intra coded MBs to be inserted in each frame by taking into account this feedback information about the rate of lost packets, as shown in Figure 1.6.

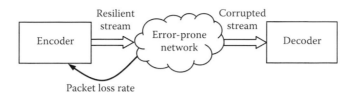

FIGURE 1.6
Network-aware video encoding architecture. (From Soares, L.D. et al., Efficient network-aware macroblock mode decision for error resilient H.264/AVC video coding, *Proceedings of the SPIE Conference on Applications of Digital Image Processing*, vol. 7073, San Diego, CA, August 2008.)

In the method presented here, the intra/inter mode decision is still based on the α_{RD} parameter, but this time α_{RD} may depend on several aspects, such as the content type, the content spatial and temporal resolutions, the coding bit rate, and the PLR of the network.

This way, by considering a mapping function f_{NMD}, it will be possible to dynamically determine the α_{RD} parameter from the following expression:

$$\alpha_{RD} = f_{NMD}(\mathbf{S}, PLR) \tag{1.9}$$

where
   PLR is the packet loss rate
   S can be an n-dimensional vector characterizing the encoding scenario, for
       example, in terms of the content motion activity and the texture coding
       complexity, the content spatial and temporal resolutions, and the coding
       bit rate

In this work, however, as it will be shown later in Section 1.4.3, the encoding scenario can be characterized solely by the encoded bit rate with a good approximation. The f_{NMD} function basically maps the encoding scenario and the network PLR into a "good" α_{RD} parameter that dynamically maximizes the average decoding video quality. Notice that, although it is not easy to obtain a general function, it can be defined for several classes of content and a discrete limited set of encoding parameters and PLRs. In this chapter, it will be shown that, by carefully designing the f_{NMD} function, significant gains can be obtained in terms of video quality regarding the reference method described in Section 1.4.4.

Therefore, the network-aware MB mode decision (NMD) method can be briefly described through the following steps in terms of encoder operation:

1. Obtain the packet loss rate through network feedback.
2. Compute the α_{RD} parameter through the mapping function given by Equation 1.9 (and detailed in the following).

3. Perform intra/inter mode decision using the α_{RD} parameter, computed in Step 2, for the next MB to be encoded, and encode the MB.

4. Check if a new network feedback report has arrived; if yes, go back to Step 1; if not, go back to Step 3.

Notice that it is out of the scope of this chapter to define when the network reports are issued, since this will depend on how the network protocols are configured and the varying characteristics of the network itself [30]. Nevertheless, in real application scenarios, it is important to design appropriate interfacing mechanisms between the codec and the underlying network, in order that both encoder and decoder can adaptively adjust their operations according to the network conditions [12].

Through Equation 1.9, the encoder is able to adjust the amount of intra refresh according to the network error conditions and the available bit rate. This intra refresh method typically increases the intra refresh for the more complex MBs, which are those typically more difficult to conceal. The main problem of this approach is that it does not guarantee that all MBs in the scene are refreshed. This is clearly illustrated in Figure 1.7 for the *Foreman* sequence, where the right image represents the relative amount of MB intra refresh along the sequence (lighter blocks mean more intra refresh). As it can be seen, with this intra refresh scheme some MBs are never refreshed, which can lead to errors propagating indefinitely along time in these MB positions (dark blocks in Figure 1.7).

1.4.3 Model for the f_{NMD} Mapping Function

In order to devise a model for the mapping function f_{NMD} defined in Equation 1.9, it is first important to see how the optimal α_{RD} parameter varies with PLR.

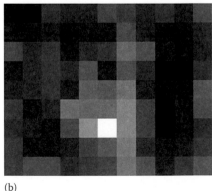

(a)   (b)

FIGURE 1.7
Relative amount of intra refresh (b) for the MBs of the *Foreman* sequence (a) (*QCIF*, 15 Hz, 128 kbit/s, and $\alpha_{RD}=1.1$). (From Nunes, P. et al., Automatic and adaptive network-aware macroblock intra refresh for error-resilient H.264/AVC video coding, *Proceedings of the IEEE International Conference on Image Processing*, Cairo, Egypt, p. 3074, November 2009. With permission. © 2009 IEEE.)

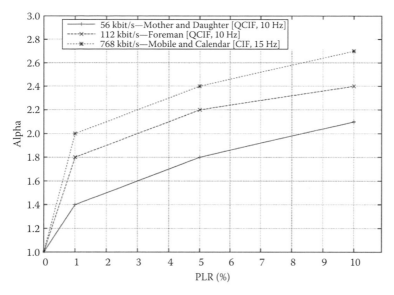

FIGURE 1.8

Example of optimal α_{RD} versus PLR for various sequences and bit rates. (From Soares, L.D. et al., Efficient network-aware macroblock mode decision for error resilient H.264/AVC video coding, *Proceedings of the SPIE Conference on Applications of Digital Image Processing, vol.* 7073, San Diego, CA, August 2008.)

This is plotted in Figure 1.8 for three different sequences (i.e., *Mother and Daughter, Foreman,* and *Mobile and Calendar*) encoded at different bit rates, and resolutions, for illustrative purposes. Each curve in Figure 1.8 corresponds to a different encoding scenario **S**, in terms of the content motion activity and the texture coding complexity, the content spatial and temporal resolutions, and the coding bit rate (see Equation 1.9). As shall be detailed later in Section 1.5, these three sequences have also been encoded at many other bit rates, and the kind of curves obtained was always similar.

As can be seen from the plots in Figure 1.8, the behavior of the optimal α_{RD} parameter versus the PLR is similar to that of a charging capacitor [31] (but starting at $\alpha_{RD} = 1.0$). Therefore, for a given sequence and for a given bit rate (i.e., a given encoding scenario **S**), it should be possible to model the behavior of the α_{RD} parameter with respect to the PLR with the following expression:

$$\alpha_{RD} = 1 + K_1 \times \left(1 - e^{-K_2 \times PLR}\right) \tag{1.10}$$

where *PLR* represents the packet loss rate, while K_1 and K_2 represent constants that are specific to the considered encoding scenario, notably the sequence characteristics and bit rate. However, the main problem in using Equation 1.10 to compute α_{RD} is that, for a given sequence, a different set of K_1 and K_2 would be needed for each of the considered bit rates, which would be extremely

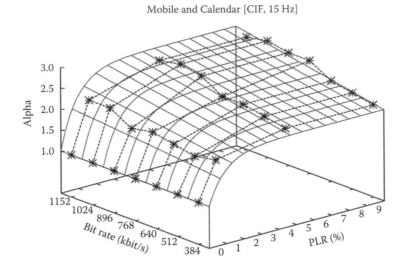

FIGURE 1.9

Optimal α_{RD} versus PLR and bit rate for the *Mobile and Calendar* sequence. (From Soares, L.D. et al., Efficient network-aware macroblock mode decision for error resilient H.264/AVC video coding, *Proceedings of the SPIE Conference on Applications of Digital Image Processing*, vol. 7073, San Diego, CA, August 2008.)

unpractical. In order to address this issue, it is important to understand how the optimal α_{RD} parameter varies when both the PLR and the bit rate vary. This variation is illustrated in Figure 1.9 for the *Mobile and Calendar* sequence.

After close inspection of Figure 1.9, it can be seen that the K_1 value, which basically dictates the value of α_{RD} toward which the curve asymptotically converges, depends linearly on the used bit rate and, therefore, it can be modeled by the following expression:

$$K_1\left(r_b\right) = a \times r_b + b \tag{1.11}$$

where r_b is the bit rate, while a and b are the parameters that need to be estimated for a given sequence.

As for the K_2 value, which dictates the growth rate of the considered exponential, it appears, after exhaustive testing, to not depend on the used bit rate. Therefore, as a first approach, it can be considered to be constant, as in

$$K_2 = c \tag{1.12}$$

This behavior was observed for the three different video sequences mentioned earlier and, therefore, makes it possible to establish a final expression which allows the video encoder to automatically select, for a given sequence, an adequate α_{RD} parameter when the PLR and the bit rate r_b are known:

$$\alpha_{RD} = f_{NMD}\left(r_b, PLR\right) = 1 + \left(a \times r_b + b\right) \cdot \left(1 - e^{-c \times PLR}\right) \tag{1.13}$$

where a, b, and c are the model parameters that need to be estimated (see Ref. [26]). After extensive experimentation, it was found that the parameters a, b, and c can be considered more or less independent of the sequence, which means that a single set of parameters could be used for three different video sequences with a low fitting error. This basically means that the encoding scenario **S**, defined in Section 1.4.2, can be well represented only by the bit rate r_b.

As explained in Ref. [26], the parameters a, b, and c could be obtained by considering four packet loss rates and two different bit rates for three different sequences, corresponding to a total of 24 $\left(r_b, PLR\right)$ pairs, with the iterative Levenberg–Marquardt method [32,33]. By following this approach, the estimated parameters are $a = 0.83 \times 10^{-6}$, $b = 0.97$, and $c = 0.90$.

1.4.4 Network-Aware Cyclic Intra Refresh

The approach presented in Section 1.4.2 can also be followed to simply adjust the number of cyclic intra refreshed MBs per frame, based on the feedback received about the network PLR, without any RD cost considerations. This is shown in Figure 1.10, where it is clear that for each PLR condition there are a number of cyclic intra refresh MBs that maximize the decoded video quality. However, when comparing the best PSNR results of Figures 1.5 and

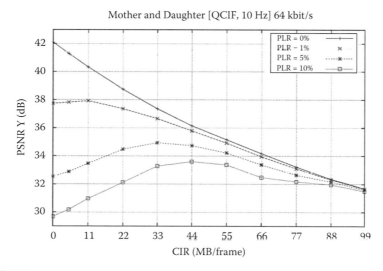

Mother and Daughter [QCIF, 10 Hz] 64 kbit/s

FIGURE 1.10
PSNR versus number of CIR MBs for various PLRs for the *Mother and Daughter* sequence. (From Soares, L.D. et al., Efficient network-aware macroblock mode decision for error resilient H.264/AVC video coding, *Proceedings of the SPIE Conference on Applications of Digital Image Processing,* vol. 7073, San Diego, CA, August 2008.)

1.10 (both obtained for the *Mother and Daughter* sequence encoded with the same spatial and temporal resolutions and the same bit rate), for a given PLR, the PSNR values obtained by varying α_{RD} are always higher. For example, for a PLR of 5%, a maximum average PSNR of 37.03 dB is achieved for $\alpha_{RD} = 1.9$ (see Figure 1.5), while a maximum PSNR of only 34.94 dB is achieved for 33 cyclically intra refreshed MBs in each frame (see Figure 1.10), a difference of approximately 2 dB. This shows that by adequately choosing the α_{RD} parameter it should be possible to achieve a higher quality than when using the optimal number of CIR MBs. This is mainly due to the fact that when simply cyclically intra refreshing some MBs in a given frame, the additional RD cost of that decision can be extremely high, penalizing the overall video quality, since the "cheap" intra MBs are not looked for as in the efficient intracoding refresh solution based on the α_{RD} parameter.

1.4.5 Intra Refresh with Network-Aware α_{RD} and CIR Selection

The main drawback of the scheme described in Section 1.4.3 of not being able to guarantee that all MBs are periodically refreshed, can be alleviated by introducing some additional CIR MBs per frame to guarantee that all MB positions are refreshed with a minimum periodicity. This requirement raises the question of how to adaptively select an adequate amount of CIR MBs that is sufficiently high to avoid long-term error propagation without penalizing too much the encoder RD performance.

A possible approach to tackle this problem is to decide the adequate α_{RD} value and the number of CIR MBs per frame separately, using a different model for each of these two error resilience parameters. For the α_{RD} selection, the model in Equation 1.9 is used. As for the selection of the number of CIR MBs, it was verified after exhaustive testing [27] that the optimal amount of CIR MBs tends to increase linearly with the bit rate r_b, for a given PLR, but tends to increase exponentially with the PLR, for a given bit rate. Based on these observations, the following model was considered for the selection of the amount of CIR MBs per frame:

$$CIR = f_{CIR}(PLR, r_b) = (a_1 \cdot r_b + b_1) \cdot e^{c_1 \cdot PLR} \tag{1.14}$$

where a_1, b_1, and c_1 are the model parameters that need to be estimated. In Ref. [27], these parameters have been determined by nonlinear curve fitting (the Levenberg–Marquardt method) of the optimal amount of CIR MBs per frame, experimentally determined for a set of representative test sequences, encoding bit rate ranges and packet loss rates. The estimated parameters were $a_1 = 12.97 \times 10^{-6}$, $b_1 = -0.13$, and $c_1 = 0.24$; these parameter values will also be considered here.

Figure 1.11 shows the proposed model as well as the experimental data for the *Mobile and Calendar* test sequence. As can be seen, a simple linear model would not have represented well the experimental data.

Mobile and Calendar [CIF, 15 Hz]

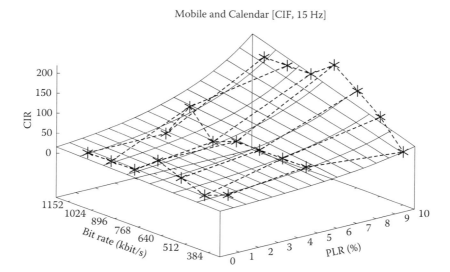

FIGURE 1.11
Optimal amount of CIR MBs per frame versus PLR and bit rate for the *Mobile and Calendar* sequence. (From Nunes, P. et al., Automatic and adaptive network-aware macroblock intra refresh for error-resilient H.264/AVC video coding, *Proceedings of the IEEE International Conference on Image Processing*, Cairo, Egypt, p. 3075, November 2009. With permission. © 2009 IEEE.)

The CIR order is randomly defined once before encoding, as described in Section 1.2.2 (and in Ref. [25]), to avoid the subjectively disturbing effect of performing sequential (e.g., raster scan) refresh. The determined order is then cyclically followed with the computed number of MBs being refreshed in each frame.

Therefore, the complete network-aware MB intracoding refresh (NIR) scheme (which was initially proposed in Ref. [27]) can be briefly described by the following steps in terms of encoder operation:

Step 1. Obtain the PLR value through network feedback.

Step 2. Compute the number of CIR MBs to be used per frame, by using the proposed f_{CIR} function defined by Equation 1.14 and rounding it to the nearest integer.

Step 3. Compute the α_{RD} value by using the f_{NMD} function defined by Equation 1.9 in Section 1.4.2.

Step 4. For each MB in a frame, check if it should be forced to intra mode according to the CIR order and the determined number of CIR MBs per frame; if not, perform intra/inter mode decision using the α_{RD} value computed in Step 3; encode the MB with selected mode.

Step 5. At the end of the frame, check if a new network feedback report has arrived; if yes, go back to Step 1; if not, go back to Step 4.

The definition of when the network reports are issued depends on how the network protocols are configured and the varying characteristics of the network itself [34].

Notice that independently selecting the α_{RD} value and the amount of CIR MBs, while they are likely interdependent, can lead to chosen values that do not correspond to the optimal (α_{RD}, *CIR*) pair. However, it has been verified after extensive experimentation that the considered independent selection process is still robust in the sense that the chosen values are typically close enough to the optimal pair and, therefore, the overall performance is not dramatically penalized.

1.5 Performance Evaluation

To evaluate the performance of the complete NIR scheme described in this chapter, it has been compared in similar conditions to a reference intra refresh scheme, which basically corresponds to the network-aware version with the cyclic intra refresh scheme of the H.264/AVC reference software [25] described in Section 1.4.4. This solution has been adopted because at the time of writing no other network-aware intra refresh techniques, which adaptively take into account the current network conditions, were known.

In the reference scheme, the optimal number of CIR MBs per frame is selected manually for the considered network conditions, while in the considered NIR solution, the selection of the amount of CIR MBs per frame and the α_{RD} parameter is done fully automatically. For the complete NIR and reference schemes, the *Mother and Daughter*, the *Foreman*, and the *Mobile and Calendar* video sequences have been encoded using the H.264/AVC Baseline Profile [25]. The used test conditions, which are representative of those currently used for personal communications over mobile networks, are summarized in Table 1.1. For QCIF, each frame was divided into three slices, while for CIF each frame was divided into six slices. In both cases, each slice

TABLE 1.1

Test Conditions

Video Test Sequence	Mother and Daughter	Foreman	Mobile and Calendar
Spatial resolution	QCIF	QCIF	CIF
Frame rate (Hz)	10	10	15
Bit rate (kbit/s)	24–64	48–128	384–1152

Source: Nunes, P., Soares, D., and Periera, F., Error resilient macroblock rate control for H.264/AVC video coding, *Proceedings of the IEEE International Conference on Image Processing*, San Diego, CA, p. 2134, October 2008. With permission. Copyright 2008 IEEE.

consists of three MB rows. After encoding, each slice was mapped to an RTP packet for network transmission [34].

For the reference scheme, the number of cyclically intra refreshed MBs per frame was chosen for each PLR and bit rate, such that the decoded video quality would be the best possible. This was done manually by performing an exhaustive set of tests using many different amounts of CIR MBs per frame and then choosing the one that leads to the highest decoded average PSNR value, obtained by averaging over 50 different error patterns. For the QCIF video sequences, the possible values for the number of cyclically intra refreshed MBs were chosen from the representative set {0, 5, 11, 22, 33, …, 99}, while for the CIF video sequences the representative set consisted of {0, 22, 44, 66, …, 396}.

To simulate the network conditions, three different PLRs were considered: 1%, 5%, and 10%. Since each slice is mapped to one RTP packet, each lost packet will correspond to a lost video slice. Packet losses are considered independent and identically distributed. For each one of the studied PLRs, each coded bitstream has been corrupted and then decoded 50 times (i.e., corresponding to 50 different error patterns or runs), while applying the default error concealment technique implemented in the H.264/AVC reference software [25,28]. The presented results correspond to PSNR averages of these 50 different runs for the luminance component (PSNR Y).

For the conditions mentioned earlier, PSNR Y results are shown in Tables 1.2 through 1.4 for the *Mother and Daughter, Foreman,* and *Mobile and Calendar* video sequences, respectively. In these tables, NIR refers to the complete network-aware intracoding refresh scheme described in this chapter, and JM refers to the reference technique (winning cases appear in bold). In addition, OPT corresponds to the manual selection of the best (α_{RD}, CIR) pair.

TABLE 1.2

PSNR Results for the *Mother and Daughter* Sequence

| Bit rate (kbit/s) | PSNR Y (dB) | | | | | | | | |
| | PLR = 1% | | | PLR = 5% | | | PLR = 10% | | |
	JM	NIR	OPT	JM	NIR	OPT	JM	NIR	OPT
24	34.67	34.62	35.21	31.91	**32.48**	33.32	30.68	**31.79**	32.23
32	36.01	**36.42**	36.65	32.59	**34.06**	34.57	31.12	**32.87**	33.16
40	36.57	**37.26**	37.36	33.44	**34.91**	35.40	32.02	**33.61**	34.03
48	36.97	**38.00**	38.14	33.92	**35.75**	35.87	32.68	**34.28**	34.47
56	37.77	**38.68**	38.69	34.49	**36.33**	36.54	33.14	**34.46**	35.14
64	37.92	**39.34**	39.34	34.94	**36.84**	37.03	33.60	**35.13**	35.55

Source: From Nunes, P., Soares, D., and Pereira, F., Automatic and adaptive network-aware macroblock intra refresh for error-resilient H.264/AVC video coding, *Proceedings of the IEEE International Conference on Image Processing,* Cairo, Egypt, p. 3076, November 2009. With permission. Copyright 2009 IEEE.

TABLE 1.3

PSNR Results for the *Foreman* Sequence

	PSNR Y (dB)								
	PLR = 1%			*PLR = 5%*			*PLR = 10%*		
Bit rate (kbit/s)	JM	NIR	OPT	JM	NIR	OPT	JM	NIR	OPT
48	30.02	29.92	30.32	25.94	**27.01**	27.01	23.89	**24.44**	24.88
64	31.06	**31.79**	31.83	26.96	**28.59**	28.59	25.14	**25.76**	26.22
80	32.05	**32.71**	32.73	**28.08**	**29.30**	29.30	26.88	26.24	26.96
96	32.63	**32.90**	33.69	**29.06**	**29.75**	29.93	28.27	27.12	28.34
112	33.34	**33.70**	34.34	**29.89**	**30.01**	30.50	29.20	27.90	29.28
128	33.74	**34.57**	35.05	**30.79**	**30.98**	31.32	30.01	28.77	30.08

Source: From Nunes, P., Soares, D., and Periera, F., Automatic and adaptive network-aware macroblock intra refresh for error-resilient H.264/AVC video coding, *Proceedings of the IEEE International Conference on Image Processing*, Cairo, Egypt, p. 3076, November 2009. With permission. Copyright 2009 IEEE.

TABLE 1.4

PSNR Results for the *Mobile and Calendar* Sequence

	PSNR Y (dB)								
	PLR = 1%			*PLR = 5%*			*PLR = 10%*		
Bit rate (kbit/s)	JM	NIR	OPT	JM	NIR	OPT	JM	NIR	OPT
384	25.42	**25.62**	25.75	21.49	**21.98**	21.98	19.17	**19.75**	19.75
512	26.32	**26.79**	26.79	22.62	**22.76**	23.04	20.41	**20.83**	20.83
640	27.12	**27.45**	27.60	23.42	**23.42**	23.86	21.40	**21.66**	21.78
768	27.84	**28.35**	28.46	24.07	**24.40**	24.50	22.23	**22.34**	22.44
896	28.37	**28.80**	29.30	**24.58**	24.46	25.14	22.71	22.45	23.00
1024	28.78	**29.27**	29.78	**25.13**	**25.21**	25.63	23.23	23.00	23.33
1152	29.25	**29.88**	30.37	**25.47**	**25.65**	26.15	23.58	23.07	23.79

Source: From Nunes, P., Soares, D., and Periera, F., Automatic and adaptive network-aware macroblock intra refresh for error-resilient H.264/AVC video coding, *Proceedings of the IEEE International Conference on Image Processing*, Cairo, Egypt, p. 3076, November 2009. With permission. Copyright 2009 IEEE.

No visual results are given here, because the direct comparison of peer frames (encoded with different coding mode selection schemes) is rather meaningless in this case; only the comparison of the total video quality for several error patterns makes sense. This is due to the fact that the generated streams for the proposed and the reference techniques are different and, even if the same error pattern is used to corrupt them, the errors will affect different parts of the data at a given time instant, causing very different artifacts.

To help the reader to better read the gains obtained with the proposed technique, the results obtained for the *Mother and Daughter* sequence are also

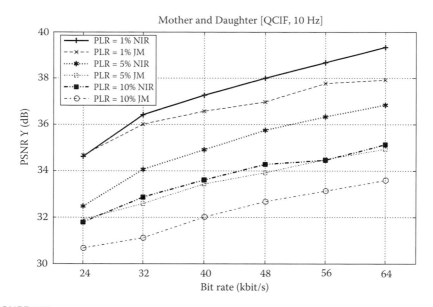

FIGURE 1.12

PSNR results for the *Mother and Daughter* sequence. (From Nunes, P., Soares, D., and Periera, F., Automatic and adaptive network-aware macroblock intra refresh for error-resilient H.264/AVC video coding, *Proceedings of the IEEE International Conference on Image Processing,* Cairo, Egypt, p. 3076, November 2009. With permission. Copyright 2009 IEEE.)

shown in a plot in Figure 1.12, for both JM and NIR. For the *Foreman* and the *Mobile and Calendar* sequences, the trends are similar.

The presented results show that, when the fully automatic NIR scheme is used, the decoded video quality is significantly improved for the vast majority of tested conditions when compared to the reference method with a manually selected amount of CIR MBs (JM). Improvements of the NIR method can be as high as 1.90 dB for the *Mother and Daughter* sequence encoded at 64 kbit/s and a PLR of 5%. The most significant exception is for the PLR of 10% and higher bit rates (see Tables 1.3 and 1.4). This exception is due to the fact that, for these PLR and bit rate values, the number of CIR MBs chosen with the proposed f_{CIR} is slightly different from the optimal values.

When comparing the NIR scheme to the one proposed in Ref. [26], which does not use CIR, the NIR PSNR Y values are most of the times higher than or equal to those achieved in Ref. [26]. The highest gains occur for the *Foreman* sequence encoded at 128 kbit/s and a PLR of 10% (0.90 dB), and for the *Mobile and Calendar* sequence encoded at 768 kbit/s and a PLR of 10% (0.60 dB). For the cases, where the NIR leads to lower PSNR Y values, the losses are never more than 0.49 dB, which happens for the *Mobile and Calendar* sequence encoded at 896 kbit/s and a PLR of 5%.

Notice, however, that the scheme in Ref. [26] cannot guarantee that all MBs will eventually be refreshed, which is a major drawback for real usage in error-prone environments, such as mobile networks. On the other hand, the one described in this chapter can, not only overcome this drawback, but it does so fully automatically, without any user intervention.

1.6 Final Remarks

This chapter describes a method to efficiently and fully automatically perform intracoding refresh, while taking into account the PLR of the underlying network and the encoded bit rate. The described method can be used to efficiently generate error-resilient H.264/AVC bitstreams that are perfectly adapted to the channel error characteristics. This is extremely important because it can mean that error-resilient video transmission will be possible in environments with varying error characteristics with an improved quality, notably, when compared to the case where the MB intracoding decisions are taken without considering the error characteristics of the network.

Acknowledgments

The authors would like to acknowledge that the work described in this chapter was developed at Instituto de Telecomunicações (Lisboa, Portugal) and was supported by FCT project PEst-OE/EEI/LA0008/2011.

References

1. A. H. Li, S. Kittitornkun, Y.-H. Hu, D.-S. Park, J. Villasenor, Data partitioning and reversible variable length codes for robust video communications, *Proceedings of the IEEE Data Compression Conference*, Snowbird, UT, pp. 460–469, March 2000.
2. G. Cote, S. Shirani, F. Kossentini, Optimal mode selection and synchronization for robust video communications over error-prone networks, *IEEE Journal on Selected Areas in Communications*, 18(6), 952–965, June 2000.
3. S. Wenger, G. D. Knorr, J. Ott, F. Kossentini, Error resilience support in H.263+, *IEEE Transactions on Circuits and Systems for Video Technology*, 8(7), 867–877, November 1998.
4. L. P. Kondi, F. Ishtiaq, A. K. Katsaggelos, Joint source-channel coding for motion-compensated DCT-based SNR scalable video, *IEEE Transactions on Image Processing*, 11(9), 1043–1052, September 2002.

5. H. M. Radha, M. van der Schaar, Y. Chen, The MPEG-4 fine-grained scalable video coding method for multimedia streaming over IP, *IEEE Transactions on Multimedia*, 3(1), 53–68, March 2001.

6. T. Schierl, T. Stockhammer, T. Wiegand, Mobile video transmission using scalable video coding, *IEEE Transactions on Circuits and Systems for Video Technology*, 17(9), 1204–1217, September 2007.

7. R. Puri, K. Ramchandran, Multiple description source coding through forward error correction codes, *Proceedings of the Asilomar Conference on Signals, Systems, and Computers*, Pacific Grove, CA, vol. 1, pp. 342–346, October 1999.

8. V. K. Goyal, Multiple description coding: Compression meets the network, *IEEE Signal Processing Magazine*, 18(5), 74–93, September 2001.

9. K. Stuhlmüller, N. Färber, M. Link, B. Girod, Analysis of video transmission over lossy channels, *IEEE Journal on Selected Areas in Communications*, 18(6), 1012–1032, June 2000.

10. L. D. Soares, F. Pereira, Error resilience and concealment performance for MPEG-4 frame-based video coding, *Signal Processing: Image Communication*, 14(6–8), 447–472, May 1999.

11. A. K. Katsaggelos, F. Ishtiaq, L. P. Kondi, M.-C. Hong, M. Banham, J. Brailean, Error resilience and concealment in video coding, *Proceedings of the European Signal Processing Conference*, Rhodes, Greece, pp. 221–228, September 1998.

12. Y. Wang, S. Wenger, J. Wen, A. Katsaggelos, Error resilient video coding techniques *IEEE Signal Processing Magazine*, 17(4), 61–82, July 2000.

13. F. Zhai, A. Katsaggelos, *Joint Source-Channel Video Transmission*, Morgan & Claypool Publishers, San Rafael, CA, 2007.

14. ISO/IEC 14496-10, Information Technology—Coding of Audio-Visual Objects—Part 10: Advanced Video Coding, 2005.

15. ISO/IEC 14496-2, Information Technology—Coding of Audio-Visual Objects—Part 2: Visual (2nd Edn.), 2001.

16. P. Haskell, D. Messerschmitt, Resynchronization of motion compensated video affected by ATM cell loss, *Proceedings of the IEEE International Conference on Acoustics, Speech and Signal Processing*, San Francisco, CA, vol. 3, pp. 545–548, March 1992.

17. G. Côté, F. Kossentini, Optimal intra coding of blocks for robust video communication over the Internet, *Signal Processing: Image Communication*, 15(1–2), 25–34, September 1999.

18. J. Y. Liao, J. D. Villasenor, Adaptive intra block update for robust transmission of H.263, *IEEE Transactions on Circuits and Systems for Video Technology*, 10(1), 30–35, February 2000.

19. P. Frossard, O. Verscheure, AMISP: A complete content-based MPEG-2 error-resilient scheme, *IEEE Transactions on Circuits and Systems for Video Technology*, 11(9), 989–998, September 2001.

20. Z. He, J. Cai, C. Chen, Joint source channel rate-distortion analysis for adaptive mode selection and rate control in wireless video coding, *IEEE Transactions on Circuits and Systems for Video Technology*, 12(6), 511–523, June 2002.

21. H. Shu, L. Chau, Intra/Inter macroblock mode decision for error-resilient transcoding, *IEEE Transactions on Multimedia*, 10(1), 97–104, January 2008.

22. H-J. Ma, F. Zhou, R.-X. Jiang, Y.-W. Chen, A network-aware error-resilient method using prioritized intra refresh for wireless video communications, *Journal of Zhejiang University - Science A*, 10(8), 1169–1176, August 2009.

23. P. Nunes, L. D. Soares, F. Pereira, Error resilient macroblock rate control for H.264/AVC video coding, *Proceedings of the IEEE International Conference on Image Processing*, San Diego, CA, pp. 2132–2135, October 2008.

24. Z. Li, F. Pan, K. Lim, G. Feng, X. Lin, S. Rahardaj, Adaptive basic unit layer rate control for JVT, *Doc. JVT-G012, 7th MPEG Meeting*, Pattaya, Thailand, March 2003.

25. ISO/MPEG & ITU-T, H.264/AVC Reference Software, Available: http://iphome.hhi.de/suehring/tml/download/

26. L. D. Soares, P. Nunes, F. Pereira, Efficient network-aware macroblock mode decision for error resilient H.264/AVC video coding, *Proceedings of the SPIE Conference on Applications of Digital Image Processing*, vol. 7073, San Diego, CA, pp. 1–12, August 2008.

27. P. Nunes, L. D. Soares, F. Pereira, Automatic and adaptive network-aware macroblock intra refresh for error-resilient H.264/AVC video coding, *Proceedings of the IEEE International Conference on Image Processing*, Cairo, Egypt, pp. 3073–3076, November 2009.

28. K.-P. Lim, G. Sullivan, T. Wiegand, Text description of joint model reference encoding methods and decoding concealment methods, *Doc. JVT-X101, ITU-T VCEG Meeting*, Geneva, Switzerland, June 2007.

29. T. Wiegand, H. Schwarz, A. Joch, F. Kossentini, G. Sullivan, Rate-constrained coder control and comparison of video coding standards, *IEEE Transactions on Circuits and Systems for Video Technology*, 13(7), 688–703, July 2003.

30. H. Schulzrinne, S. Casner, R. Frederick, V. Jacobson, RTP: A transport protocol for real-time applications, *Internet Engineering Task Force, RFC 1889*, January 1996.

31. R. C. Dorf, J. A. Svoboda, *Introduction to Electric Circuits*, 5th Edition, Wiley, New York, 2001.

32. K. Levenberg, A method for the solution of certain non-linear problems in least squares, *Quarterly of Applied Mathematics*, 2(2), 164–168, July 1944.

33. D. Marquardt, An algorithm for the least-squares estimation of nonlinear parameters, *SIAM Journal of Applied Mathematics*, 11(2), 431–441, June 1963.

34. S. Wenger, H.264/AVC over IP, *IEEE Transactions on Circuits and Systems for Video Technology*, 13(7), 645–656, July 2003.

2

Distributed Video Coding: Principles and Challenges

Jürgen Slowack and Rik Van de Walle

CONTENTS

2.1 Introduction

A video compression system consists of an encoder that converts uncompressed video sequences into a compact format suitable for transmission or storage, and a decoder that performs the opposite operations to facilitate video display.

Compression is typically achieved by exploiting similarities between frames (temporal direction), as well as similarities between pixels within the same frame (spatial direction). The conventional way is to exploit these similarities at the encoder. Using already-coded information, the encoder generates a prediction of the information still to be coded. Next, the difference between the information to be coded and the prediction is further processed and compressed through entropy coding.

The accuracy of the prediction determines the compression performance, in the sense that more accurate predictions will lead to smaller residuals and better compression. As a consequence, computationally complex algorithms have been developed to search for the best predictor. This has led to a complexity imbalance, in which the encoder is significantly more complex than the decoder.

A radically different approach to video coding—called distributed video coding (DVC)—has emerged during the past decade. In DVC, the prediction is generated at the decoder instead of at the encoder. As this prediction—called side information—typically contains errors, additional information is sent from the encoder to the decoder to allow correcting the side information. Generating the prediction signal at the decoder shifts the computational burden from the encoder to the decoder side. This facilitates applications in which encoding devices are relatively cheap, small, and/or power-friendly. Some examples of these applications include wireless sensor networks, wireless video surveillance, and videoconferencing using mobile devices [44].

Many publications covering DVC have appeared (including a book on distributed source coding [DSC] [16]). The objective of this chapter is therefore to provide a comprehensive overview of the basic principles behind DVC and illustrate these principles with examples from the current state-of-the-art. Based on this description, the main future challenges will be identified and discussed.

2.2 Theoretical Foundations

Before describing the different DVC building blocks in detail we start by highlighting some of the most important theoretical results. This includes a discussion on the Slepian–Wolf and Wyner–Ziv (WZ) theorems, which are generally regarded as providing a fundamental information–theoretical basis for DVC. It should be remarked that these results apply to DSC in general and that DVC is only a special case.

2.2.1 Lossless Distributed Source Coding (Slepian–Wolf)

David Slepian and Jack K. Wolf considered the configuration depicted in Figure 2.1, in which two sources X and Y generate correlated sequences of

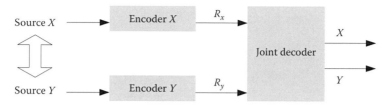

FIGURE 2.1
Slepian and Wolf consider the setup in which two correlated sources X and Y are coded independently, but decoded jointly.

information symbols [51]. Each of these sequences is compressed by a separate encoder, namely, one for X and one for Y. The encoder of each source is constrained to operate without knowledge of the other source, explaining the term DSC. The decoder, on the other hand, receives both coded streams as input and should be able to exploit the correlation between the sources X and Y for decoding the information symbols.

Surprisingly, Slepian and Wolf proved that the compression bound for this configuration is the same as in the case where the two encoders are allowed to communicate. More precisely, they proved that the rates R_X and R_Y of the coded streams satisfy the following set of equations:

$$R_X + R_Y \geq H(X,Y),$$

$$R_X \geq H(X \mid Y), \tag{2.1}$$

$$R_Y \geq H(Y \mid X),$$

where $H(.)$ denotes the entropy. These conditions can be represented graphically, as a so-called admissible or achievable rate region, as depicted in Figure 2.2.

While any point on the line $H(X,Y)$ is equivalent from a compression point of view, special attention goes to the corner points of the achievable rate region. For example, the point $(H(X \mid Y), H(Y))$ corresponds to the special case of source coding with side information available at the decoder, as depicted in Figure 2.3. This case is of particular interest in the context of current DVC solutions, where side information Y is generated at the decoder and used to decode X. According to the Slepian–Wolf theorem, the minimal rate required in this case is the conditional entropy $H(X \mid Y)$.

2.2.2 Lossy Compression with Receiver Side Information (Wyner–Ziv)

The work of Slepian and Wolf relates to lossless compression. These results were extended to lossy compression by Aaron D. Wyner and Jacob Ziv [65]. Although introducing quality loss seems undesirable at first thought, it is

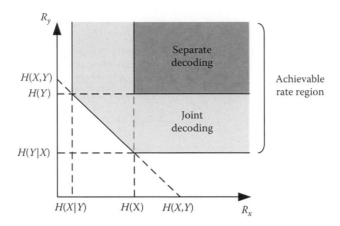

FIGURE 2.2
Graphical representation of the achievable rate region.

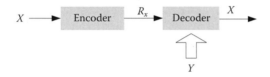

FIGURE 2.3
(Lossless) source coding with side information available at the decoder.

often necessary to allow some loss of quality at the output of the decoder in order to achieve even higher compression ratios (i.e., lower bit rates).

Denote the acceptable distortion between the original signal X and the decoded signal X' as $D = E[d(X, X')]$, where d is a specific distortion metric (such as the mean-squared error). Two cases are considered for compression with side information available at the decoder. In the first case, the side information Y is not available at the encoder. The rate of the compressed stream for this case is denoted $R_{X|Y}^{WZ}(D)$. In the second case, Y is made available to the encoder as well, resulting in a rate denoted $R_{X|Y}(D)$. With these notations, Wyner and Ziv proved that

$$R_{X|Y}^{WZ}(D) - R_{X|Y}(D) \geq 0 \qquad (2.2)$$

In other words, not having the side information available at the encoder results in a rate loss greater than or equal to zero, for a particular distortion D. Interestingly, the rate loss has been proved to be zero in the case of Gaussian memoryless sources and a mean-squared error (MSE) distortion metric.

The results of Wyner and Ziv were further extended by other research-ers, for example, proving that the equality also holds in case X is equal to the sum of arbitrarily distributed Y and independent Gaussian noise [46]. In addition, Zamir showed that the rate loss for sources with general statistics is less than 0.5 bits per sample when using the MSE as a distortion metric [68].

2.3 General Concept

The theorems of Slepian–Wolf and Wyner–Ziv apply to DSC, and therefore also to the specific case of DVC. Basically, the theorems indicate that a DVC system should be able to achieve the same compression performance as a conventional video compression system. However, the proofs do not provide insights on how to actually construct such a system. As a result, the first DVC systems have appeared in the scientific literature only about 30 years later.

The common approach in the design of a DVC system is to consider Y as being a corrupted version of X. This way, the proposed setup becomes highly similar to a channel-coding scenario. In the latter, a sequence of information symbols X could be sent across an error-prone communication channel, so that Y has been received instead of X. To enable successful recovery of X at the receiver's end, the sender could include additional error-correcting infor-mation calculated on X, such as turbo or low-density parity-check (LDPC) codes [33].

The difference between such a channel-coding scenario and the setup depicted in Figure 2.3 is that in our case Y is already available at the decoder. In other words, the encoder should only send the error-correcting informa-tion to allow recovery of X (or X' in the lossy case). Since Y is already avail-able at the decoder instead of being communicated by the encoder, the errors in Y are said to be induced by virtual noise (also called correlation noise) on a virtual communication channel.

2.4 Use-Case Scenarios in the Context of Wireless Networks

By generating Y itself at the decoder side as a prediction of the original X at the encoder, the complexity balance between the encoder and the decoder becomes totally different from a conventional video compression system such as H.264/AVC [64]. While conventional systems feature an encoder that is significantly more complex than the decoder, in DVC the complexity bal-ance is completely the opposite.

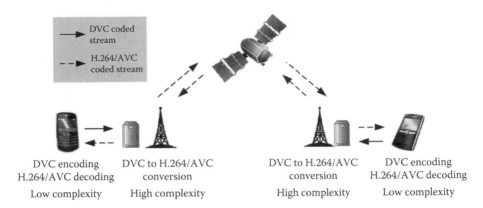

FIGURE 2.4
In the context of videoconferencing using mobile devices, DVC can be used in combination with conventional video coding techniques (such as H.264/AVC), which allows to assign computationally less complex steps to mobile devices, while performing computationally complex operations in the network.

This complexity distribution offers a significant advantage in the context of wireless video applications as they often require cheap, small, and/or power-friendly encoders. For example, DVC techniques can be used to realize compression in the case of videoconferencing using mobile devices [44]. In such a scenario, as illustrated with an example in Figure 2.4, the computationally less demanding operations such as DVC encoding and H.264/AVC decoding can be executed at the mobile devices. This lowers the devices' energy consumption and computational requirements, which is a significant advantage in practice. On the other hand, more complex operations such as converting the DVC stream to an H.264/AVC coded stream [42] can be performed by network nodes, with less strict constraints on computational complexity and energy consumption.

A second important application scenario for DVC includes wireless visual sensor networks [31]. These networks typically consist of a large set of sensors monitoring the environment to enable surveillance or environmental tracking, delivering their information over wireless channels to a control unit for further processing. Since sensor nodes are typically battery-powered, the reduced energy consumption of the encoding step in the case of DVC is a great advantage.

Another example is wireless capsule endoscopy [15]. In this scenario, a capsule-sized camera is given to a patient and swallowed in order to obtain visual information about possible anomalies in the patient's gastrointestinal tract. The use of a wireless device, in the form of a capsule, not only offers advantages for the comfort of the patient, but it also enables to examine parts of the small intestine, which cannot be examined using conventional wired endoscopy. To reduce the size of the endoscopic capsule only relatively simple video processing techniques can be incorporated, making DVC an ideal candidate.

Other scenarios and advantages of using DVC can be found further in this chapter, after we provide details about the different building blocks of a typical DVC system.

2.5 DVC Architectures and Components

Two DVC architectures can be considered the pioneers in the field, namely, the PRISM system developed at Berkeley [48] and the system developed at Stanford [1]. Since then, many researchers have proposed extensions and improvements. Most of these extensions were originally based on the Stanford architecture, and therefore we will first introduce this architecture briefly. Based on this discussion we will explore the details of the different building blocks in separate subsections, providing references to the literature to illustrate different approaches.

Figure 2.5 depicts the DVC system initially proposed by Aaron et al. [2]. Some labels are chosen slightly more general (e.g., "transformation" instead of "DCT") for the sake of this overview chapter. First, we will explain the operation of the encoder, followed by a description of the decoder.

At the encoder, the video sequence is first partitioned into key frames and WZ frames. Both types of frames will be coded using different coding modes. The key frames are coded using conventional intracoding techniques (such as H.263+ or H.264/AVC intracoding). The WZ frames are partitioned into nonoverlapping blocks of a certain size (e.g., 4×4), and each of these blocks is transformed using the discrete cosine transformation (DCT). Transformation coefficients at corresponding locations within one WZ frame are grouped together, into the so-called coefficient bands. For example, one

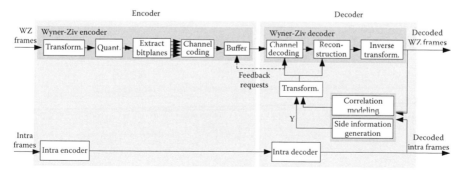

FIGURE 2.5
Typical architecture of a transform-domain distributed video codec.

coefficient band will contain all first coefficients of each block. Next, the coefficients in one coefficient band are quantized using a scalar quantizer. From the quantized coefficients in one coefficient band, bits at corresponding positions are extracted to form one bitplane. For example, all most significant bits of the third coefficient band will be grouped to form one bitplane. Next, each of the bitplanes is coded by a channel coder. The resulting channel codes are temporarily stored in a buffer and sent to the decoder in portions (through a process called puncturing) upon the decoder's request.

At the decoder, key frames are decoded independently from other frames, using conventional intra decoding techniques. For each WZ frame, the decoder generates side information using already decoded intra and/or WZ frames as references. The side information is transformed, and its correlation with the original is estimated. This allows extracting reliabilities needed by the channel decoder. The latter requests bits from the encoder's buffer until reliable decoding of the quantized transformation coefficients is achieved. After decoding, given the decoded quantized coefficients and the side information, the decoder estimates the value of the original coefficients before quantization. This process is called reconstruction. After this step, the decoded WZ frame is obtained by applying the inverse transformation. The decoded frame can then be used as a reference frame for generating side information for WZ frames to be decoded in the future.

Now that we have a basic understanding of how a particular DVC system works, in the following subsections we will discuss each of the components in more detail, providing references to the literature to illustrate alternative solutions. We will discuss the components in a slightly different order, as it will allow motivating the goal and logic behind these elements more clearly.

2.5.1 Side Information Generation

The side information plays a crucial role in the performance of any DVC system. This is clearly demonstrated by the Slepian–Wolf theorem, which states that in the lossless case the minimum rate is given by the conditional entropy $H(X|Y)$. In other words, the higher the correlation between X and Y, the lower the rate can be.

To construct side information Y showing strong resemblance with the original X at the encoder, the decoder should exploit as much information as possible. This could involve analyzing frames that have already been decoded, analyzing frames that are partially decoded, and/or exploiting any additional information provided by the encoder such as hash and/or low-resolution information.

The available literature on side information generation is quite large. We will therefore cluster these algorithms according to the techniques used and/or information exploited. For each of the clusters, we will identify the possible advantages and disadvantages.

2.5.1.1 Frame-Level Interpolation Strategies

Most systems in the literature use already decoded frames to generate one side information frame through interpolation. This strategy requires changing the decoding order so that at least one past and one future frame (in display order) are available at the time of generating the side information. For example, the frame sequence $I_1 - W_2 - W_3 - W_4 - I_5$, with key frames denoted I and WZ frames denoted W, can be decoded in the following order: $I_1 - I_5 - W_3 - W_2 - W_4 - \ldots$ In this example, the side information for W_2 can be generated using one decoded frame from the past (in display order), that is, I_1, and two from the future, namely W_3 and I_5.

Most approaches use only one past and one future frame (typically the closest ones) for the generation of the side information. The simplest approach is to calculate the pixel-by-pixel average of both reference frames [1]. However, in most cases this will not result in a very accurate prediction. Better performance can be achieved by tracking the motion between the reference frames and creating the side information by interpolation along this motion path. For example, in a typical block-based approach the decoder partitions the future reference frame into nonoverlapping blocks of a certain size, and for each of these blocks it determines the best matching block in the past reference frame. Further, assuming that the motion is uniform between the reference frames, the decoder performs linear interpolation of the motion to create the side information [2].

The obtained motion vectors are often noisy when compared to the true motion, and additional filtering and refinement steps have been proposed by several researchers to improve the result. For example, the well-known codec called DISCOVER includes sub-pixel motion vector refinement and spatial smoothing through vector median filtering [5]. Figure 2.6 provides an example of the side information generated using these techniques, showing inaccuracies in the side information, particularly for moving regions, as illustrated by the residual. Alternatively, instead of using a block-based approach, other solutions such as mesh-based approaches can be considered [30].

In the previous techniques, motion interpolation is performed by assuming that the motion between the reference frames is approximately uniform. This may not be true in all cases. Higher accuracy can be obtained by analyzing the motion between more than two reference frames and performing interpolation in a nonlinear fashion [4,45].

Using interpolation techniques for generating side information has two disadvantages. The first problem is that coding frames in a different order than the display order introduces delay. This may cause problems for scenarios in which delay is of critical importance (e.g., videoconferencing). A second problem is that it requires inserting key frames, which have relatively poor compression performance.

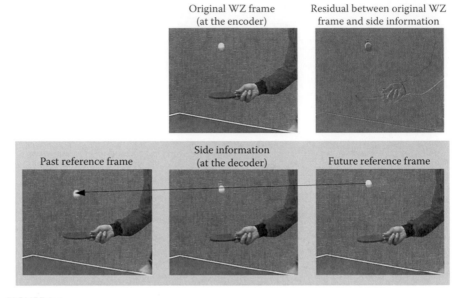

FIGURE 2.6
Example of side information generated using block-based techniques.

2.5.1.2 Frame-Level Extrapolation Strategies

In an extrapolation strategy, side information is generated using reference frames from the past only. As a result, the (de)coding order and display order can remain the same, solving the problem of increased delay as in the case of interpolation.

Some of the first methods on side information generation through extrapolation in DVC make use of block-based motion search between the two past reference frames. The obtained motion field is extrapolated by reusing the obtained motion vectors for the current frame [32]. Further improvements on these techniques include using more advanced search strategies [11] as well as using advanced filtering and extrapolation steps [60].

While there are definitely practical advantages for extrapolation compared to interpolation, the downside of extrapolation is that the compression performance of the WZ frames is typically lower than that in the case of interpolation. The main reason is that using information about the past alone is in general less accurate than using information about both the past and the future. This is intuitive, since, for example, new objects entering the scene could be present in future frames but not in past frames.

Because of the performance loss of extrapolation compared to interpolation, techniques based on interpolation are far more common in the literature. The techniques described in the remainder of this chapter will therefore apply to interpolation unless explicitly stated otherwise.

2.5.1.3 Encoder-Aided Techniques

Estimating the true motion between the reference frames and performing accurate interpolation or extrapolation of this motion has been shown to be a difficult task. Specifically for sequences with complex motion characteristics (e.g., involving rapid motion, deformation, occlusion) the side information can often be of poor quality, reducing significantly the compression performance.

To aid the decoder in generating better side information, techniques have been developed in which the encoder sends additional information to the decoder. This information can be a hash containing a subset of the frequency information of the frame for which side information needs to be generated [3,8], or a low-resolution version coded using a regular coder [15,36]. The additional information sent by the encoder can be used by the decoder to make a better decision about which motion vectors correspond to the true motion and/or how to improve the interpolation or the extrapolation process. This leads to a more accurate side information.

Although the additional information provided by the encoder can be of great benefit, there are two important issues that need to be dealt with. First of all, the additional complexity at the encoder for calculating the hash should remain limited, given the DVC use-case scenario of simple encoders and complex decoders. Second, the hash introduces additional rate overhead, which must be compensated for by improved side information accuracy. The problem is that the optimal amount of hash information is difficult to determine, since the encoder does not know the level of improvement a particular hash could bring to the side information at the decoder. The solution provided by most existing techniques is therefore to determine a good hash size based on training on several video sequences [8].

2.5.1.4 Partitioning and Iterative Refinement

So far, the decoder only exploited past and/or future reference frames for generating the side information, possibly using additional hash information provided by the encoder. There is still additional information that can be exploited at the decoder to improve the accuracy of the side information even further. That is, due to the fact that the WZ frames are typically partitioned and coded as different segments, each already decoded segment can be used to improve the side information for the remaining segments to be decoded. A schematic illustration of this concept is provided in Figure 2.7.

For example, merely all DVC systems apply coefficient band and bitplane-level partitioning as in the Stanford architecture. Already decoded coefficient bands can provide an indication of whether the current candidate for a particular side information block is reliable or not. When considered unreliable, an update of the side information can be performed by replacing the block with a more suitable candidate. This improves the compression performance

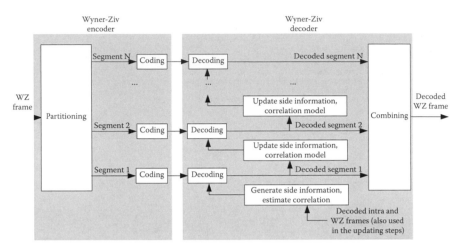

FIGURE 2.7
WZ frames are typically split up into different segments (e.g., coefficient bands or bitplanes) that are coded separately. (At the decoder, knowledge about already decoded segments can be used to improve the decoding of the remaining segments.)

of the remaining coefficient bands to be decoded [37]. Alternatives to this strategy have been proposed by other researchers as well [25,67]. Also, the data partitioning strategy can be modified to facilitate side information refinement. For example, different spatial resolutions of the same frame can be created, which allows generating side information for high-resolution layers using the already decoded lower-resolution layers [20]. An analytical study on using lower-resolution information for refining the motion estimation process has been provided by Liu et al. [35].

Incrementally refining the side information with each decoded segment has been shown to greatly improve the compression performance. There are not many disadvantages to this approach, apart from the fact that the complexity of the decoder is increased (but this is not violating the DVC use-case of simple encoders and complex decoders). Perhaps, a disadvantage is that it forces decoding of the segments to be sequential, which could increase delay and make parallelization of decoder operations more difficult.

2.5.2 Correlation Noise Estimation

Although the aforementioned techniques may allow generating an accurate prediction of the original available at the encoder, inevitably, often there are still errors that need to be corrected to obtain a sufficient level of quality. To correct errors in the side information, highly efficient channel-coding techniques such as LDPC or turbo coding can be used. However, these techniques require decoder-side knowledge about the error probabilities of the (virtual) communication channel. Failure to provide such information accurately has been shown to result in severe performance degradation of the channel code [55].

The error probabilities can be obtained by modeling the correlation between the original at the encoder and the side information at the decoder. Obviously, this correlation cannot be measured since X and Y are never simultaneously available at the encoder or decoder. Hence, the correlation needs to be estimated by the decoder. This estimation process is often referred to as correlation noise estimation or virtual noise estimation.

The correlation between the original X and the side information Y is commonly modeled through a conditional density function. Apart from a limited number of contributions proposing different models [36,38], most researchers use a Laplacian conditional density function to model the correlation, that is,

$$f_{X|Y}(x \mid y) = \frac{\alpha}{2} e^{-\alpha|x-y|}, \tag{2.3}$$

where the relation between α and the distribution's variance σ^2 is given by $\sigma^2 = 2/\alpha^2$.

At the decoder, the realization y of the side information is available, meaning that estimating the correlation noise only involves estimating the distribution's scale parameter α.

In the early DVC systems, the decoder often obtained a suitable value for α through an offline training stage [2], or even by granting the decoder access to the original [57]. However, these approaches are impractical or lack flexibility. Current techniques try to model the fact that the reliability of the side information varies across the frame, in the sense that some spatial regions may be well predicted while the accuracy of predicting other regions could be rather poor. For example, in Figure 2.8, the accuracy of predicting the person's face is rather poor while the quality of predicting the background is much higher. These differences can be estimated by analyzing the process of side information generation. In the common block-based interpolation approach, a particular spatial block in the side information is generated using one past and one future reference block. For regions with complex motion (involving deformation, occlusion, etc.), the difference between past and future reference blocks is typically large. On the other hand, small differences between past and future reference blocks often correspond to regions that are well predicted. In other words, there is a relation between the motion-compensated residual R (i.e., the difference between past and future reference blocks) and the virtual noise N. This relation between N and R—illustrated with an example in Figure 2.8—can be exploited by the decoder to estimate a suitable value for α based on R [13].

Several variants to this approach have been proposed, for example, estimating the virtual noise in the pixel-domain first before transforming it to

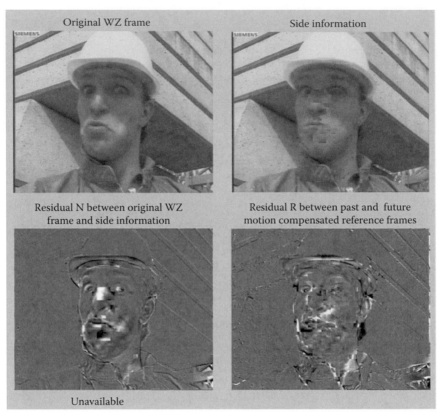

FIGURE 2.8

(See color insert.) The motion-compensated residual (bottom-right) resembles the true residual between the original and the side information (bottom-left). (Since the true residual is unavailable during (de)coding, the motion-compensated residual can be used to estimate α.)

the DCT domain [21,63]. Also, the relation between the virtual noise and the motion-compensated residual that appears to be influenced by quantization should be taken into account [53,61]. Also, similar to the refinement techniques in the context of side information generation, exploiting information about the already decoded segments (such as coefficient bands) can improve the accuracy of α for the remaining segments to be decoded [26].

Instead of exploiting knowledge about the motion-compensated residual, a more generic approach can be used in which the α parameter is progressively learned as the WZ frames are decoded. Such a technique has the advantage of being independent from the way side information is generated [20].

The correlation model allows extracting probabilities for the quantized symbols and/or bitplanes. This step is explained in more detail using the example in Figure 2.9. In this example, each quantization bin has been assigned a unique three-bit binary string $b_0b_1b_2$. Using the bin labels and the

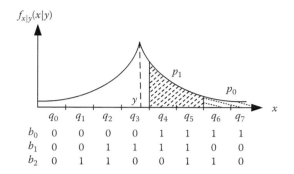

FIGURE 2.9
Reliabilities for the channel decoder can be obtained by integration over the estimated conditional distribution.

correlation model it is then straightforward to calculate the bit probabilities through integration. For example, in Figure 2.9, the probability of b_0 being 1 is given by the sum of the indicated regions p_0 and p_1.

Bitplanes are often decoded one after the other. By doing so, information from the already decoded bitplanes can be used to refine the probabilities for the next bitplane. For example, in Figure 2.9, if b_0 has been successfully decoded as being equal to 1, then the probability for b_1 being 1 is given by $P(b_1 = 1 | b_0 = 1, y) = p_1/(p_0 + p_1)$.

Although from a theoretical perspective the bin labeling strategy should not have an influence on the entropy, in practice there can be some differences, for example, with regard to encoder-side rate estimation [14]. For these reasons, Gray mapping is often used (as in Figure 2.9) since it has the advantage that neighboring bins only differ in one bit, which improves the correlation between the quantized original and side information.

A second remark is that, although the order in which the bitplanes are decoded does not affect the entropy, in practice there seems to be some impact on compression performance [59]. This is because the channel decoder often does not guarantee that the decoded bitstream is correct. Instead, it tries to guarantee a particular level of fidelity.

2.5.3 Channel Coding

The symbol/bit reliabilities obtained in the previous section are required to initialize the decoding process of high performing channel-coding solutions such as punctured turbo and LDPC codes [33]. Turbo codes have been used particularly in the early DVC systems, while currently more and more systems are using LDPC codes—mainly the so-called LDPC Accumulate (LDPCA) codes [58]—because of their superior performance [43].

One of the downsides of using LDPC and turbo codes is that the computational complexity of the channel decoding process is quite high. This can be dealt with by simplifying the decoding strategy or by exploiting parallelization [56]. In addition to these solutions, techniques can be incorporated that avoid executing more decoding iterations than strictly necessary, by stopping the decoding process when the probability of improving the result through additional iterations is very low [9,62].

Another downside is that the compression performance of these codes typically decreases as the length of the words used as input decreases. In the case of short word lengths simpler techniques may be considered, for example, using BCH block codes as in the PRISM system [47].

A new and interesting alternative to conventional channel codes is to use distributed arithmetic codes [23]. The main idea here is to adjust the common arithmetic coding process so that the subintervals partially overlap. To discriminate the correct interval the decoder uses the side information. This new technique promises better compression at small block lengths, better compression at similar complexity, and on-the-fly estimation of probabilities based on past symbols.

After channel decoding has terminated, the decoder needs to analyze whether the output of the channel decoder is reliable or not. Initial approaches adopted offline strategies in which the decoder compares the output with the original [12]. If the average bit error is larger than a threshold (typically 10^{-3}), channel decoding was considered successfully terminated. Otherwise, decoding had failed. In more practical solutions, the encoder provides a cyclic redundancy check (CRC) of the original to aid the decoder in discriminating between successful and unsuccessful decoding [5,47]. Alternatively, the decoder can estimate whether decoding is successful or not by analyzing the probability updating process during channel decoding [62].

Obviously, in the ideal case the encoder will have provided enough information to enable successful decoding, but this is difficult to guarantee, as explained in the following section.

2.5.4 Determining the WZ Rate

The amount of information (often called WZ bits) that needs to be provided by the channel encoder to enable successful decoding depends on several factors. The most important factor is the accuracy of the side information and the correlation model. If the side information has poor quality and/or if the correlation between the side information and the original is inaccurately modeled, the channel decoder will need more WZ bits from the encoder to enable successful decoding. On the other hand, fewer bits are needed if the side information and correlation model are accurate.

However, how can the encoder know the accuracy of the side information and the correlation model at the decoder? After all, sending more than the necessary amount of information results in a waste of bits (and hence, reduced compression performance), while sending not enough information will result in unsuccessful decoding and decreased output quality.

In the literature, there are two approaches for solving this problem. The most popular and simplest approach is to use a feedback channel from the decoder to the encoder, as in the pioneering Stanford architecture [2,5]. Basically, it is up to the decoder to inform the encoder whether decoding is successful or not. When decoding is unsuccessful, the encoder sends more information so that the decoding process can be restarted.

The advantage of having a feedback channel is that the decoder can avoid WZ rate overhead by approximating the optimal rate from below. However, this strategy limits the usefulness of the system since a feedback channel cannot be supported in storage applications. Also, performing multiple request-and-decoding passes introduces possibly large delays, which may not be tolerated in many streaming applications.

To overcome these issues, several researchers have proposed solutions in which a feedback channel is not present. Obviously, if the encoder does not receive any information about the outcome of the channel decoding process, then it needs to determine the WZ rate by itself. An essential problem is that—to support the DVC use-case of a simple encoder but complex decoder—the additional complexity added to the encoder to make a decision about the number of WZ bits to be sent should remain very limited. Therefore, in most feedback-free DVC systems, the encoder uses very simple techniques to generate a coarse estimation of the side information. Based on this approximation, the required number of WZ bits is estimated. For example, in PRISM, the WZ rate is determined by comparing the current pixel block with the colocated block in the previous frame [48]. In more recent approaches, the WZ rate is estimated based on the average of the two adjacent key frames [39,50], or by using fast motion estimation techniques [14,22].

Feedback free DVC systems are definitely more practical, but on the other hand their compression performance is typically lower due to occasional underestimation or overestimation of the WZ rate. Also, although limited in most cases, removing the feedback channel comes at the price of increased encoding complexity.

Instead of removing the feedback channel completely, hybrid techniques have been proposed in which the encoder performs an initial estimation of the WZ rate. After sending this amount, the decoder is allowed to request for more information using the conventional feedback strategy [18,28]. While these solutions do not eliminate the feedback channel, they do decrease its usage.

2.5.5 Transformation and Quantization

The transformation and quantization steps applied in most systems are highly similar to the techniques used in conventional video compression. In most cases a DCT is applied, although occasionally wavelet-based systems have been proposed as well [24]. For the quantization step, uniform scalar quantization is applied using standard quantization matrices [2]. Only the backward quantization step, called reconstruction, is different from the approaches in most conventional video compression systems. This step is explained now.

The process of reconstruction tries to estimate the values at the input of the quantizer (at the encoder), given the decoded quantized value at the decoder. To improve the accuracy of this estimation, the decoder can exploit various types of information.

The simplest approach is to exploit knowledge about the side information. Suppose the decoded quantized value q relates to a range from A to B, then the decoder could simply select the value within this range that is closest to the side information y [6]:

$$x' = \begin{cases} A, & y < A \\ y, & y \\ B, & y \geq B \end{cases} \tag{2.4}$$

The decoder could also exploit information provided by the correlation model to improve the accuracy of the reconstruction process. For example, calculating the centroid over the quantization bin minimizes the distortion in the mean-squared error sense [29]:

$$x' = E[x \mid x \in q, y] = \frac{\displaystyle\int_A^B x f_{X|y}(x)dx}{\displaystyle\int_A^B f_{X|y}(x)dx} \tag{2.5}$$

Slightly modified solutions account for the fact that bitplanes may be unsuccessfully decoded [52]. Even more advanced techniques improve the reconstruction process, for example, through an additional constraint on the spatial smoothness of the decoded frame [69].

2.5.6 Mode Decision

The previous sections illustrated that using a WZ approach requires estimating various parameters. First of all, the decoder needs to generate an approximation of the original available at the encoder. Next it needs to model the correlation noise between the side information and the original. There is an additional difficulty in feedback-free DVC systems, since the encoder needs to estimate the number of WZ bits to be sent to the decoder. Inaccuracies in each of these steps will penalize the compression performance of the system, even up to the point where using a WZ approach becomes less interesting than other low-complexity alternatives such as intracoding and skip. To deal with such cases of inefficient WZ coding, researchers have incorporated different coding modes. These modes have been introduced on different hierarchy levels, for example, at the frame, block, coefficient band, and bitplane levels.

At frame level, most systems discriminate between key frames and WZ frames. The main reason for including key frames is that it allows generating side information through interpolation. However, the compression performance of the key frames is quite low, due to the fact that these frames are typically intracoded, not exploiting temporal correlations. To lower the overhead introduced by the key frames, one could increase the period of these frames (e.g., to one out of four frames instead of one out of two). However, this will increase the distance between the reference frames used for generating side information, which negatively impacts the compression performance of the WZ frames. As a solution, instead of using a fixed intra period, more intelligent techniques will search for the right balance between key frames and WZ frames [7,66]. Additionally, instead of intracoding the key frames more advanced techniques can be developed so that temporal correlations are exploited for the key frames as well [19].

Block-level mode decision accounts for spatial differences in the quality of the side information. Often, due to inaccuracies in the side information and correlation model, spatial regions with complex motion characteristics can be better coded through intracoding instead of WZ coding. On the other hand, for regions with simple motion characteristics (such as static background), the side information could already be a sufficiently close approximation, so that no additional WZ rate is required for these regions. Such considerations have motivated researchers to introduce intra and skip modes at the block level. For example, in PRISM, the decision between skip, WZ, and intracoding is made based on a comparison between the current pixel block and the colocated block in the previous frame [48]. In other techniques, error-correcting information is still calculated on the entire frame, but instead, skip and intracoding are used to improve the quality of the side information [10] and/or improve the efficiency of channel decoding [41].

In addition to the fact that WZ coding is sometimes less efficient for coding particular spatial regions, WZ coding can be less efficient for coding particular frequency bands and/or bitplanes. Often, low frequencies can be rather well predicted while the quality of predicting higher frequencies is poor. Therefore, intracoding these higher frequencies could be a better solution [48]. Instead of making this decision at the encoder side, in feedback-based DVC systems, the decoder could be responsible for performing mode decision, sending the modes back to the encoder through the feedback channel [52].

2.6　Evaluation of DVC Compression Performance

In this section we will evaluate the compression performance of two DVC systems by comparing them with the state of the art in conventional video compression, that is, H.264/AVC [64].

A first DVC system under evaluation is the DISCOVER codec [5], due to its importance in the literature as well as the fact that binaries for this system are available online at www.discoverdvc.org. The functional diagram of the system is similar to the one provided in Figure 2.5, featuring frame intracoding using H.264/AVC, channel coding using LDPC codes, and side information generation through frame-level interpolation. No additional modes are incorporated in this system.

The second DVC system is the codec developed by the authors of this chapter [52]. The general operation of this codec is similar to DISCOVER, but it includes bitplane-level coding modes (skip, WZ, and intracoding) and improved techniques for correlation noise modeling.

Two configurations of the H.264/AVC codec (JM 14.1, extended profile, one slice per picture) are included in the comparison: one applying intracoding only and the second one with all features enabled.

Tests for different sequences were conducted. All have CIF resolution (i.e., 352×288 pixels) and a frame rate of 30 frames per second. One out of four frames is coded using intracoding. Only the luma component is coded due to the restrictions imposed by the DISCOVER software.

The results provided in Figure 2.10 indicate that there is a significant performance gap between H.264/AVC and the two DVC systems under evaluation. The main reason for this gap is that encoder-side motion estimation and mode decision as performed in H.264/AVC are much more effective, since the original can be used for these computations. In DVC, motion estimation and mode decision are performed without using the original frame, and so confidence measures need to be used instead. As a result of this essential difference, the highly efficient block partitioning strategies and coding modes incorporated in H.264/AVC have not found their match yet in current DVC systems.

When comparing the two DVC systems, we conclude that using additional coding modes such as bitplane intracoding has a considerably positive effect on compression performance, particularly for sequences with complex motion characteristics such as *"Bus"* (Figure 2.10). Due to these improvements, our configuration is able to outperform H.264/AVC intracoding for both sequences, in contrast to DISCOVER.

2.7 Other DVC Architectures and Scenarios

So far we have focused on using DVC for the compression of one video sequence. However, apart from this scenario, DVC techniques can be used in other contexts as well.

DVC can be used to jointly code video sequences captured from a single scene but from different viewpoints (e.g., in the context of wireless visual

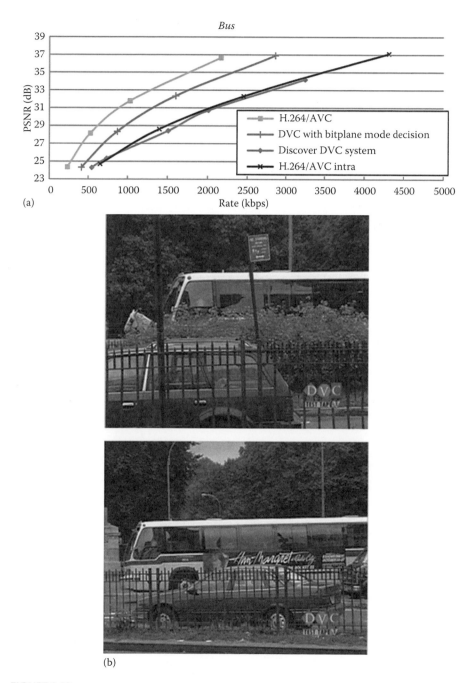

FIGURE 2.10
PSNR versus bit rate for the high-motion *"Bus"* sequence.

(continued)

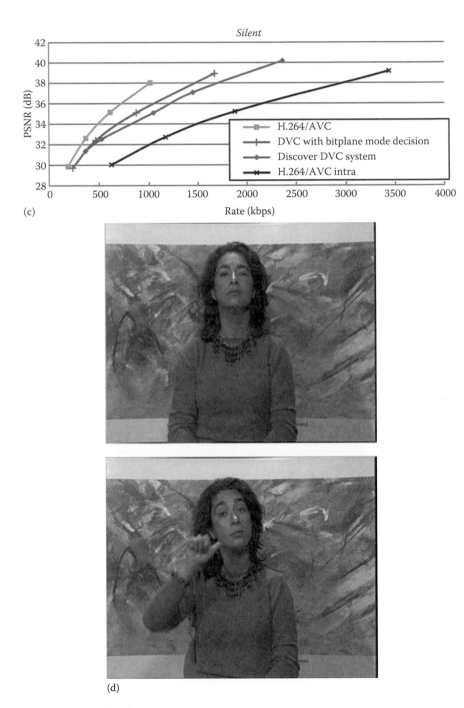

(c)

(d)

FIGURE 2.10 (continued)
PSNR versus bit rate for the relatively low-motion sequence called *"Silent."*

sensor networks). To generate side information for a particular frame, the decoder can exploit correlations between frames part of the same view as well as correlations between frames of different views [24]. The advantage of using a DVC approach for this scenario compared to adopting a conventional approach based on the existing multiview video coding standard (MVC) is that there is no communication necessary between the encoders.

Using channel codes for video compression as in DVC has the advantage that the system is more robust to transmission errors compared to a conventional system applying source coding and separate channel coding (such as H.264/AVC). Therefore, several contributions in the literature focus on applying DVC techniques for improved error resilience in various video coding scenarios [49,70]. This is an additional argument for using DVC in wireless environments, where transmission errors occur more frequently.

A different scenario relates to flexible distribution of complexity between the encoder and the decoder. As explained earlier, conventional video compression features a complex encoder but simple decoder, while the complexity distribution in DVC is the reverse. By combining techniques from both approaches, the distribution of complexity can be made dynamic. For example, based on the available computational resources on both devices, complexity can be shifted from the encoder to the decoder and vice versa. This can be particularly interesting in situations where available computational complexity is nonstatic, for example, in the case of multitasking or battery-constrained devices [54].

DVC techniques can also be used to improve the compression performance of conventional video coding techniques such as H.264/AVC video coding. If the encoder generates the side information as well, it can exploit this knowledge to improve compression, for example, by including the side information as a mutual encoder–decoder reference frame for motion estimation [40].

Finally, DVC techniques can be used in the context of authentication and biometrics [17,34]. In the latter, the side information is considered as a noisy version of the original, and authentication is given in case applying the error-correcting information leads to recovery of the original.

2.8 Future Challenges and Research Directions

Despite significant research efforts in the context of DVC there are still important challenges requiring further research. One of these challenges is how to improve the compression performance further. A first approach could be to improve the existing techniques for side information generation, correlation modeling, mode decision, etc. For example, since iterative refinement approaches are shown to provide gains without considerable drawbacks, extending these approaches could be an interesting research direction.

Another interesting observation is that the way the data are partitioned (e.g., in coefficient bands and bitplanes) has a strong impact on the techniques that can be used for side information generation, correlation modeling, etc. Therefore, careful design of the partitioning algorithm is of great importance. In essence, the questions that need to be answered are: what data need to be sent to the decoder at a given point in time and how should they be coded? What is the most efficient strategy? Sending a hash of the original frame, sending intracoded high frequencies, or WZ-coded low frequencies? Answering these questions is not trivial, but designing an efficient partitioning strategy may be the key to improve compression performance.

A different category of challenges includes how to make DVC systems more practical. Currently, most systems employ a feedback channel from the decoder to the encoder. Obviously, issuing a high number of requests per WZ frame results in potentially large delays, which could not be acceptable for the application at hand. The alternative, that is, feedback-free DVC architectures, feature a delicate trade-off between additional encoder-side computational complexity and compression performance. Given that side information generation is likely to become more and more complex in the future, an important challenge in the context of feedback-free DVC architectures is therefore to keep on designing suitable encoder-side rate estimation algorithms with low complexity.

A second issue for practical deployment of DVC systems is the absence of a rate control scheme. Most conventional video compression systems incorporate rate control mechanisms so that the bit rate at the output of the encoder can be controlled, for example, with respect to a given constraint on network bandwidth. In DVC, such algorithms are very rare (e.g. [27]), and further research is required to incorporate rate control techniques in other DVC architectures as well.

Finally, given the advantages of using DVC techniques in other contexts than the monoview scenario, new scenarios can be proposed and explored.

References

1. A. Aaron and B. Girod. Compression with side information using turbo codes. In *Proceedings of the IEEE Data Compression Conference (DCC)*, Snowbird, UT, pp. 252–261, April 2002.
2. A. Aaron, S. Rane, E. Setton, and B. Girod. Transform-domain Wyner-Ziv codec for video. In *Proceedings of the SPIE Visual Communications and Image Processing*, San Jose, CA, vol. 5308, pp. 520–528, January 2004.
3. A. Aaron, R. Zhang, and B. Girod. Wyner-Ziv video coding with hash-based motion compensation at the receiver. In *Proceedings of the IEEE International Conference on Image Processing (ICIP)*, Singapore, pp. 3097–3100, October 2004.

4. M. Akinola, L. Dooley, and P. Wong. Wyner-ziv side information generation using a higher order piecewise trajectory temporal interpolation algorithm. In *Proceedings of the International Conference on Graphic and Image Processing*, Manila, Philippines, December 2010.

5. X. Artigas, J. Ascenso, M. Dalai, S. Klomp, D. Kubasov, and M. Ouaret. The DISCOVER codec: Architecture, techniques and evaluation. In *Proceedings of the Picture Coding Symposium (PCS)*, Lisbon, Portugal, November 2007.

6. J. Ascenso, C. Brites, and F. Pereira. Motion compensated refinement for low complexity pixel based Distributed Video Coding. In *IEEE International Conference on Advanced Video and Signal Based Surveillance*, Como, Italy, pp. 593–598, September 2005.

7. J. Ascenso, C. Brites, and F. Pereira. Content adaptive Wyner-Ziv video coding driven by motion activity. In *Proceedings of the IEEE International Conference on Image Processing (ICIP)*, Atlanta, GA, pp. 605–608, October 2006.

8. J. Ascenso and F. Pereira. Adaptive hash-based side information exploitation for efficient Wyner-Ziv coding. In *Proceedings of the IEEE International Conference on Image Processing (ICIP)*, San Antonio, TX, pp. III–29–III–32, September 2007.

9. J. Ascenso and F. Pereira. Complexity efficient stopping criterion for LDPC based distributed video coding. In *Proceedings of the Mobimedia*, Kingston, UK, pp. 28:1–28:7, September 2009.

10. J. Ascenso and F. Pereira. Low complexity intra mode selection for efficient distributed video coding. In *Proceedings of the International Conference on Multimedia and Expo (ICME)*, New York, NY, pp. 101–104, July 2009.

11. S. Borchert, R. P. Westerlaken, R. Klein Gunnewiek, and R. L. Lagendijk. On extrapolating side information in distributed video coding. In *Proceedings of Picture Coding Symposium (PCS)*, Lisbon, Portugal, November 2007.

12. C. Brites, J. Ascenso, and F. Pereira. Feedback channel in pixel domain Wyner-Ziv video coding: Myths and realities. In *14th European Signal Processing Conference*, Florence, Italy, September 2006.

13. C. Brites and F. Pereira. Correlation noise modeling for efficient pixel and transform domain Wyner-Ziv video coding. *IEEE Transactions on Circuits and Systems for Video Technology*, 18:1177–1190, September 2008.

14. C. Brites and F. Pereira. An efficient encoder rate control solution for transform domain Wyner-Ziv video coding. *IEEE Transactions on Circuits and Systems for Video Technology*, 21:1278–1292, September 2011.

15. N. Deligiannis, F. Verbist, J. Barbarien, J. Slowack, R. Van de Walle, P. Schelkens, and A. Munteanu. Distributed coding of endoscopic video. In *Proceedings IEEE International Conference on Image Processing (ICIP)*, Brussels, Belgium, pp. 1853–1856, September 2011.

16. P. L. Dragotti and M. Gastpar. *Distributed Source Coding: Theory, Algorithms, and Applications*. Academic Press, Waltham, MA, 2009.

17. S. C. Draper, A. Khisti, E. Martinian, A. Vetro, and J. S. Yedidia. Using distributed source coding to secure fingerprint biometrics. In *Proceedings of the IEEE International Conference on Acoustics, Speech and Signal Processing*, Honolulu, Hawai, vol. 2, pp. II–129–II–132, April 2007.

18. A. Elamin, V. Jeoti, and S. Belhouari. Feed back channel usage reduction in distributed video coding. In *Proceedings of the International Conference on Intelligent and Advanced Systems (ICIAS)*, Kuala Lumpur, Malaysia, pp. 1–4, June 2010.

19. G. R. Esmaili and P. C. Cosman. Wyner-Ziv video coding with classified correlation noise estimation and key frame coding mode selection. *IEEE Transactions on Image Processing*, 20(9):2463–2474, September 2011.
20. X. Fan, O. C. Au, and N. M. Cheung. Transform-domain adaptive correlation estimation (TRACE) for Wyner-Ziv video coding. *IEEE Transactions on Circuits and Systems for Video Technology*, 20(11):1423–1436, November 2010.
21. X. Fan, O. C. Au, and N. M. Cheung. Adaptive correlation estimation for general Wyner-Ziv video coding. In *IEEE International Conference on Image Processing (ICIP)*, Cairo, Egypt, pp. 1409–1412, November 2009.
22. C. Fu and J. Kim. Encoder rate control for block-based distributed video coding. In *IEEE International Workshop on Multimedia Signal Processing (MMSP)*, Saint Malo, France, pp. 333–338, October 2010.
23. M. Grangetto, E. Magli, and G. Olmo. Distributed arithmetic coding for the slepian-wolf problem. *IEEE Transactions on Signal Processing*, 57(6):2245–2257, June 2009.
24. X. Guo, Y. Lu, F. Wu, D. Zhao, and W. Gao. Wyner-Ziv-based multiview video coding. *IEEE Transactions on Circuits and Systems for Video Technology*, 18(6):713–724, June 2008.
25. X. Huang and S. Forchhammer. Improved virtual channel noise model for transform domain Wyner-Ziv video coding. In *Proceedings of the IEEE International Conference on Acoustics, Speech and Signal Processing (ICASSP)*, Taipei, Taiwan, pp. 921–924, April 2009.
26. X. Huang and S. Forchhammer. Transform domain Wyner-Ziv video coding with refinement of noise residue and side information. In *Proceedings of SPIE*, Huangshan, China, vol. 7744, pp. 774418-1–774418-9, July 2010.
27. M. Jakubowski. Constant rate control algorithm for Wyner-Ziv video codec. In *Proceedings of the SPIE*, Wilga, Poland, vol. 7502, pp. 75020A-1–75020A-10, May 2009.
28. D. Kubasov, K. Lajnef, and C. Guillemot. A hybrid encoder/decoder rate control for a Wyner-Ziv video codec with a feedback channel. In *IEEE MultiMedia Signal Processing Workshop*, Chania, Crete, pp. 251–254, October 2007.
29. D. Kubasov, J. Nayak, and C. Guillemot. Optimal reconstruction in Wyner-Ziv video coding with multiple side information. In *IEEE MultiMedia Signal Processing Workshop*, Chania, Crete, pp. 183–186, October 2007.
30. D. Kubasov and C. Guillemot. Mesh-based motion-compensated interpolation for side information extraction in Distributed Video Coding. In *Proceedings of the IEEE International Conference on Image Processing (ICIP)*, Atlanta, GA, pp. 261–264, October 2006.
31. C. Li, J. Zou, H. Xiong, and C. W. Chen. Joint coding/routing optimization for distributed video sources in wireless visual sensor networks. *IEEE Transactions on Circuits and Systems for Video Technology*, 21(2):141–155, February 2011.
32. Z. Li, L. Liu, and E. J. Delp. Rate distortion analysis of motion side estimation in Wyner-Ziv video coding. *IEEE Transactions on Image Processing*, 16(1):98–113, January 2007.
33. S. Lin and D. J. Costello. *Error Control Coding*, 2nd edn. Prentice Hall, Upper Saddle River, NJ, 2004.
34. Y.-C. Lin, D. Varodayan, and B. Girod. Image authentication and tampering localization using distributed source coding. In *IEEE Workshop on Multimedia Signal Processing*, Chania, Crete, pp. 393–396, October 2007.

35. W. Liu, L. Dong, and W. Zeng. Motion refinement based progressive side-information estimation for Wyner-Ziv video coding. *IEEE Transactions on Circuits and Systems for Video Technology*, 20:1863–1875, December 2010.

36. B. Macchiavello, D. Mukherjee, and R. L. Querioz. Iterative side-information generation in a mixed resolution Wyner-Ziv framework. *IEEE Transactions on Circuits and Systems for Video Technology*, 19(10):1409–1423, October 2009.

37. R. Martins, C. Brites, J. Ascenso, and F. Pereira. Refining side information for improved transform domain Wyner-Ziv video coding. *IEEE Transactions on Circuits and Systems for Video Technology*, 19(9):1327–1341, September 2009.

38. T. Maugey, J. Gauthier, B. Pesquet-Popescu, and C. Guillemot. Using an exponential power model for Wyner-Ziv video coding. In *International Conference on Acoustics, Speech and Signal Processing*, Dallas, TX, pp. 2338–2341, March 2010.

39. M. Morbée, J. Prades-Nebot, A. Pizurica, and W. Philips. Feedback channel suppression in pixel-domain Distributed Video Coding. In *Annual Workshop on Circuits, Systems and Signal Processing (ProRISC)*, Veldhoven, the Netherlands, pp. 154–157, November 2006.

40. M. Munderloh, S. Klomp, and J. Ostermann. Mesh-based decoder-side motion estimation. In *Proceedings of the IEEE International Conference on Image Processing (ICIP)*, Hong Kong, pp. 2049–2052, September 2010.

41. S. Mys, J. Slowack, J. Škorupa, P. Lambert, and R. Van de Walle. Introducing skip mode in distributed video coding. *Signal Processing: Image Communication*, 24(3):200–213, 2009.

42. E. Peixoto, R. L. de Queiroz, and D. Mukherjee. A Wyner-Ziv video transcoder. *IEEE Transactions on Circuits and Systems for Video Technology*, 20(2):189–200, February 2010.

43. F. Pereira, J. Ascenso, and C. Brites. Studying the gop size impact on the performance of a feedback channel-based Wyner-Ziv video codec. In *IEEE Pacific-Rim Symposium on Image and Video Technology*, Santiago, Chile, pp. 801–815, December 2007.

44. F. Pereira, L. Torres, C. Guillemot, T. Ebrahimi, R. Leonardi, and S. Klomp. Distributed video coding: Selecting the most promising application scenarios. *Signal Processing: Image Communication*, 23:339–352, 2008.

45. G. Petrazzuoli, M. Cagnazzo, and B. Pesquet-Popescu. High order motion interpolation for side information improvement in DVC. In *Proceedings of the International Conference on Acoustics, Speech and Signal Processing*, Dallas, TX, pp. 2342–2345, March 2010.

46. S. S. Pradhan, J. Chou, and K. Ramchandran. Duality between source coding and channel coding and its extension to the side information case. *IEEE Transactions on Information Theory*, 49(5):1181–1203, May 2003.

47. R. Puri, A. Majumdar, and K. Ramchandran. PRISM: A video coding paradigm with motion estimation at the decoder. *IEEE Transactions on Image Processing*, 16(10):2436–2448, October 2007.

48. R. Puri and K. Ramchandran. PRISM: A new robust video coding architecture based on distributed compression principles. In *Proceedings of the Allerton Conference on Communication, Control and Computing*, Allerton, Italy, October 2002.

49. S. Rane, P. Baccichet, and B. Girod. Systematic lossy error protection of video signals. *IEEE Transactions on Circuits and Systems for Video Technology*, 18(10):1347–1360, October 2008.

50. A. Rehman, H. Chen, and E. Steinbach. Addressing the uncertainty in critical rate estimation for pixel-domain Wyner-Ziv video coding. In *Proceedings of the SPIE Conference on Visual Communications and Image Processing (VCIP)*, July 2010.

51. D. Slepian and J. K. Wolf. Noiseless coding of correlated information sources. *IEEE Transactions on Information Theory*, 19(4):471–480, 1973.

52. J. Slowack, S. Mys, J. Škorupa, N. Deligiannis, P. Lambert, A. Munteanu, and R. Van de Walle. Rate-distortion driven decoder-side bitplane mode decision for distributed video coding. *Signal Processing: Image Communication*, 25(9):660–673, October 2010.

53. J. Slowack, S. Mys, J. Škorupa, P. Lambert, C. Grecos, and R. Van de Walle. Accounting for quantization noise in online correlation noise estimation for distributed video coding. In *Proceedings of the Picture Coding Symposium (PCS)*, May 2009.

54. J. Slowack, J. Škorupa, S. Mys, P. Lambert, C. Grecos, and R. Van de Walle. Flexible distribution of complexity by hybrid predictive-distributed video coding. *Signal Processing: Image Communication*, 25(2):94–110, February 2010.

55. T. A. Summers and S. G. Wilson. SNR mismatch and online estimation in turbo decoding. *IEEE Transactions on Communications*, 46:421–423, April 1998.

56. Y. Tonomura, T. Nakachi, and T. Fuji. Efficient index assignment by improved bit probability estimation for parallel processing of distributed video coding. In *Proceedings of the International Conference on Acoustics, Speech and Signal Processing*, Las Vegas, NV, pp. 701–704, March 2008.

57. A. Trapanese, M. Tagliasacchi, S. Tubaro, J. Ascenso, C. Brites, and F. Pereira. Improved correlation noise statistics modeling in frame-based pixel domain Wyner-Ziv video coding. *International Workshop on Very Low Bitrate Video*, Sardinia, Italy, pp. 16–21, September 2005.

58. D. Varodayan, A. Aaron, and B. Girod. Rate-adaptive codes for distributed source coding. *EURASIP Signal Processing Journal*, Special Issue on Distributed Source Coding, 86(11):3123–3130, November 2006.

59. Y. Vatis, S. Klomp, and J. Ostermann. Inverse bit plane decoding order for turbo code based distributed video coding. In *Proceedings of the IEEE International Conference on Image Processing (ICIP)*, San Antonio, TX, pages II-1–II-4, October 2007.

60. J. Škorupa, J. Slowack, S. Mys, N. Deligiannis, J. De Cock, P. Lambert, C. Grecos, A. Munteanu, and R. Van de Walle. Efficient low-delay distributed video coding. *IEEE Transactions on Circuits and Systems for Video Technology*, vol. 22, no. 4, pp. 530–544, April 2012.

61. J. Škorupa, J. Slowack, S. Mys, N. Deligiannis, J. De Cock, P. Lambert, A. Munteanu, and R. Van de Walle. Exploiting quantization and spatial correlation in virtual-noise modeling for distributed video coding. *Signal Processing: Image Communication*, 25(9):674–686, 2010.

62. J. Škorupa, J. Slowack, S. Mys, P. Lambert, C. Grecos, and R. Van de Walle. Stopping criterions for turbo coding in a Wyner-Ziv video codec. In *Proceedings of the Picture Coding Symposium (PCS)*, Chicago, IL, May 2009.

63. J. Škorupa, J. Slowack, S. Mys, P. Lambert, and R. Van de Walle. Accurate correlation modeling for transform-domain Wyner-Ziv video coding. In *Proceedings of the Pacific-Rim Conference on Multimedia (PCM)*, Tainan, Taiwan, pp. 1–10, December 2008.

64. T. Wiegand, G. J. Sullivan, G. Bjøntegaard, and A. Luthra. Overview of the H.264/AVC video coding standard. *IEEE Transactions on Circuits and Systems for Video Technology*, 13:560–576, July 2003.
65. A. D. Wyner. On source coding with side information at the decoder. *IEEE Transactions on Information Theory*, 21(3):294–300, 1975.
66. C. Yaacoub, J. Farah, and B. Pesquet-Popescu. Content adaptive GOP size control with feedback channel suppression in distributed video coding. In *Proceedings of the IEEE International Conference on Image Processing (ICIP)*, Cairo, Egypt, pp. 1397–1400, November 2009.
67. S. Ye, M. Ouaret, F. Dufaux, and T. Ebrahimi. Improved side information generation with iterative decoding and frame interpolation for distributed video coding. In *IEEE International Conference on Image Processing (ICIP)*, pp. 2228–2231, October 2008.
68. R. Zamir. The rate loss in the Wyner-Ziv problem. *IEEE Transactions on Information Theory*, 42(6):2073–2084, 1996.
69. Y. Zhang, H. Xiong, Z. He, and S. Yu. Reconstruction for distributed video coding: A Markov random field approach with context-adaptive smoothness prior. In *Proceedings of SPIE*, Huangshan, China, pp. 774417-1–774417-10, July 2010.
70. Y. Zhang, C. Zhu, and K.-H. Yap. A joint source-channel video coding scheme based on distributed source coding. *IEEE Transactions on Multimedia*, 10(8):1648 –1656, December 2008.

3

Computer Vision–Aided Video Coding

Manoranjan Paul and Weisi Lin

CONTENTS

3.1 Introduction

The latest advanced video coding standard H.264/AVC [1] improves rate-distortion performance significantly compared to its predecessors (e.g., H.263) and competitors (e.g., MPEG-2/4) by introducing a number of innovative ideas in Intra- and Inter-frame coding [2,3]. Major performance improvement has taken place by means of motion estimation (ME) and motion compensation (MC) using variable block size, sub-pixel search, and multiple reference frames (MRFs) techniques [4–8]. It has been demonstrated that MRFs facilitate better predictions than using just one reference frame, for video with repetitive motion, uncovered background, noninteger pixel displacement, lighting change, etc. The requirement of index codes (to identify the particular reference frame used), computational time in ME and MC (which increases almost linearly with the number of reference frames), and memory buffer size (to store decoded frames in both encoder and decoder for referencing) limits the number of reference frames used in practical applications. The optimal number of MRFs depends on the content of the video sequences. Typically, the number of reference frames varies from one to five. If the cycle of repetitive motion, exposing uncovered background, noninteger pixel displacement, or lighting change exceeds the number of reference frames used in MRFs coding system, there will not be any improvement and therefore, the related computation (mainly that of ME) and bits for index codes are wasted.

To tackle with the major problem of MRFs, a number of techniques [5–8] have been developed for reducing the computation associated with it. Huang et al. [5] searched either the previous or every reference frame based upon the result of the intra-prediction and ME from the previous frame. This approach can reduce 76%–96% of computational complexity by avoiding unnecessary search for reference frames. Moreover, this approach is orthogonal to conventional fast block matching algorithms, and they can be easily combined to achieve further efficient implementation. Shen et al. [6] proposed an adaptive and fast MRF selection algorithm based on the hypothesis that homogeneous areas of video sequences probably belong to the same video object, move together as well, and thus have the same optimal reference frame. Simulation results show that this algorithm deducts 56%–74% of computation time in ME. Kuo et al. [7] proposed a fast MRF selection algorithm based on the initial search results using 8×8-pixel block. Hachicha et al. [8] used *Markov random fields* algorithm relying on robust moving pixel segmentation, and saved 35% of coding time by reducing the number of reference frames to three instead of five without image quality loss.

Most of the fast MRF selection algorithms including the aforementioned techniques used one reference frame (in the best case) when their assumptions on the correlation of the MRFs selection procedure are satisfied, or five reference frames (in the worst case) when their assumptions completely failed. But, it is obvious that in terms of rate-distortion performance, these techniques cannot outperform the H.264 with five reference frames, which is considered as optimal. Moreover, due to the limited number of reference frames (the maximum is five in practical implementations), uncovered background may not be encoded efficiently using the existing techniques. Some algorithms [9–11] determined and exploited uncovered background using preprocessing and/or postprocessing and computationally expensive video segmentation for coding. Uncovered background can also be efficiently encoded using sprite coding through object segmentation. Most of the video coding applications could not tolerate inaccurate video/object segmentations and expensive computational complexity incurred by segmentation algorithms. Ding et al. [12] used a background frame for video coding. The background frame is made up of blocks which keep unchanged (based on zero motion vectors) in a certain number of continuous frames. Due to the dependency on block-based motion vectors and lack of adaptability in multimodal backgrounds for dynamic environment, this background frame could not perform well.

Recently in the computer vision field, a number of dynamic background generation algorithms based on the dynamic background modeling (DBM) [13–16] using the Gaussian mixture model (GMM) have been introduced for robust and real-time object detection from dynamic environment where ground-truth background is unavailable. A static background model does not remain valid due to illumination variation over time, intentional or unintentional camera displacement, shadow/reflection of foreground objects, and intrinsic background motions (e.g., waving tree leaves, clouds, etc.) [15]. Object can be detected more accurately by subtracting background frame (generated from the background model) from the current frame. First, a dynamic frame known as a most common frame in a scene (McFIS) is generated from video frames using DBM, and then used for improving video coding efficiency. A McFIS can be effective in the following ways:

- McFIS can be used for video compression as a reference frame for referencing static and uncovered areas because of its capability of capturing a whole cycle of repetitive motion, exposing uncovered background, noninteger pixel displacement, or lighting change.

- If a McFIS is generated from encoded frames, intrinsically it has better error recovery capacity for error-prone channel transmission as McFIS model has already contained pixel intensity history of previously encoded frames.

- A simple mechanism for adaptive group of picture (AGOP) determination and scene changed detection (SCD) is possible using McFIS as it represents common background of a scene. Thus, any mechanism for SCD and AGOP determination by comparing the difference between McFIS and the current frame is more effective. In fact, the SCD (therefore AGOP) is integrated with reference frame generation.

- A computationally efficient video encoder can be developed using McFIS as a long-term reference frame against multiple frames to achieve same or better rate-distortion performance. Further computational time can be reduced if we can use small search length for ME and MC while McFIS is referenced for static areas that have no or little motion.

- In video coding, an intra (I-) frame is used as an anchor frame for referencing the subsequent frames, as well as error propagation prevention, indexing, etc. To get better rate-distortion performance, a frame should have the best similarity with the frames in a group of picture (GOP), so that when it is used as a reference frame for a frame in the GOP we need the least bits to achieve the desired image quality, minimize the temporal fluctuation of quality, and also maintain a more consistent bit count per frame. McFIS has the quality to be an I-frame.

The rest of the chapter is organized as follows. Section 3.2 describes the dynamic background modeling with McFIS generation strategies. Section 3.3 describes two advanced video coding techniques. Section 3.4 explains a number of coding strategies where McFIS can be used to improve rate-distortion efficiency, while Section 3.5 concludes the chapter.

3.2 Dynamic Background Modeling

Moving foreground consisting of people, vehicles, or other active objects is at the center of all interactions and events in the real world. In terms of coding points of view, active object consumes most of the bits for encoding and it is a central point for quality judgments. Encoding the moving entities is thus the primary focus of any video coding technology to construct high-quality video within the constrained bit rates. Foreground detection is usually performed by maintaining a model of the scene background, which is subtracted from each incoming video frame, known as background subtraction. The simplest background model could be a static background image captured without any moving entity [13]. However, this simplest model is not suitable for real-world scenarios, as the background can change over time due to illumination variation, camera jitter, intrinsic background motion (e.g., waving tree leaves, clouds), and shadow/reflection. To deal with these challenges, an adaptive background model, capable of updating itself over time to reflect changing circumstances, has been proposed in [13,14]. Apart from the aforementioned challenges, a foreground detection technique is also expected to meet some operational requirements for video coding application such as (1) low complexity, (2) high responsiveness, (3) high stability, and (4) low specificity (environment invariant).

From the aforementioned Gaussian models, background and foreground are determined using different techniques. Stauffer et al. [13] used a user-defined threshold based on the background and foreground ratio. A predefined threshold does not perform well in object/background detection because the ratio of background and foreground varies from video to video. Lee et al. [14] used two parameters (instead of a threshold used in Ref. [13]) of a sigmoid function by modeling the posterior probability of a Gaussian to be background.

This method also depends on the proportion by which a pixel is going to be observed as background. Moreover, the generated background has delayed response because of using the weighted mean of all the background models.

3.2.1 McFIS Generation

Traditional dynamic background modeling techniques [13,14] primarily focus on object detection; thus, it has less concern about real-time processing as well as rate-distortion performance optimization in video compression. Moreover, the generated background has delay response because of using the weighted mean of all the background models. To avoid *mean effect* (mean is considered as an *artificially* generated value and sometimes far from the original value) and delay response, Haque et al. [15] used a parameter called *recentVal*, m to store recent pixel intensity value when a pixel satisfies a model in the Gaussian mixture. They used *classical* background subtraction method, which identifies an object if the value of the current intensity differs from m of the *best* background model by a well-studied threshold. This method reduces not only delay response but also learning rates, which are sometimes desirable criteria for real-time object detection.

Generally, each pixel position of a scene is modeled independently by a mixture of K Gaussian distributions [13–15] in DBM techniques. A pixel position may be occupied by different objects and backgrounds in different frames. Each Gaussian model represents the intensity distribution of one of the different components, for example, objects, background, shadow, illumination, surround changes (like clouds in an outdoor scene), etc., observed by the pixel position in different frames. A Gaussian model is represented by the recent pixel intensity, mean of pixel intensity, pixel intensity variance, and weight (i.e., how many times this model is satisfied by the incoming pixel intensity). The system starts with an empty set of models and initial parameters. If the maximum number of models allowable for a pixel is three, we can get a maximum of $3 \times H \times W$ models for each video scene, where $H \times W$ is the resolution of a frame.

For a pixel, the McFIS generation algorithm is shown in Figure 3.1, which takes the recent pixel intensity (i.e., pixel intensity at the current time, X^t) and existing model for that pixel position (if any) as inputs and returns background pixel intensity (i.e., McFIS) and the background model. For every new observation X^t at the current time t, it is first matched against the existing models in order to find one such that the difference between the newly arrived pixel intensity and the mean of the model is within 2.5 times of the standard deviation (STD) of that model. If such a model exists, its associated parameters are updated with a learning rate parameter. The recent pixel intensity of the model is replaced by the newly arrived pixel intensity. If such a model does not exist, a new Gaussian is introduced with the intensity as a mean (μ), a high STD (σ), recent pixel value (γ), and a

Algorithm [Ψ^t, Ω^t] = McMISgeneration (X^t, Ω^{t-1}) *Parameters:* X^t *is the pixel intensity at time t;* Ω *is the structure of K Gaussian mixture models at time t-1 where each model contains mean, STD, weight, and recent value i.e.,* $\{\mu_k^{t-1}, \sigma_k^{t-1}, \omega_k^{t-1}, \gamma_k^{t-1}\}$; Ψ^t *is the* background pixel intensity i.e., McFIS at time *t*.

For the first time $\Omega_1^t = \{X^t, 30, 0.001, X^t\}$; $\Psi^t = X^t$;

otherwise

    IF $(|X^t - \mu_k^{t-1}| \leq 2.5\sigma_k^{t-1})$ for any $k \leq K$

        Update $\mu_k^t, \sigma_k^{t^2}, \omega_k^t$, according to Equations (2),

        (3), and (4); $\gamma_k^t = X^t$;

    ELSE

        Find the maximum number of models, τ in Ω;

        IF $(\tau < K)$

            $\mu_{\tau+1}^t = X^t; \sigma_{\tau+1}^t = 30; \omega_{\tau+1}^\tau = 0.001$;

            $\gamma_{\tau+1}^\tau = X^t$;

        ELSE

            $\mu_\tau^t = X^t; \sigma_\tau^t = 30; \omega_\tau^t = 0.001; \gamma_\tau^t = X^t$;

        ENDIF

    ENDIF

Normalize all ω_k^t so that $\sum_{\forall k} \omega_k^t = 1$;

$\Omega_k^t = \{\mu_k^t, \sigma_k^t, \omega_k^t, \gamma_k^t\}$ for all k;

Sort Ω_k^t based on ω^t/σ^t in descending order;

$\Psi^t = \mu_1^t$;

FIGURE 3.1
Pseudocode for McFIS generation algorithm.

low weight (ω), and the least probable model is evicted. The least probable model is determined based on the lowest value of w/σ among the models. We have fixed the initial parameters in this implementation as follows: maximum number of models for a pixel $K = 3$, learning rate $\alpha = 0.1$, weight $\omega = 0.001$, and variance $\sigma = 30$.

Assume that kth Gaussian at time t represents a pixel intensity with mean μ_k^t, STD σ_k^t, the recent value γ_k^t, and the weight ω_k^t such that $\sum_{\forall k} \omega_k^t = 1$. The learning parameter α is used to balance the contribution between the current and past values of parameters such as weight, STD, mean, etc. After initialization, for every new observation X_t at the current time t, it is first matched against the existing models in order to find one (e.g., kth model) such that $|X^t - \mu_k^{t-1}| \leq 2.5 \sigma_k^{t-1}$. If such a model exists, the corresponding recent value parameter γ_k^t is updated with X^t. Other associated parameters are updated with learning rates as follows:

$$\mu_k^t = (1-\alpha)\mu_k^{t-1} + \alpha X^t \tag{3.1}$$

$$\sigma_k^{t^2} = (1-\alpha)\sigma_k^{t-1^2} + \alpha(X^t - \mu_k^t)(X^t - \mu_k^t) \tag{3.2}$$

$$\omega_k^t = (1-\alpha)\omega_k^{t-1} + \alpha \qquad\qquad (3.3a)$$

and the weights of the remaining Gaussians (i.e., l where $l \neq k$) are updated as

$$\omega_l^t = (1-\alpha)\omega_l^{t-1} \qquad\qquad (3.3b)$$

The weights are then renormalized. If such a model does not exist, a new model is introduced with $\gamma = \mu = X^t$, $\sigma = 30$, and $\omega = 0.001$ by evicting the Kth (i.e., the third model based on w/σ in descending order) model if it exists.

To get the background pixel intensity from the DBM technique for a particular pixel, one can take the mean value of the background model that has the highest value of w/σ. In this way, a background frame (comprising background pixels) as the McFIS can be made. One example of McFIS is shown in Figure 3.2 using the first 50 original frames of *Silent* video sequence. Figure 3.2a show the 50th frame of video, and Figure 3.2b shows McFIS. The circle in Figure 3.2b indicates the uncovered background captured by the corresponding McFIS. To capture the uncovered background by any single frame is impossible unless this uncovered background is visible for one.

Paul et al. [16] observed that mean and recent intensity values are two extreme cases to generate true background intensity (i.e., McFIS) for better video coding while encoded video frames are used in background generation. The mean is too generalized for pixel intensities over the time and the recent intensity value is too biased to only the recent pixel intensity. Thus, a weighting between mean and recent intensity is a compromise. It also reduces the delay response (due to mean) and speeds up the learning rates (due to recent pixel intensity). Note that normally three models are used for a pixel in DBM as one represents foreground objects, another represents static

(a)   (b)

FIGURE 3.2
Examples of McFIS and uncovered background (inside the circle) using *Silent* video sequence: (a) original 50th frame and (b) corresponding McFIS using first 50 frames of *Silent* video sequence.

First frame   Test frame   Ideal result   Scheme in (13) Scheme in (14) Scheme in (15)

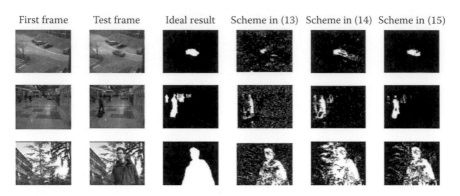

FIGURE 3.3
Visual comparison results in fast learning rate ($\alpha = 0.1$) using three techniques with the test sequences: (a) *PETS2000*; (b) *PETS2006-S7-T6-B1*; (c) *Moved Object*; and (d) *Waving Trees*. (From Haque, M. et al., Improved Gaussian mixtures for robust object detection by adaptive multi-background generation, in *IEEE International Conference on Pattern Recognition*, 2008, pp. 1–4.)

backgrounds, and the third one represents different transitional object/ background. Paul et al. [17] also observed that recent pixel intensity can be effective in generating McFIS while original frames (i.e., without any distortion due to quantization) are used.

3.2.2 Object Detection Using McFIS

Visual comparison results are presented in Figure 3.3 using various methods [13–15] for object detection. Standard test video sequences such as *PETS2000*, *PETS2006-S7-T6-B1*, *Moving Object*, and *Waving Trees* are used for comparison. Experimental results show that background generation method using recent value [15] outperforms other methods in terms of clear object detection at a faster rate.

3.2.3 Scene Change Detection Using McFIS

A number of algorithms [4,18–21] are proposed in the literature for AGOP and SCD. Dimou et al. [18] used dynamic threshold based on the mean and standard deviation of the previous frames for SCD. Their reported accuracy is 94% on average. Alfonso et al. [19] used ME and MC to find the SCD. To avoid repetitive scene change, they imposed lower limit of scene change as four frames. The success rate of this method is 96% with 7.5%–15% more compression and 0.2 dB quality loss. Matsuoka et al. [20] proposed a combined SCD and AGOP method based on fixed thresholds generated from the accumulated difference of luminance pixel components. They used the number of the intensive pixels (NIP) to investigate the frame characteristics. A pixel of a frame is considered as an intensive one if the luminance pixel difference between the adjacent frames is more than 100. If NIP exceeds a predefined threshold between two frames, then an I-frame is inserted at that position assuming the occurrence of SCD; otherwise they restricted GOP size to either 8 or 32 based on the NIP and another threshold. Song et al. [21] proposed another scene change detection method focusing on the Hierarchical B-picture structure [22]. Recently, Ding et al. [4] combined AGOP and SCD for better coding efficiency based on the motion vectors and the sum of absolute transformed differences (SATD) using 4×4 pixels block. This method ensured 98% accuracy of SCD with 0.63 dB image quality improvement.

Most of the existing methods used some metrics computed using already processed frames and the current frames. McFIS is the most similar frame comprising stable portion of the scene (mainly background) compared to the individual frame in a scene. Thus, the SCD is determined by a *simple* metric computed using the McFIS and the current frame. For SCD using McFIS, Paul et al. [16] randomly selected 50% of the pixel position of a frame and determined the sum of absolute difference (SAD) between McFIS and the current frame. If the SAD for the current frame is greater than that of the previous frame by 1.7, then they considered that SCD occurs and inserted an I-frame, otherwise continued intercoding (no other AGOP). This would be effective compared to the existing algorithms as the McFIS is *equivalent* to a group of already processed frames. Moreover, a scene change means the change of background of a video sequence. As the McFIS has the *history* of the scene we do not need a *rigorous* process (like Ding's algorithm) for SCD.

To see the effectiveness of the McFIS in SCD, one mixed video sequence has been created: Mixed video of 700 frames comprising 11 different standard video sequences. *Mixed A* video sequence comprises the first 50/100 frames of the specified QCIF videos, as shown in Table 3.1. From the table,

TABLE 3.1

Mixed Video Sequences for SCD and AGOP

Mixed A (QCIF)	Frames	Frames in Mixed Sequence
Akiyo	100	1–100
Miss America	50	101–150
Claire	50	151–200
Car phone	50	201–250
Hall Monitor	100	251–350
News	50	351–400
Salesman	100	401–500
Grandma	50	501–550
Mother	50	551–600
Suzie	50	601–650
Foreman	50	651–700

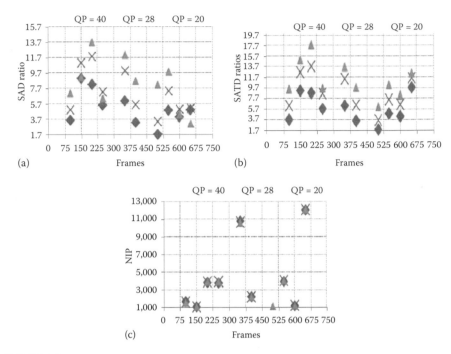

FIGURE 3.4
Scene change detection using McFIS, Ding's, and Matsuoka's methods for mixed video comprising 11 QCIF video sequences. (a) SCD using McFIS for mixed video (b) Ding's method for mixed video (c) Matsuoka's method for mixed video.

it is clear that for mixed sequence, totally 10 scene changes occurred at 101, 151, 201, 251, 351, 401, 501, 551, 601, and 651-th frame. SCD technique using McFIS is compared against two most recent and effective AGOP and SCD algorithms [4,20] for efficient video coding.

Figure 3.4 shows the SCD results by the techniques McFIS, Ding's, and Matsuoka's approaches using three $QPs = \{40, 28, \text{ and } 20\}$ for *Mixed* video sequence. SAD ratio (Figure 3.4a), SATD ratio (Figure 3.4b), and NIP (Figure 3.4c) are plotted using McFIS, Ding's, and Matsuoka's algorithms, respectively. As we mentioned earlier, for the McFIS, a SCD occurs if the SAD ratio is above 1.7 (i.e., the SAD for the current frame is 70% greater than that of the previous frame). For Ding's algorithm a SCD occurs if the SATD ratio is more than 1.7 [4]. For Matsuoka's algorithm, a SCD occurs if the NIP is more than 1000 [20] for QCIF sequences. Thus, it is clear from the figure that for each of the SCD positions (i.e., 101, 151, 201, 251, 351, 401, 501, 551, 601, and 651-th frame), the McFIS-based technique and Ding's methods successfully detect all scene changes. On the other hand, Matsuoka's method successfully detects all scene changes except at the 501-th frame for $QP = 40$ and 28 due to the similarity in background between the *salesman* and *grandma* video sequences.

3.3 Efficient Video Coding Techniques

H.264 introduces variable block-sized ME and MC where a 16×16-pixel mac-roblock (MB) is partitioned into several small rectangular or square-shaped blocks. ME and MC are carried out for all possible combinations, and the ultimate block size is selected based on the Lagrangian optimization [23,24] using the bits and distortions of the corresponding blocks. The real-world objects, by nature, may be in any arbitrary shapes; thus, the ME and MC using only rectangular/square-shaped blocks may roughly approximate the real shape, and thus the coding gain would not be satisfactory. A number of research works are conducted using nonrectangular block partitioning [25–27], called geometric shape partitioning, motion-based implicit block partitioning, and L-shaped partitioning. Requirement of excessively high computational complexity in the segmentation process and the marginal improvement over H.264 makes them less effective for real-time applications. Moreover, the requirement of precious extra bits for encoding the area covering almost static background makes the aforementioned algorithms inefficient in terms of rate-distortion performance [3]. Exploiting nonrectangular block partition-ing and partial block skipping, an advanced video coding technique named pattern-based video coding (PVC) technique [3,28,29] outperforms the H.264. Details of the PVC techniques are discussed in following subsection.

H.264 has one thing in common compared with all other video coding schemes, that is, all the coding schemes encode a video one image frame by one image frame, where the image frame is formed in the spatial domain and with some physical meaning (e.g., a natural scene). It is reasonable since video is usually first (before coding) captured by a sensor (e.g., a camera) and finally (after decoding) displayed in the form of one image frame by one image frame. However, in the sense of data structure, video is nothing more than a three-dimensional (3D) (direction) data matrix; to distinguish between X, Y, and T is meaningless, where X and Y are the spatial directions and T is the temporal direction of a video. One such coding paradigm is 3D transform-based video coding [30,31]; this coding paradigm is not compat-ible with H.264 and therefore could not be adopted in the existing video coding systems. Recently, Liu et al. [32] proposed a new framework of video coding, which is H.264-compatible and takes the advantage of both H.264 and 3D transform-based coding schemes. Similarly, with the 3D transform-based video coding, they ignored the physical meaning of X, Y, and T axes rather than explicitly distinguishing T-axis as temporal axis as in H.264. The detail of the method is discussed in Section 3.3.2.

3.3.1 Pattern-Based Video Coding

To exploit the nonrectangular block partitioning and partial block skipping for static background area in an MB, the PVC [3,28,29] schemes partition the

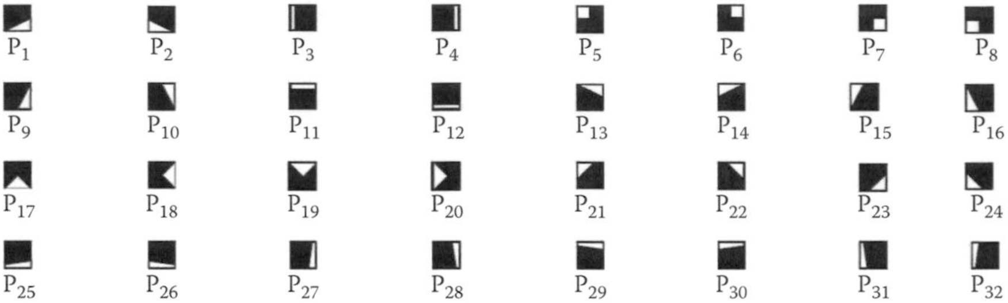

P_1 P_2 P_3 P_4 P_5 P_6 P_7 P_8

P_9 P_{10} P_{11} P_{12} P_{13} P_{14} P_{15} P_{16}

P_{17} P_{18} P_{19} P_{20} P_{21} P_{22} P_{23} P_{24}

P_{25} P_{26} P_{27} P_{28} P_{29} P_{30} P_{31} P_{32}

FIGURE 3.5
The pattern codebook of 32 regular-shaped, 64-pixel patterns, defined in 16×16 blocks, where the white region represents "1" (motion) and the black region represents "0" (no motion). (From Paul, M. et al., *IEEE Trans. Circ. Syst. Video Technol.*, 15(6), 753, 2005.)

MBs via a simplified segmentation process that avoids handling the exact shape of the moving objects, so that the popular MB-based ME could be applied. The PVC algorithm focuses on the moving regions (MRs) of the MBs, through the use of a set of regular 64-pixel pattern templates (Figure 3.5). The MR is defined as the difference between the current MB and the colo-cated MB of the reference frame. The pattern templates were designed using "1"s in 64 pixel positions and "0"s in the remaining 192 pixel positions in a 16×16-pixel MB. The MR of an MB is defined as a region comprising a collection of pixel positions where pixel intensity differs from its reference MB. Using some similarity measures, if the MR of a MB is found well covered by a particular pattern, then the MB can be classified as a region-active MB and coded by considering only the 64 pixels of the pattern, with the remaining 192 pixels being skipped as static background. Embedding PVC in the H.264 standard as an extra mode provides higher compression as larger segment with static background is coded with the partial skipped mode.

3.3.2 Optimal Compression Plane for Video Coding

In the optimal compression plane (OCP) scheme, the frames are allowed to be formed in a non-XY plane. OCP determination process is to be used as a preprocessing step prior to any standard video coding scheme. The essence of the scheme is to form the frames in the plane formed by two axes (among X, Y, and T) corresponding to signal correlation evaluation, which enables better prediction (therefore, better compression). The preprocessing approach for video coding is different from the existing paradigms by exploring the information redundancy in a fuller extent. Rather than explicitly distinguishing T-axis as a temporal axis, the OCP scheme ignores the physical meaning of X, Y, and T axes (somewhat similar with 3D transform) and focuses on the amount of video redundancy along each axis (more specifically, on the correlation coefficient [CC] along each axis). OCP-based video

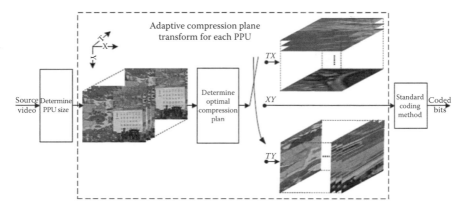

FIGURE 3.6
Block diagram of the proposed scheme (illustrated with XY and non-XY frames of "Mobile" video sequence for better visual impression). (From Liu, A. et al., *IEEE Trans. Image Process.*, 20(10), 2788, 2011.)

coding framework consists of selection of appropriate preprocessing unit (PPU; its size, i.e., number of frames in the XY plane is denoted as N), adaptive OCP decision, and video coding with a standard compression method. In each PPU, it is possible to form frames in the XY, TX, or TY plane. It is obvious that it needs 2 bits of overhead to represent the three possible coding planes for each PPU, and this bit overhead is included in our rate calculation throughout this work.

The insight observation for the OCP framework is that the sampling processing is as usual, but it ignores the physical meaning of a video during the video coding processing for better coder performance. People deem video as successive natural image frames, which are formed in the spatial domain and with clear physical meaning (e.g., a natural scene). However, in the sense of data structure, video is nothing more than a 3D data matrix, and the distinction among X (a spatial dimension), Y (the other spatial dimension), and T (the temporal dimension) is not absolutely necessary, in the viewpoint of compression. The adaptive plane selection according to video content makes sense and the coding system benefits from it. A block diagram of the OCP scheme is shown in Figure 3.6.

3.4 McFIS in Video Coding

Computer vision tool such as dynamic background modeling, that is, McFIS can be used for different purposes of video coding. In the following sections we will explore the applications of the McFIS in video coding.

3.4.1 McFIS as a Better I-Frame

H.264 as well as other modern standards uses intra (I-) and inter (predicted [P]- and bi-directional [B]-) frames for improved video coding. An I-frame is encoded using only its own information and thus can be used for error propagation prevention, fast backward/forward play, random access, indexing, etc. On the other hand, a P- or B-frame is coded with the help of previously encoded I- or P-frame(s) for efficient coding. In the H.264 standard, frames are coded as a GOP comprising one I-frame with subsequent inter-frames. The number of I-frames is fewer compared to the inter-frames because an I-frame typically requires several times more bits compared to its intercoded counterpart for the same image quality. An I-frame is used as an anchor frame for referencing the subsequent inter-frames of a GOP directly or indirectly. Thus, encoding error (due to the quantization) of an I-frame is propagated and accumulated toward the end of the frames of a GOP. As a result the image quality degrades and the bits requirement increases toward the end of the GOP. When another I-frame is inserted for the next GOP, better image quality (with the cost of more bits) is recovered, and then again quality degrades toward the end of GOP. As a result, the farther an inter-frame is from the I-frame, the lower the quality becomes. The fluctuation of image quality (or bits per frame) is not desirable for perceptual quality (or bit rate control). By selecting the first frame as an I-frame without verifying its suitability to be an I-frame, we sacrifice (1) overall rate-distortion performance because of poor selection of an I-frame, and (2) perceptual image quality by introducing image quality fluctuation. A frame being the first frame of a GOP is not automatically the best I-frame. An ideal I-frame should have some qualities: the best similarity with the frames in a GOP, so that when it is used as a reference frame for inter-frames in the GOP it needs fewer bits to achieve the desired image quality for better rate-distortion performance and perceptual image quality.

McFIS can be generated using DBM with the first several original frames of a scene in a video and encoding it as an I-frame with finer quantization. All frames of the scene are coded as inter-frames using two reference frames: one is the immediate previous frame and another is the McFIS assuming that moving regions and the background regions of the current frame will be referenced using the immediate previous frames and the McFIS, respectively. As all frames are coded as inter-frames using direct referencing from the McFIS, this provides less fluctuation in PSNR and bit count for the entire scene. The McFIS has higher *similarity* to all the frames of the scene and thus can be a better I-frame. One can continue to use the current McFIS as a second reference frame unless SCD occurs. If SCD occurs, a McFIS has to be generated again using the first several frames from the new scene and encoded as an I-frame. All of the frames of the new scene are encoded as inter-frames unless SCD occurs again. A joint SCD and AGOP technique can be developed to make the McFIS relevant to the potential referencing for the inter-frames of each new scene [17].

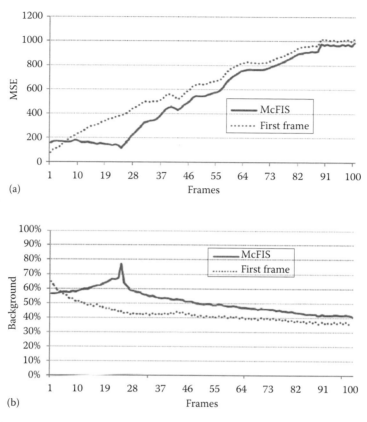

FIGURE 3.7

Effectiveness of McFIS as an I-frame compared to the first frame: (a) mean square error for the first frame and the McFIS with the rest 100 frames for indication of dissimilarity; (b) percentages of background generated by the first frame and the McFIS with the rest 100 frames (from 2 to 101 frames). (From Paul, M. et al., *IEEE Trans. Circ. Syst. Video Technol.*, 21(9), 1242, 2011.)

Figure 3.7 shows two cases of evidence to demonstrate the effectiveness of McFIS compared to the first frame as an I-frame. As mentioned earlier, an I-frame should have higher similarity with the rest of the frames. To check this, the mean square error (MSE) of a frame is calculated in a video sequence evaluated with the first frame and the McFIS, respectively. Obviously, the higher MSE value indicates more dissimilarity. Figure 3.7a shows the average results of MSE using first 100 frames of eight video sequences, namely, *Hall Monitor, News, Salesman, Silent, Paris, Bridge close, Susie,* and *Popple.* The figure shows that McFIS results in less MSE than the first frame, and this indicates that McFIS is more similar to the rest of the frames than the first frame. As a result, we need fewer bits and achieve better quality if one uses McFIS (instead of the first frame) as an I-frame and direct reference frame.

From another angle, Figure 3.7b also demonstrates the effectiveness of McFIS for improving coding performance compared to the first frame as an I-frame. The subfigure shows average percentages of "background" for those video sequences using the McFIS and the first frame, respectively. A pixel is defined as a background pixel if that pixel has not more than one level (in 0–255 scale) difference with the colocated pixel in the McFIS (or first frame). The subfigure shows that there are more background pixels in McFIS than the first frame. This confirms that McFIS represents more background regions by capturing the most common features in the video compared to that of the first frame. This leads to more referencing from the McFIS for uncovered/normal background area to improve video coding performance. Note that there is a dip and a peak with McFIS at the 25th frame in both subfigures of Figure 3.7, respectively. These are due to the most *similarity* of the McFIS with the 25th frame as we generate the McFIS using the first 25 frames where the latest (i.e., the 25th) frame has the highest impact (due to the weight and *recent pixel intensity*) on the McFIS generation.

3.4.2 McFIS as a Reference Frame

McFIS can be used as an extra reference frame to refer static and uncovered background areas in different video coding schemes. In the following sections we will discuss the applications of McFIS in different coding paradigms such as pattern-based video coding, optimal compression plane, hierarchical bi-predictive pictures, computational time reductions, and bit rate and PSNR fluctuation reductions.

3.4.2.1 PVC Using McFIS

For pattern matching in PVC coding approach, MR needs to be generated for current MB. The MR generated from the difference of the current MB and the colocated MB from the *traditional* reference frames (i.e., the immediate previous frame or any frame which is previously encoded) may contain moving object and uncovered background (UCB) (Figure 3.8). The ME and MC using pattern-covered MR for UCB would not be accurate if there is no similar region in the reference frames. As a result no coding gain can be achieved for the UCB using the PVC. Similar issues occur for any other H.264 variable size block modes due to the lack of suitable matching region in the reference frames. Thus, we need a reference frame where we will find the UCB for the current MB if that region was evidenced. Only a *true* background of a scene can be the best choice to be the reference frame for UCB. Moreover, an MR generated from the true background against the current frame represents only moving object instead of both the moving object and the UCB. Thus, the selection of the best matched pattern against the newly generated MR is the best approximation of the object/partial object in an MB. The ME and MC using the best matched pattern carried out on the

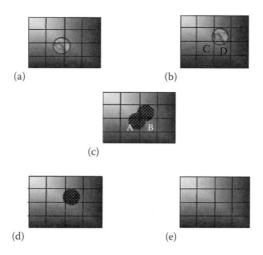

FIGURE 3.8
Motion estimation and compensation problem using blocks or patterns when there is occlusion: (a) reference frame, (b) current frame, (c) MR without McF, (d) MR using McFIS, and (e) true background.

immediate previous frame will provide more accurate motion vector and thus minimum residual errors for the object/partial object of the MB. The rest of the area (which is not covered by the pattern) is copied from the true background frame. The immediate previous frame is used for ME and MC assuming that the object is visible in the immediate previous frame. The other modes of H.264 can also use true background as well as the immediate previous frame (in the multiple reference frames technique) as two separate reference frames. The Lagrangian optimization will pick the optimal reference frame. The experimental results reveal that this approach improves the rate-distortion performance.

3.4.2.2 OCP Using McFIS

Treating a video as a 3D data tube and rearranging the video frames in other two directions revolutionizes the way to treat the traditional video features such as background, motion, object, panning, zooming, etc. Due to the rearrangement of the traditional XY plane images into the TX or TY plane images, object and/or camera motions of a traditional video can be transformed into simplified motions or simple background; for example, horizontal motions and vertical motions can be transformed into a static background in the TX or TY plane images, respectively, or a heterogeneous object can be transformed into a smooth object in TX or TY plane. Besides this, camera motions such as zooming and panning could not be effectively estimated by the traditional translational motion estimation adopted into the H.264. Camera motions can also be transformed into a simplified motion/background in the TX or TY

plane; thus any existing coding technique can encode them more efficiently using the TX or TY plane.

The McFIS is used as a second reference frame for encoding the current frame assuming that the motion part of the current frame would be referenced using the immediate previous frame and the static background part would be referenced using the McFIS. The ultimate reference is selected at the block and sub-block levels using the Lagrangian multiplier. The McFIS is used as a long-term reference frame in the dual reference frames concept which is a subset of the concept of the MRFs. Paul et al. [33] exploit the newly introduced background in the TX or TY plane through the McFIS. In the method, they first determine the OCP and then encode the video toward the optimal plane with the McFIS. The experimental results reveal that the proposed technique improves a significant video quality compared to the existing OCP technique and the H.264 video coding standard with comparable computational complexity.

Using the aforementioned procedure, different McFISes are generated using the silent videos of $VXYT, VTYX, VTXY$, respectively. Figure 3.9 shows (a) original 50th frame, (b) the 50th McFIS using XY plane, (c) the 50th McFIS using TX plane, and (d) the 50th McFIS using TY plane. If we compare the

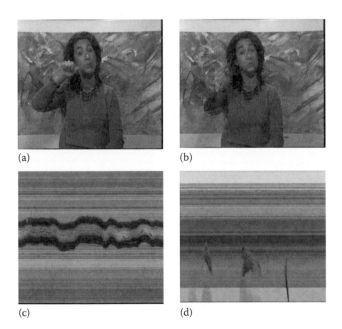

(a)   (b)

(c)   (d)

FIGURE 3.9
McFISes at different planes: (a) original 50th frame, (b) McFIS at 50th frame using V_{XYT} video, (c) McFIS at 50th frame using V_{TXY} video, and (d) McFIS at 50th frame using V_{TYX} video where first 288 frames of *Silent* video sequence are used. (From Paul, M. and Lin, W., Efficient video coding considering a video as 3D data cube, in *IEEE International conference on Digital Image Computing: Techniques and Applications (IEEE DICTA-11)*, 2011.)

McFISes in Figure 3.9c and d against the McFIS in Figure 3.9b, we can easily differentiate them in terms of smoothness. As we mentioned earlier, smoother objects or motion can be obtained after transforming the video into different planes. The McFISes in Figure 3.9c and d provide smother images compared to the McFIS in Figure 3.9b. This result also indicates that a better rate-distortion performance can be achieved using the smother McFIS (Figure 3.9c or d) compared to the rough McFIS (Figure 3.9b), as the primary goal of the McFIS is to capture more background.

3.4.2.3 Hierarchical Bipredictive Pictures

Figure 3.10a shows a popular dyadic hierarchical bipredictive picture (HBP) prediction structure with encoding image types, coding, and display order of a GOP (comprises 16 frames) where two bidirectional reference frames (solid arrows only) are used. To get the better coding performance of the HBP structure compared to the other structure (e.g., IPP or IBBP), different *quantization parameters* are used for different hierarchy levels. Normally, finer quantization is applied for the frames that are more frequently used as reference frames for the other frames directly or indirectly. For example, Frame 9 (according to display order) in Figure 3.10a is used more frequently (four times directly for frames 5, 7, 11, and 13, and 10 times indirectly for frames 2, 3, 4, 6, 8, 10, 12, 14, 15, and 16) compared to any other B-frames as a reference frame.

The MRFs technique can be applied on the HBP. Intuitively, a triple frame referencing technique can be made under the dyadic HBP structure, which is shown in Figure 3.10a based on the closeness and the availability of the reference frames for a frame. To encode a frame, two solid arrows come from two frames and one dotted arrow comes from the third reference frame. For example, to encode Frame 5, Frame 1 and Frame 9 are used as two bidirectional frames, and Frame 17 is used as the third reference frame. On the other hand, to encode Frame 2, Frame 1 and Frame 3 are used as bidirectional frames, and Frame 5 is used as the third reference frame. Both examples demonstrate that the MRFs technique in the HBP is not uniform in terms of the distance from the encoding frame and the reference frames. Thus, sometimes it is difficult to exploit the advantages of the MRF features from the close vicinity frames. Moreover, it could not ensure the implicit background/foreground referencing, and thus it has no physical meaning for referencing. To overcome the limitation of effectiveness using the MRFs technique in the dyadic HBP structure (Figure 3.10a) and to exploit the implicit background/foreground referencing, Paul et al. [22] proposed a new HBP scheme using the McFIS as a third reference frame. By this they assumed that motion areas of the current frame would be referenced from the two B-frames and static/uncovered background areas would be referenced from the McFIS. Experimental results reveal that the technique improves the rate-distortion performance significantly.

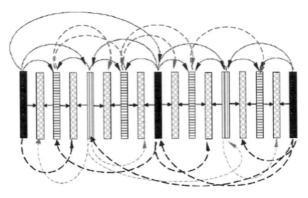

Image type: I₁ B₃ B₂ B₃ B₁ B₃ B₂ B₃ P₁ B₃ B₂ B₃ B₁ B₃ B₂ B₃ I₂
Display order: 1 2 3 4 5 6 7 8 9 10 11 12 13 14 15 16 17
(a) Coding order: 1 6 5 7 4 9 8 10 3 13 12 14 11 16 15 17 2

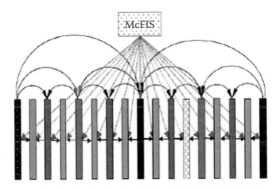

Image type: P₁ B₃ B₂ B₃ B₁ B₃ B₂ B₃ P₂ B₃ B₂ B₃ B₁ B₃ B₂ B₃ P₃
Display order: 1 2 3 4 5 6 7 8 9 10 11 12 13 14 15 16 17
(b)   Coding order: 2 7 6 8 5 10 9 11 4 14 13 15 12 17 16 18 3

FIGURE 3.10
(See color insert.) (a) Dyadic hierarchical B-picture prediction structure using two frames and three frames including the third frame (dotted arrows) and (b) proposed triple frame referencing with the McFIS as the third frame.

3.4.2.4 Computational Time Reduction in Video Coding Using McFIS

To see the amount of computational reduction of McFIS-based approach, we have used the H.264 with fixed GOP and five reference frames. In video coding, major portion of entire computational time is used for ME and MC. Although the proposed scheme needs some extra computational time to generate McFIS and encode it as I-frame, this extra time is not significant in comparison with the ME reduction. To compare the experimental results we have implemented the proposed and the H.264 schemes adapted from

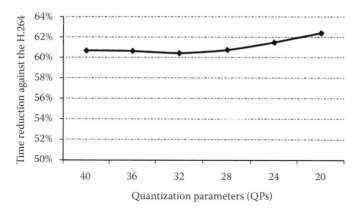

FIGURE 3.11
Average computational time reduction by the proposed scheme against the H.264 with fixed GOP and five reference frames using different standard video sequences (*Silent, Bridge close, Paris, Hall Monitor, News, Salesman, News, Susie,* and *Popple*). (From Paul, M. et al., *IEEE Trans. Circ. Syst. Video Technol.*, 21(9), 1242, 2011.)

JM 10.1 H.264/AVC reference software on a PC with Intel(R) Core$^{(TM)}$2 CPU 6600@2.40 GHz, 2.39 GHz, and 3.50 GB of RAM. Figure 3.11 shows experimental results of computational reduction of the McFIS-based scheme (where McFIS is used as an I-frame and second reference frame) against the H.264 with five reference frames, using a number of video sequences (*Mixed video, Silent, Paris, Bridge close, Hall Monitor, Salesman, News, Susie,* and *Popple*) over different QPs, that is, 40, 36, 32, 28, 24, and 20. The computational complexity is calculated based on the overall encoding time including processing operations and accessing to the data. This figure confirms that the McFIS-based approach reduces 61% of the computation on average.

The McFIS-based technique requires extra computational time for the generation and encoding of the McFIS. This extra time is not more than 3% of the overall encoding time of a scene if we assume that a scene length is 100 frames, ME search length is 15, and single reference frame is used. The experimental results suggest that the proposed scheme saves −43%, 17%, and 58% computational time, reduces 22%, 20%, and 19% of bit rates, and improves 1.53, 1.47, and 1.45 dB image quality against the H.264 with one, two, and five reference frames, respectively, for *News* video sequence on average. As can be seen, the proposed method is more efficient even in comparison with the H.264 using two reference frames. The McFIS generation and encoding time is fixed and does not depend on the number of reference frames. Thus, when any fast motion estimation (such as Unsymmetrical–cross Multi-Hexagon-grid Search (UMHexagonS) [34,35]) is used, the percentage of time saving is lower compared to that when we use exhaustive search. For example, when we turn on the UMHexgonS for both the proposed scheme and the H.264 with five reference frames, the computational time saving is around 50%, which is significant as well. When we turn on the fast skip mode [3,29] for both the proposed scheme and the H.264 with five reference frames, the computational complexity is even better for the proposed scheme, as the proposed scheme produces more skip modes using the McFIS.

3.4.2.5 Consistency of Image Quality and Bits per Frame Using McFIS

Encoding the first frame as an I-frame and referencing in the conventional way, errors (due to the quantization) are propagated and accumulated toward the end of the GOP. Figure 3.12 shows bits per frame and PSNRs at frame level by the H.264 and the McFIS-based scheme using the first 256 frames of *News*

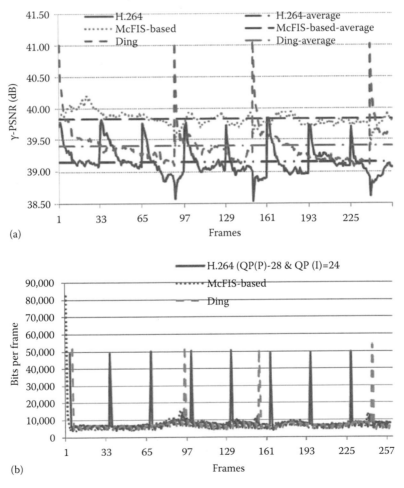

FIGURE 3.12
Fluctuations of PSNR (a) and bits per frame (b) by the H.264, Ding's, and the McFIS-based schemes using first 256 frames of *News* sequence.

video sequence. The figure demonstrates that the proposed scheme provides not only better PSNR, that is, 39.83 dB using 200 kbps bit rate but also consistent PSNR and bits per frame over the scene compared to the H.264 (i.e., 39.15 dB using 214 kbps) and Ding's algorithm (39.41 dB using 200 kbps). Note that McFIS bits are considered at bit rate calculations. Figure 3.12b shows that only one McFIS is required for *News* video sequence as there is no scene change or there is no significant drop of referencing using the McFIS within 256 frames. Thus, for this sequence the fluctuation of bits is less compared to that of other methods. The standard deviations of the PSNR using the McFIS-based algorithm, Ding's algorithm, and the H.264 are 0.1122, 0.255, and 0.2343, respectively. The PSNR fluctuations using the McFIS-based approach, Ding's algorithm, and the H.264 are 0.8, 2.0, and 1.5 dB, respectively.

The McFIS-based scheme can provide more consistent image quality and bits per frame because a common frame McFIS is used as a reference frame directly (thus, no error propagation toward the end of the scene) for all interframes in a scene. As McFIS is encoded with finer quantization, there is less error due to the quantization. Encoding an I-frame in the conventional coding scheme with that level of fine quantization requires enormous number of bits (due to the regular insertion of I-frame at the beginning of a GOP for fixed size of GOP where the GOP size has to be small in order to cater for possible scene changes). Naturally, the generated McFIS enables lower PSNR fluctuation because it represents the most common and stable features in the video segment (on the contrary, the first frame only represents itself). A new McFIS needs to be generated and encoded if there is a significant drop of the PSNR of an image or the percentage of referencing drops significantly compared to the other frames of a scene. Two thresholds as 2.0 dB and 3%, respectively, are used in Ref. [16].

Figure 3.13 shows reference mapping using *Silent* and *Paris* video sequences by the McFIS-based scheme and Ding's algorithm [4]. A scattered referencing takes place using Ding's algorithm for the immediate previous and second previous frames. For the proposed method, moving object areas (black regions in Figure 3.13e and f) are referenced using the immediate previous frame, whereas the background regions are referenced using McFIS (normal area in Figure 3.13e and f). A large number of areas (normal regions in Figure 3.13e and f) are referenced by the McFIS, and this indicates the effectiveness of the McFIS for improving coding performance.

3.4.2.6 R D Performance Improvement

Figure 3.14 shows the overall rate-distortion curves using the McFIs-based technique, Ding's, Matsuoka's, and the H.264 (with fixed GOP and five reference frames) algorithms for a number of video sequences. The experimental results confirm that the McFIS-based scheme outperforms the H.264 as well as other two existing algorithms in the most cases even with the video sequences (e.g., *Tennis*, *Trevor*, *Bus*, etc.) with camera motions. McFIS-based

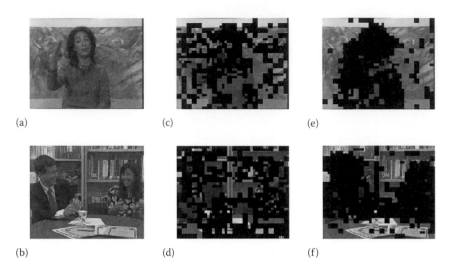

FIGURE 3.13
Frame-level reference maps by the proposed and Ding's methods for *Silent* and *Paris* video sequences: (a) and (b) are the decoded 31st frames of *Silent* and *Paris* videos; (c) and (d) are the reference maps by Ding's algorithm; and (e) and (f) are the reference maps by the proposed algorithm where black regions are referenced from the immediate previous frame while other regions are referenced from the McFIS (for the proposed) or the second previous frame (for Ding's).

scheme as well as the other two techniques (Ding and Matsuoka) with two reference frames could not outperform the H.264 with five reference frames for the video sequences (e.g., *Tempete*, *Mobile*, and *Foreman*) with camera motions. It is due to the fact that the proposed, Ding's, and Matsuoka's techniques are not explicitly designed for camera motions. The performance improvement by the McFIS-based scheme is relatively high for *News*, *Salesman*, *Silent*, and *Hall Monitor* video sequences compared to the other sequences. This is due to the relatively larger background areas in these cases, and hence a larger number of references are selected from the McFIS.

3.5 Conclusions

In this chapter, the issue of effective, dynamic I-frame insertion, and reference frame (termed as the most common frame in scene, or McFIS for short) generation in video coding has been tackled simultaneously with a Gaussian mixture based model for dynamic background. To be more specific, the McFIS-based method used the generated McFIS' inherent capability of scene change detection and adaptive GOP determination for integrated decision for efficient video coding. The McFIS is generated using real-time Gaussian

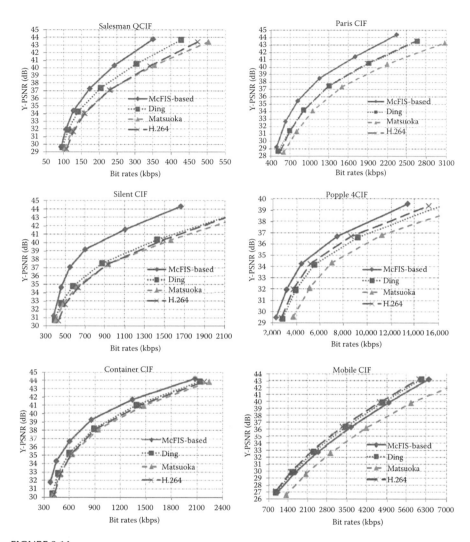

FIGURE 3.14
Rate-distortion performance comparison using different techniques such as H.264 with five reference frames, McFIS-based technique, Ding's technique, and Matsuoka's technique for different video sequences such as *Salesman, Paris, Silent, Popple, Container,* and *Mobile.*

mixture model. McFIS can be used as the second reference frame for efficient encoding of background. In essence, the scheme allows moving object areas being referenced with the immediate previous frame while background regions are being referenced with McFIS.

A dynamic background modeling is also discussed using decoded or distorted frames instead of original frames. This allows wider scope of use with dynamic background modeling because raw video feeds (without any lossy compression) are usually not available and noise/error is inevitable especially in the case of wireless transmission. By foreground and background referencing, McFIS can improve rate-distortion performance in the uncovered background region, which is almost impossible by the traditional multiple reference schemes. The McFIS-based scheme effectively reduces computational complexity by limiting the reference frames into only two without sacrificing rate-distortion performance (actually it improves, compared to the relevant existing algorithms). By introducing McFIS as a reference frame, one can avoid the complication of selecting long-term reference frame.

Two advanced video coding techniques, namely, PVC and OCP, are also discussed in the chapter. McFIS is used in PVC to determine the moving region which is used in pattern-matching criteria. McFIS is also used to be referenced for newly generated static and smooth areas by the OCP technique. Improved rate-distortion performance can be achieved using McFIS in the PVC and OCP video coding techniques.

The video coding technique using McFIS outperforms the existing relevant schemes, in terms of rate-distortion and computational requirement. The experimental results show that the technique detects scene changes more effectively compared to the two state-of-the-art algorithms, and outperforms them by 0.5–2.0 dB PSNR for coding quality. The technique outperforms the H.264 with fixed GOP and five reference frames by 0.8–2.0 dB in PSNR and by around 60% of reduced computational time.

References

1. ITU-T Recommendation H.264: Advanced video coding for generic audiovisual services, 03/2009.
2. Wiegand, T., G. J. Sullivan, G. Bjøntegaard, and A. Luthra, 2003. Overview of the H.264/AVC video coding standard. *IEEE Transactions on Circuits and Systems for Video Technology*, 13(7), 560–576.
3. Paul, M. and M. Murshed, 2010. video coding focusing on block partitioning and occlusions. *IEEE Transactions on Image Processing*, 19(3), 691–701.
4. Ding J.-R. and J.-F. Yang, 2008. Adaptive group-of-pictures and scene change detection methods based on existing H.264 advanced video coding information. *IET Image Processing*, 2(2), 85–94.
5. Huang, Y.-W., B.-Y. Hsieh, S.-Y. Chien, S.-Y. Ma, and L.-G. Chen, 2006. Analysis and complexity reduction of multiple reference frames motion estimation in H.264/AVC. *IEEE Transactions on Circuits and Systems for Video Technology*, 16(4), 507–522.
6. Shen, L., Z. Liu, Z. Zhang, and G. Wang, 2007. An adaptive and fast multi frame selection algorithm for H.264 Video coding. *IEEE Signal Processing Letters*, 14(11), 836–839.
7. Kuo, T.-Y. and H.-J. Lu, 2008. Efficient reference frame selector for H.264. *IEEE Transactions on Circuits and Systems for Video Technology*, 18(3), 400–405.

8. Hachicha, K., D. Faura, O. Romain, and P. Garda, 2009. Accelerating the multiple reference frames compensation in the H.264 video coder. *Journal of Real-time Image Processing*, 4(1), 55–65.

9. Hepper, D., 1990. Efficiency analysis and application of uncovered background prediction in a low bit rate image coder. *IEEE Transactions on Communication*, 38, 1578–1584.

10. Chien, S.-Y., S.-Y. Ma, and L.-G. Chen, 2002. Efficient moving object segmentation algorithm using background registration technique. *IEEE Transactions on Circuits and Systems for Video Technology*, 12(7), 577–586.

11. Totozafiny, T., O. Patrouix, F. Luthon, and J.-M. Coutellier, 2006. Dynamic background segmentation for remote reference image updating within motion detection JPEG 2000. *IEEE International Symposium on Industrial Electronics*, Montreal, Quebec, Canada, pp. 505–510.

12. Ding, R., Q. Dai, W. Xu, D. Zhu, and H. Yin, 2004. Background-frame based motion compensation for video compression. *IEEE International Conference on Multimedia and Expo (ICME)*, Taipei, Vol. 2, pp. 1487–1490.

13. Stauffer C. and W. E. L. Grimson, 1999. Adaptive background mixture models for real-time tracking. *IEEE Conference on Computer Vision and Pattern Recognition*, Vol. 2, pp. 246–252.

14. Lee, D.-S., 2005. Effective Gaussian mixture learning for video background subtraction. *IEEE Transactions on Pattern Analysis and Machine Intelligence*, 27(5), 827–832.

15. Haque, M., M. Murshed, and M. Paul, 2008. Improved Gaussian mixtures for robust object detection by adaptive multi-background generation. *IEEE International Conference on Pattern Recognition*, Tampa, FL, pp. 1–4.

16. Paul, M., W. Lin, C. T. Lau, and B.-S. Lee, 2010. Video coding using the most common frame in scene. *IEEE International Conference on Acoustics, Speech, and Signal Processing (IEEE ICASSP-10)*, Dallas, TX, pp. 734–737.

17. Paul, M., W. Lin, C. T. Lau, and B.-S. Lee, 2011. Explore and model better I-frame for video coding. *IEEE Transaction on Circuits and Systems for Video Technology*, 21(9), 1242–1254.

18. Dimou, A., O. Nemethova, and M. Rupp, 2005. Scene change detection for H.264 using dynamic threshold techniques. *EURASIP Conference on Speech and Image Processing, Multimedia Communications and Service*, Smolenice, Slovak Republic, pp. 222–227.

19. Alfonso, D., B. Biffi, and L. Pezzoni, 2006. Adaptive gop size control in H.264/AVC encoding based on scene change detection. *Signal Processing Symposium*, Reykjavik, Iceland, pp. 86–89.

20. Matsuoka, S., Y. Morigami, T. Song, and T. Shimamoto, 2008. Coding efficiency improvement with adaptive gop size selection for H.264/SVC. *International Conference on Innovative Computing Information and Control, (ICICIC)*, Dalian, Liaoning, pp. 356–359.

21. Song, T., S. Matsuoka, Y. Morigami, and T. Shimamoto, 2009. Coding efficiency improvement with adaptive gop selection for H.264/SVC. *International Journal of Innovative Computing, Information and Control*, 5(11), 4155–4165.

22. Paul, M., W. Lin, C. T. Lau, and B.-S. Lee, 2011, McFIS in hierarchical bipredictive picture-based video coding for referencing the stable area in a scene. *IEEE International Conference on Image Processing (IEEE ICIP-11)*, Brussels, Belgium, pp. 3521–3524.

23. Paul, M., M. Frater, and J. Arnold, 2009. An efficient mode selection prior to the actual encoding for H.264/AVC encoder. *IEEE Transactions on Multimedia*, 11(4), 581–588.
24. Paul, M., W. Lin, C. T. Lau, and B.-S. Lee, 2011. Direct intermode selection for H.264 video coding using phase correlation. *IEEE Transaction on Image Processing*, 20(2), 461–473.
25. Escoda, O. D., P. Yin, C. Dai, and X. Li, 2007. Geometry-adaptive block partitioning for video coding. *IEEE International Conference on Acoustics, Speech, and Signal Processing (ICASSP-07)*, Honolulu, Hawaii, pp. I-657–660.
26. Kim, J. H., A. Ortega, P. Yin, P. Pandit, and C. Gomila, 2008. Motion compensation based on implicit block segmentation, *IEEE International Conference on Image Processing (ICIP-08)*, San Diego, CA, pp. 2452–2455.
27. Chen, S., Q. Sun, X. Wu, and L. Yu, 2008. L-shaped segmentations in motion-compensated prediction of H.264. *IEEE International Conference on Circuits and Systems*, Seattle, WA, pp. 1620–1623.
28. Paul, M., M. Murshed, and L. Dooley, 2005. A real-time pattern selection algorithm for very low bit-rate video coding using relevance and similarity metrics. *IEEE Transactions on Circuits and Systems for Video Technology*, 15(6), 753–761.
29. Wong, K.-W., K.-M. Lam, and W.-C. Siu, 2001. An efficient low bit-rate video-coding algorithm focusing on moving regions. *IEEE Transactions on Circuits and System for Video Technology*, 11(10), 1128–1134.
30. Liu, Y., F. Wu, and K. N. Ngan, 2007. 3-D object-based scalable wavelet video coding with boundary effect suppression. *IEEE Transactions on Circuits and System for Video Technology*, 17(5), 639–644.
31. Seran, V. and L. P. Kondi, 2006. New scaling coefficients for bi-orthogonal filter to control distortion variation in 3D wavelet based video coding. In *Proceedings of the International Conference on Image Processing*, Atlanta, GA, pp. 1873–1876.
32. Liu, A., W. Lin, M. Paul, and F. Zhang, 2011. Optimal compression plane determination for video coding. *IEEE Transaction on Image Processing*, 20(10), 2788–2799.
33. Paul, M. and W. Lin, 2011. Efficient video coding considering a video as 3D data cube. *IEEE International Conference on Digital Image Computing: Techniques and Applications (IEEE DICTA-11)*, Noosa, Queensland, Australia, pp. 170–174
34. Chen, Z., P. Zhou, Y. He, and J. Zheng, 2006. Fast integer-PEL and fractional-PEL motion estimation for H.264/AVC. *Journal of Visual Communication and Image Representation*, 17(2), 264–290.
35. Rahman, C. A. and W. Badawy, 2005. Umhexagons algorithm based motion estimation architecture for H.264/AVC. *Fifth International Workshop on System-on-Chip for Real-time Applications (IWSOC'05)*, Banff, Alberta, Canada, pp. 207–210.

4

Macroblock Classification Method for Computation Control Video Coding and Other Video Applications Involving Motions

Weiyao Lin, Bing Zhou, Dong Jiang, and Chongyang Zhang

CONTENTS

4.1 Introduction and Related Work

Complexity-scalable video coding (CSVC) (or computational-scalable/power-aware video coding) is of increasing importance to many applications (Burleson et al., 2001; Chen et al., 2006; He et al., 2005; Huang et al., 2005; Kim et al., 2006; Lin et al., 2008; Tai et al., 2003; Yang et al., 2005; Yi and Ling, 2005), such as video communication over mobile devices with limited power budget as well as real-time video systems that require coding the video below a fixed number of processor computation cycles.

The target of the CSVC research is to find an efficient way to allocate the available computation budget for different video parts (e.g., group of pictures [GOPs], frames, and macroblocks [MBs]) and different coding modules (e.g., motion estimation [ME], discrete cosine transform [DCT], and entropy coding) so that the resulting video quality is kept as high as possible under the given computation budget. Since the available computation budget may vary, the CSVC algorithm should be able to perform video coding under different budget levels.

Normally, in a power-rich condition, a high-quality CSVC strategy is preferred in spite of higher power consumption. On the contrary, more power-efficient CSVC strategies need to be introduced with lower power consumption to prolong the service time. In Figure 4.1, the battery discharging effects of three kinds of video coding strategies are shown (Chen et al., 2009). The power-aware CSVC strategy B can have more than twice the battery lifetime of the traditional encoder A. We can further extend the battery lifetime with a more power-efficient CSVC strategy C by gradually stepping down the power dissipation such that the battery capacity can be fully utilized with a lower loading.

Several researches have been proposed on CSVC. He et al. (2005) analyzed the video encoding system and proposed a computation-bit rate-PSNR model for the entire video encoding system. However, since they aimed at modeling

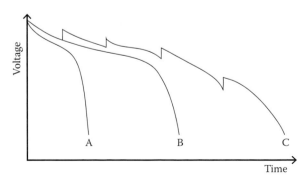

FIGURE 4.1
Battery discharging effects of three types of CSVC encoders.

the whole system, their proposed CSVC algorithm for the ME module is very simple. Besides, Burleson et al. (2001) modeled the computational-scalable algorithm for DCT module. Vanum et al. (2007) as well as Vanum et al. (2009) tried to control the coding complexity by adaptively tuning the encoder parameters. However, these methods either fail to provide an efficient computation control strategy or do not address the computation control for the important ME part.

Since ME occupies the major portion of the whole coding complexity (Wiegand et al., 2003; Zhang and He, 2003), we will focus on the computation allocation for the ME part in this chapter (i.e., computation control motion estimation [CCME]). Furthermore, since the computation often can be roughly measured by the number of search points (SPs) in ME, we will use the term *SP* and *Computation* interchangeably. Therefore, the CCME target for this chapter can be described by: given the total number of ME computation budget (e.g., number of SAD computation or SP), trying to find efficient ways to allocate them into different video parts (frames, GOPs, and MBs) such that the video coding performance is kept as high as possible under the available computation, as shown in Figure 4.2.

Many algorithms have been proposed for CCME (Chen et al., 2006; He et al., 2005; Huang et al., 2005; Kim et al., 2006; Tai et al., 2003; Yang et al., 2005). They can be evaluated by two key parts of CCME: (a) *the computation allocation,* and (b) *the MB importance measure.* They are described as follows:

1. *The computation allocation order.* Two approaches can be used for allocating the computations: one-pass flow and multi-pass flow. Most previous CCME methods (Kim et al., 2006; Tai et al., 2003; Yang et al., 2005) allocate computation in a multi-pass flow, where MBs in one frame are processed in a step-by-step fashion based on a table that measures the MB importance. At each step, the computation is allocated to the MB that is measured as *the most important* among all the MBs in the whole frame. The table is updated after each step. Since the multi-pass methods use a table for all MBs in the frame, they can

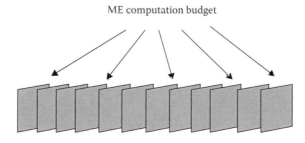

ME computation budget

FIGURE 4.2
The target for CCME: Allocate the available ME computation budget efficiently within the whole video sequence.

have a global view of the whole frame while allocating computation. However, they do not follow the regular coding order and require the ME process to jump between MBs, which is less desirable for hardware implementations. Furthermore, since the multi-pass methods do not follow the regular coding order, the neighboring MB information cannot be used for prediction to achieve better performance. Compared to the multi-pass flow approach, one-pass methods (Chen et al., 2006; Huang et al., 2005) allocate computation and perform ME in the regular video coding order. They are more favorable for hardware implementation and can also utilize the information from neighboring MBs. However, it is more difficult to develop a good one-pass method since (a) a one-pass method lacks a global view of the entire frame and may allocate unbalanced computations to different areas of the frame, and (b) it is more difficult to find a suitable method to measure the importance of MBs.

2. *The MB importance measure.* In order to allocate computation efficiently to different MBs, it is important to measure the importance of the MBs for the coding performance, so that more computation will be allocated to the more *important* MBs (i.e., MBs with larger importance measure values). Tai et al. (2003) use the current *sum of absolute difference (SAD)* value for the MB importance measure. Their assumption is that MBs with large matching costs will have more room to improve, and thus more SPs will be allocated to these MBs. Chen et al. (2006) as well as Huang et al. (2005) use a similar measure in their one-pass method. However, the assumption that larger current SAD will lead to bigger SAD decrease is not always guaranteed, which makes the allocation less accurate. Yang et al. (2005) use the ratio between the *SAD decrease* and *the number of SPs* at the previous ME step to measure the MB importance. Kim et al. (2006) use a similar measure except that they use *rate-distortion cost decrease* instead of the *SAD decrease*. However, their methods can only be used in multi-pass methods where the allocation is performed in a step-by-step fashion and cannot be applied to one-pass methods.

In this chapter, a new one-pass CCME method is proposed. We first propose a class-based MB importance measure (CIM) method where MBs are classified into different classes based on their properties. The importance of each MB is measured by combining its class information as well as its initial matching cost value. Based on the CIM method, a complete CCME framework is then proposed, which first divides the total computation budget into independent sub-budgets for different MB classes and then allocates the computation from the class budget to each step of the ME process. Furthermore, the proposed method performs ME in a one-pass flow, which is more desirable for hardware implementation. Furthermore, we also propose algorithms using our macroblock classification method for different video processing applications, including shot change detection, motion discontinuity detection, and outlier rejection for global motion estimation (GME). Experimental results demonstrate that the proposed method can allocate computation more accurately than previous methods while maintaining good quality.

The rest of the chapter is organized as follows: Section 4.2 describes our proposed CIM method. Based on the CIM method, Section 4.3 describes the proposed CCME algorithm in detail. The experimental results are given in Section 4.4. Section 4.5 describes extending our proposed macroblock classification method for other video processing applications, and Section 4.6 concludes the chapter.

4.2 The Class-Based MB Importance Measure

In this section, we discuss some statistics of ME and describe our CIM method in detail. For convenience, we use COST (Lin et al., 2008) as the ME matching cost in the rest of the chapter. The COST (Lin et al., 2008) is defined as

$$COST = SAD + \lambda_{MOTION} \cdot R(MV) \tag{4.1}$$

where
  SAD is the sum of absolute difference for the block matching error
  $R(MV)$ is the number of bits to code the *motion vector* (*MV*)
  λ_{MOTION} is the lagrange multiplier (Weigand et al., 2003)

In this chapter, the CIM method and the proposed CCME algorithm are described based on the simplified hexagon search (SHS) algorithm (Yi et al., 2005). However, our algorithms are general and can easily be extended to other ME algorithms (Li et al., 1994; Lin et al., 2008; Po and Ma, 1996; Zhou et al., 2004; Zhu and Ma, 2000).

The SHS is a newly developed ME algorithm, which can achieve performance close to full search (FS) with comparatively low SPs. The SHS process can be described as in Figure 4.3.

Before the ME process, the SHS algorithm first checks the *init_COST*, which is defined as

$$init\_COST = \min\left(COST_{(0,0)}, COST_{PMV}\right) \tag{4.2}$$

where
  $COST_{(0,0)}$ is the COST of the (0,0) MV
  $COST_{PMV}$ is the COST of the *predictive MV* (*PMV*) (Yi et al., 2005)

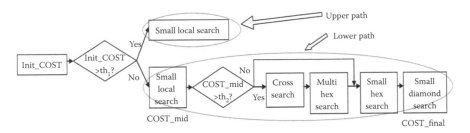

FIGURE 4.3
The SHS process. (From Lin et al., A computation control motion estimate method for complexity-scalable video coding, *IEEE Trans. Circuits Syst. Video Technol.*, 20(11), 1533, Fig. 1. With permission.)

If *init_COST* is smaller than a threshold th_1, the SHS algorithm will stop after performing a *small local search* (search four points around the position of the *init_COST*), which we call *the Upper Path*. If *init_COST* is larger than the threshold, the SHS algorithm will proceed to the steps of *Small Local Search, Cross Search, Multi Hexagon Search, Small Hexagon Search,* and *Small Diamond Search* (Yi et al., 2005), which we call *the Lower Path*. Inside the lower path, another threshold th_2 is used to decide whether or not to skip the steps of *Cross Search* and *Multi Hexagon Search*.

4.2.1 Analysis of Motion Estimation Statistics

In order to analyze the relationship between the COST value and the number of SPs, we define two more COSTs: *COST_mid* (the COST value right after the *Small Local Search* step in the *Lower Path*) and *COST_final* (the COST value after going through the entire ME process), as in Figure 4.3. Three MB classes are defined as

$$Class_{cur\_MB} = \begin{cases} 2 & \text{if } init\_COST < th_1 \\ 2 & \text{if } init\_COST \geq th_1 \quad \text{and} \mid COST\_mid - COST\_final \mid > c \\ 3 & \text{if } init\_COST \geq th_1 \quad \text{and} \mid COST\_mid - COST\_final \mid \leq c \end{cases}$$

(4.3)

where
cur_MB is the current MB
th_1 is the threshold defined in the SHS algorithm (Yi et al., 2005) to decide whether the *init_COST* is large or small (Yi et al., 2005)
c is another threshold to decide the significance of the cost improvement between *COST_mid* and *COST_final*

MBs in Class 1 are MBs with small current COST values. Class 2 represents MBs with large current COST values where additional searches can yield significant improvement. Class 3 represents MBs with large current COST values but where further searches do not produce significant improvement. If we can predict Class 3 MBs, we can save computation by skipping further searches for the Class 3 MBs. It should be noted that since we cannot get *COST_final* before actually going through the *Lower Path*, the classification method of Equation 4.3 is only used for statistical analysis. A practical classification method will be proposed later in this section. Furthermore, since MBs in Class 1 have small current COST value, their MB importance measure can be easily defined. Therefore, we will focus on the analysis of Class 2 and Class 3 MBs.

Table 4.1 lists the percentage of Class 1, Class 2, and Class 3 MBs over the total MBs for sequences of different resolutions and under different quantization parameter (QP) values where c of Equation 4.3 is set to be different values of 0, 2% *of COST_mid*, and 4% *of COST_mid*. It should be noted that 0 is the smallest possible value for c. We can see from Table 4.1 that the number of Class 3 MBs will become even larger if c is relaxed to larger values.

Figure 4.4 shows the COST value distribution of Class 2 MBs and Class 3 MBs where c of Equation 4.3 is set to be 0. We only show results for *Foreman_Qcif* with $QP = 28$ in Figure 4.4. Similar results can be observed for other sequences and other QP values. In Figure 4.4, 20 frames are coded. The experimental setting is the same as that described in Section 4.5. In order to have a complete observation, all the three COST values are displayed in Figure 4.4, where Figures 4.4a through c show the distributions of *init_COST*, *COST_mid*, and *COST_final*, respectively.

From Figure 4.4 and Table 4.1, we can observe that (a) a large portion of MBs with large current COST values can be classified as Class 3 where only a few SPs are needed and additional SPs do not produce significant improvement, and (b) the distribution of all the three COSTs for Class 2 and Class 3 is quite similar. This implies that Class 2 or Class 3 cannot be differentiated based on their COST value only.

Based on these observations, we can draw several conclusions for the computation allocation as follows:

1. The number of SPs needed for keeping the performance for each MB is not always related to its current COST value. Therefore, using the COST value alone as the MB importance measure, which has been used by many previous methods (Chen et al., 2006; Huang et al., 2005; Yang et al., 2005), may not allocate SPs efficiently.

2. Further experiments show that for Class 2 MBs, the number of SPs needed for keeping the performance is roughly proportional to their *init_COST* value (although it is not true if Class 2 and Class 3 MBs are put together).

These imply that we can have a better MB importance measure if we use the class and COST information together.

As mentioned earlier, since we cannot get *COST_final* before going through the Lower Path, Class 2 and Class 3 cannot be differentiated by their definition in Equation 4.3 in practice. Furthermore, since the COST distribution of Class 2 and Class 3 is similar, the current COST value cannot differentiate

TABLE 4.1

Percentage of Class 1, Class 2, and Class 3 MBs over the Total MBs (100 Frames for Qcif and 50 Frames for Cif and SD)

Sequence			QP=23			QP=28			QP=33		
			Class 1 MB (%)	Class 2 MB (%)	Class 3 MB (%)	Class 1 MB (%)	Class 2 MB (%)	Class 3 MB (%)	Class 1 MB (%)	Class 2 MB (%)	Class 3 MB (%)
Qcif (176×144)	Foreman_Qcif (c=0)		50	5.5	44.4	33.8	6.7	59.4	14.9	8.2	76.7
	Akiyo_Qcif (c=0)		96	0	4	89	0	10	68.7	0	31.2
	Mobile_Qcif (c=0)		6.9	0.7	92.2	1.5	0.8	97.6	0.6	0.8	98.4
Cif (352×288)	Bus_Cif	c=0	21.6	21.8	56.8	14.6	22.2	63.1	4.2	25.7	70
		c=2 Cost_mid	21.6	20.5	57.9	14.6	20.8	64.6	4.2	22.9	72.8
		c=4 Cost_mid	21.6	19.5	58.9	14.6	19.4	66	4.2	20.6	75.1
	Football_Cif (c=0)		22.4	53.1	24.5	15.3	54.1	30.5	2.3	58	39.7
	Container_Cif (c=0)		90.6	0	9.3	65.6	0.2	34.2	48.8	2.6	48.6
	Mobile_Cif	c=0	11	8.1	80.9	7.2	8.5	84.3	4.3	9.7	86
		c=2 Cost_mid	11	7.3	81.7	7.2	7.7	85.1	4.3	8.4	87.3
		c=4 Cost_mid	11	6.6	82.4	7.2	6.8	86	4.3	7.3	88.4
SD (720×576)	Foreman_Cif (c=0)		61.6	12	26.4	51.5	13.3	35.2	32.9	17.1	50
	Mobile_SD (c=0)		37.6	7.4	55	22.5	7.9	69.6	12	9	79
	Football_SD (c=0)		41.7	29.4	28.9	32	30	38	20.1	32.1	47.8
	Flower_SD (c=0)		28.7	8.7	62.6	25.1	9.6	65.3	22.7	11.4	65.9

Source: From Lin et al., A computation control motion estimate method for complexity-scalable video coding, *IEEE Trans. Circuits Syst. Video Technol.*, 20(11), 1533, Table 1, With permission.

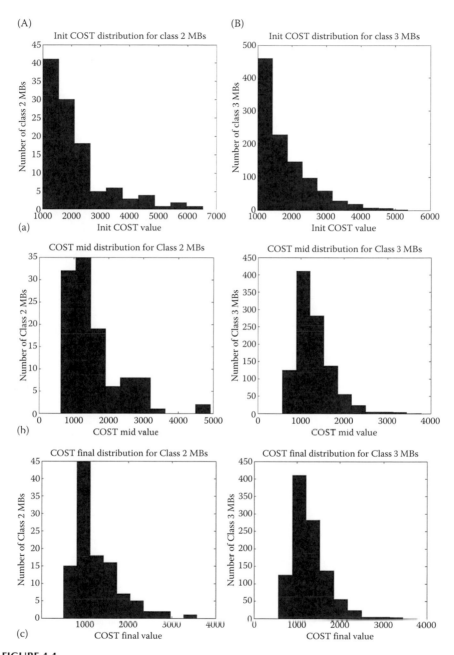

FIGURE 4.4

COST value distribution for Class 2 and Class 3 MBs for *Foreman_qcif* sequence ((A) Class 2, (B) Class 3). (a) *Init_COST* distribution comparison; (b) *COST_mid* distribution comparison; (c) *COST_final* distribution comparison. (From Lin et al., A computation control motion estimate method for complexity-scalable video coding, *IEEE Trans. Circuits Syst. Video Technol.*, 20(11), 1533, Fig. 2. With permission.)

between these two classes. Therefore, before describing our MB Importance Measure method, we first propose a practical MB classification method, which we call the predictive-MV-accuracy-based classification (PAC) algorithm. The PAC algorithm will be described in the following section.

4.2.2 The Predictive-MV-Accuracy-Based Classification Algorithm

The proposed PAC algorithm converts the definition of Class 2 and Class 3 from the *COST value* point of view to the *predictive MV accuracy* point of view. The basic idea of the PAC algorithm is described as follows:

1. If the motion pattern of a MB can be predicted accurately (i.e., if PMV is accurate), then only a small local search is needed to find the *final MV* (i.e., the MV of *COST_final*). In this case, no matter how large the COST is, additional SPs after the small local search are not needed because the *final MV* has already been found by the small local search. This corresponds to Class 3 MBs.

2. On the other hand, if the motion pattern of a MB cannot be accurately predicted, a small local search will not be able to find the *final MV*. In this case, a large area search (i.e., the *Lower Path*) after the small local search is needed to find the *final MV* with a lower COST value. This corresponds to Class 2 MBs.

Since the *MV_final* (MV for *COST_final*) cannot be obtained before going through the *Lower Path,* the final MV of the colocated MB in the previous frame is used instead to measure the accuracy of motion pattern prediction. Therefore, the proposed PAC algorithm can be described as

$$Class_{cur\_MB} = \begin{cases} 1 & \text{if } init\_COST < th \\ 2 & \text{if } init\_COST \geq th \quad \text{and} \, | \, PMV_{cur\_MB} - MV_{pre\_final} \, | > Th \\ 3 & \text{if } init\_COST \geq th \quad \text{and} \, | \, PMV_{cur\_MB} - MV_{pre\_final} \, | \leq Th \end{cases} \quad (4.4)$$

where
$| \, PMV_{cur\_MB} - MV_{pre\_final} \, |$ is the measure of the motion pattern prediction accuracy
PMV_{cur_MB} is the PMV (Yi et al., 2005) of the current MB
MV_{pre_final} is the final MV of the colocated MB in the previous frame
Th is the threshold to check whether the *PMV* is accurate or not

Th can be defined based on different small local search patterns. In the case of SHS, *Th* can be set as 1 in integer pixel resolution. According to Equation 4.4, Class 1 includes MBs that can find good matches from the previous frames. MBs with irregular or unpredictable motion patterns will be classified as Class 2. Class 3 MBs will include areas with complex textures but similar motion patterns to the previous frames.

TABLE 4.2

The Detection Rates of the PAC Algorithm

Sequence	Class 2 Detection Rate (%)	Class 3 Detection Rate (%)
Mobile Qcif	80	82
Football_Cif	71	90
Foreman_Qcif	75	76

Source: From Lin et al., A computation control motion estimate method for complexity-scalable video coding, *IEEE Trans. Circuits Syst. Video Technol.*, 20(11), 1533, Table 2, With permission.

It should be noted that the classification using Equation 4.4 is very tight (in our case, any MV difference larger than 1 integer pixel will be classified as Class 2 and a large area search will be performed). Furthermore, by including MV_{pre_final} for classification, we also take the advantage of including the temporal motion smoothness information when measuring motion pattern prediction accuracy. Therefore, it is reasonable to use MV_{pre_final} to take the place of *MV_final*. This will be demonstrated in Table 4.2 and Figure 4.5 and will be further demonstrated in the experimental results.

Table 4.2 shows the detection rates for Class 2 and Class 3 MBs with our PAC algorithm for some sequences, where the class definition in Equation 4.3 is used as the ground truth and c in Equation 4.3 is set to be 0. Table 4.2 shows that our PAC algorithm has high MB classification accuracy.

Figure 4.5 shows the distribution of MBs for each class of two example frames by using our PAC algorithm. Figures 4.5a and e are the original frames. Blocks labeled gray in (b) and (f) are MBs belonging to Class 1. Blocks labeled black in (c) and (g) and blocks labeled white in (d) and (h) are MBs belonging to Class 2 and Class 3, respectively.

Figure 4.5 shows the reasonableness of the proposed PAC algorithm. From Figure 4.5, we can see that most Class 1 MBs include backgrounds or flat areas that can find good matches in the previous frames ((b) and (f)). Areas with irregular or unpredictable motion patterns are classified as Class 2 (e.g., the *edge between the calendar and the background* as well as *the bottom circling ball* in (c), and *the running bus* as well as *the down-right logo* in (g)). Most complex-texture areas are classified as Class 3, such as the complex background and calendar in (d) as well as the flower area in (h).

4.2.3 The MB Importance Measure

Based on the discussion in the previous sections and the definition of MB classes in Equation 4.4, we can describe our proposed CIM method as follows:

1. MBs in Class 1 will always be allocated a fixed small number of SPs.
2. MBs in Class 2 will have high importance. They will be allocated more SPs, and each Class 2 MB will have a guaranteed minimum SPs

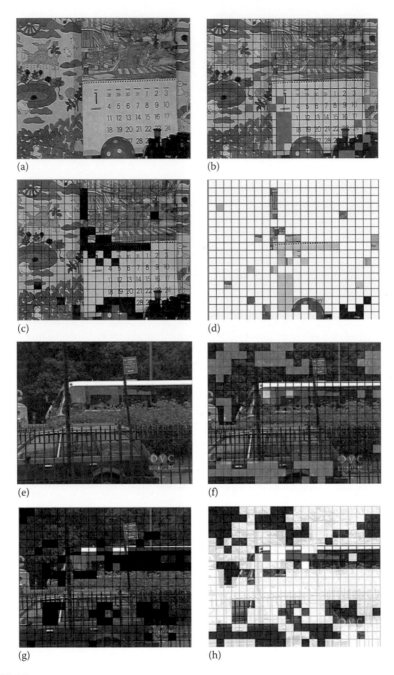

FIGURE 4.5
The original frames (a, e) and the distributions of Class 1 (b, f), Class 2 (c, g), and Class 3 (d, h) MBs for *Mobile_Cif* and *Bus_Cif*. (From Lin et al., A computation control motion estimate method for complexity-scalable video coding, *IEEE Trans. Circuits Syst. Video Technol.*, 20(11), 1533, Fig. 3. With permission.)

for coding performance purposes. If two MBs both belong to Class 2, their comparative importance is proportional to their *init_COST* value and the SPs will be allocated accordingly.

3. MBs in Class 3 will have lower importance than MBs in Class 2. Similar to Class 2, we make the comparative importance of MBs within Class 3 also proportional to their *init_COST* value. By allowing some Class 3 MBs to have more SPs rather than fixing the SPs for each MB, the possible performance decrease due to the misclassification of MBs from Equation 4.4 can be avoided. This will be demonstrated in the experimental results.

With the CIM method, we can have a more accurate MB importance measure by differentiating MBs into classes and combining the class and the COST information. Based on the CIM method, we can develop a more efficient CCME algorithm. The proposed CCME algorithm will be described in detail in the following section.

4.3 The CCME Algorithm

The framework of the proposed CCME algorithm is described in Figure 4.6. From Figure 4.6, the proposed CCME algorithm has four steps:

1. *Frame-level computation allocation (FLA).* Given the available total computation budget for the whole video sequence, FLA allocates a computation budget to each frame.

2. *Class-level computation allocation (CLA).* After one frame is allocated a computation budget, CLA further divides the computation into

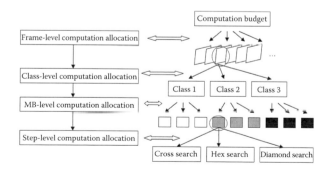

FIGURE 4.6
The framework for the proposed CCME algorithm. (From Lin et al., A computation control motion estimate method for complexity-scalable video coding, *IEEE Trans. Circuits Syst. Video Technol.*, 20(11), 1533, Fig. 4. With permission.)

three independent sub-budgets (or class budgets), with one budget for each class defined in Equation 4.4.

3. *MB-level computation allocation (MLA)*. When performing ME, each MB will first be classified into one of the three classes according to Equation 4.4. MLA then allocates the computation to the MB from its corresponding class budget.

4. *Step-level computation allocation (SLA)*. After an MB is allocated a computation budget, SLA allocates these computations into each ME step.

It should be noted that the CLA step and the MLA step are the key steps of the proposed CCME algorithm where our proposed CIM method is implemented. Furthermore, we also investigated two strategies for computation allocation for CLA and MLA steps: *the tight strategy* and *the loose strategy*. For the tight strategy, the actual computation used in the current frame *must be lower* than the computation allocated to this frame. Due to this property, the FLA step is sometimes not necessary for the tight strategy. In some applications, we can simply set the budget for all frames as a fixed number for performing the tight strategy. For the *loose strategy*, the actual computation used *for some frames* can exceed the computation allocated to these frames but the total computation used for the *whole sequence* must be lower than the budget. Since the loose strategy allows frames to borrow computation from others, the FLA step is needed to guarantee that the total computation used for *the whole sequence* will not exceed the available budget.

Since the performances of the loose-strategy algorithm and the tight-strategy algorithm are similar based on our experiments, we will only describe our algorithm based on the tight strategy in this chapter. It should be noted that since the basic ideas of the CLA and MLA processes are similar for both the tight and loose strategies, a loose-strategy algorithm can be easily derived from the description in this chapter. Furthermore, as mentioned, the FLA step is sometimes unnecessary for the tight strategy. In order to prevent the effect of FLA and to have a fair comparison with other methods, we also skip the FLA step by simply fixing the target computation budget for each frame in this chapter. In practice, various FLA methods (Chen et al., 2006; Kim et al., 2006; Tai et al., 2003; Yang et al., 2005) can be easily incorporated into our algorithm.

4.3.1 Class-Level Computation Allocation

The basic ideas of the CLA process can be summarized as follows:

1. In the CLA step, the computation budget for the whole frame C_F is divided into three independent class budgets (i.e., $C_{Class(1)}$, $C_{Class(2)}$, and $C_{Class(3)}$). MBs from different classes will be allocated computation from their corresponding class budget and will not affect each other.

2. Since the CLA step is based on the tight strategy in this chapter, the basic layer $BL_{Class(i)}$ is first allocated to guarantee that each MB has a minimum number of SPs. The remaining SPs are then allocated to the additional layer $AL_{Class(i)}$. The total budget for each class consists of the basic layer plus the additional layer. Furthermore, since the MBs in class 1 only performs a local search, the budget for class 1 only contains the basic layer (i.e., $C_{Class(1)} = BL_{Class(1)}$ and $AL_{Class(1)} = 0$).

3. The actual computation used for each class in the previous frame ($CA_{Class(i)}^{pre}$) is used as the ratio parameter for class budget allocation for the additional layer.

Therefore, the CLA process can be described as in Equation 4.5 and Figure 4.7:

$$C_{Class(i)} = BL_{Class(i)} + AL_{Class(i)} \quad i = 1,2,3 \tag{4.5}$$

where

$$BL_{Class(i)} = BL_{MB\_Class(i)} \cdot NM_{Class(i)}^{pre}$$

$$AL_{Class(i)} \begin{cases} 0 & \text{if } i = 1 \\ \min\left(AL_F \cdot \dfrac{CA_{Class(2)}^{per}}{CA_{Class(2)}^{per} + CA_{Class(3)}^{per}}, AL_{MB\_max\_Class(2)} \cdot NM_{Class(i)}^{per} \right) & \text{if } i = 2 \\ AL_F - AL_{Class(2)} & \text{if } i = 3 \end{cases}$$

$$BL_F = (BL_{Class(1)} + BL_{Class(2)} + BL_{Class(3)})$$

$$AL_F = C_F - BL_F$$

$C_{Class(i)}$ is the computation allocated to class i, and $BL_{Class(i)}$ and $AL_{Class(i)}$ represent the computation allocation for the class i basic layer and additional layer,

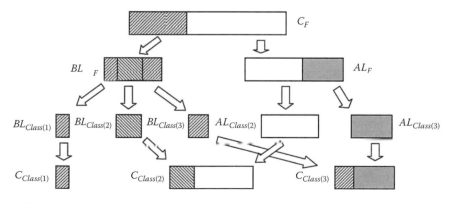

FIGURE 4.7

The tight-strategy-based CLA process. (From Lin et al., A computation control motion estimate method for complexity-scalable video coding, *IEEE Trans. Circuits Syst. Video Technol.*, 20(11), 1533, Fig. 5. With permission.)

respectively. C_F is the total computation budget for the whole frame, and BL_F and AL_F represent the basic layer computation and the additional layer computation for the whole frame, respectively. $NM_{Class(i)}^{pre}$ is the total number of MBs belonging to Class i in the previous frame and $CA_{Class(i)}^{pre}$ is the number of computation actually used for the Class i in the previous frame. $BL_{MB\_Class(i)}$ is the minimum number of computations guaranteed for each MB in the basic layer. In the case of SHS, we set $BL_{MB\_Class(1)} = BL_{MB\_Class(3)} = 6$ SPs for Class 1 and Class 3, and $BL_{MB\_Class(2)} = 25$ SPs for Class 2. As mentioned, since Class 2 MBs have higher importance in our CIM method, we guarantee them a higher minimum SP. Furthermore, in order to avoid too many useless SPs allocated to Class 2 MBs, a maximum number of SPs ($AL_{MB\_max\_Class(2)}$) is set. SPs larger than $AL_{MB\_max\_Class(2)}$ are likely wasted and therefore are allocated to Class 3 MBs (AL_F–$AL_{Class(2)}$).

From Equation 4.5 and Figure 4.7, we can summarize several features of our CLA process as follows:

1. Since *Class* is newly defined in this chapter, the CLA step is unique in our CCME method and is not included in the previous CCME algorithms (Chen et al., 2006; He et al., 2005; Huang et al., 2005; Kim et al., 2006; Tai et al., 2003; Yang et al., 2005).

2. When performing CLA, the information from the previous frame ($NM_{Class(i)}^{pre}$ and $CA_{Class(i)}^{pre}$) is used. $NM_{Class(i)}^{pre}$ provides a global view estimation of the MB class distribution for the current frame, and $CA_{Class(i)}^{pre}$ is used as a ratio parameter for class budget allocation for the additional layer.

3. The CIM method is implemented in the CLA process where (a) the CA for Class 2 is normally larger than other classes, and (b) Class 2 MBs have a larger guaranteed minimum number of SPs (i.e., $BL_{MB\_Class(2)}$ in the tight-SLA).

4.3.2 MB-Level Computation Allocation

The MLA process can be described in Equation 4.6. Similar to the CLA process, a basic layer (BL_{MB}) and an additional layer (AL_{MB}) are set. When allocating the additional layer computation, the initial COST of the current MB ($COST_{cur\_MB}^{init}$) is used as a parameter to decide the number of computations allocated. The MLA process for Class 2 or Class 3 MBs is described as in Figure 4.8:

$$C_{cur\_MB} = BL_{cur\_MB} + AL_{cur\_MB} \qquad (4.6)$$

where

$$BL_{cur\_MB} = \begin{cases} BL_{MB\_Class(1)} & \text{if } class_{cur\_MB} = 1 \\ BLC_{MB\_Class(2)} & \text{if } class_{cur\_MB} = 2 \\ BL_{MB\_Class(3)} & \text{if } class_{cur\_MB} = 3 \end{cases}$$

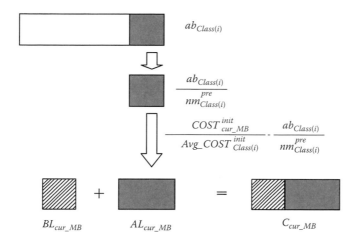

$$ab_{Class(i)}$$

$$\frac{ab_{Class(i)}}{nm^{pre}_{Class(i)}}$$

$$\frac{COST^{init}_{cur\_MB}}{Avg\_COST^{init}_{Class(i)}} \cdot \frac{ab_{Class(i)}}{nm^{pre}_{Class(i)}}$$

$$BL_{cur\_MB} \quad + \quad AL_{cur\_MB} \quad = \quad C_{cur\_MB}$$

FIGURE 4.8

The tight-MLA process for Class 2 and Class 3 MBs. (From Lin et al., A computation control motion estimate method for complexity-scalable video coding, *IEEE Trans. Circuits Syst. Video Technol.*, 20(11), 1533, Fig. 6. With permission.)

$$AL_{cur\_MB} = \begin{cases} 0 & \text{if } Class_{cur\_MB} = 1 \\ \min\left(\max\left(\dfrac{COST^{init}_{cur\_MB}}{Avg\_COST^{init}_{Class(2)}} \cdot \dfrac{ab_{Class\,(2)}}{nm^{pre}_{Class\,(2)}}, 0\right), AL_{MB\_max\_Class(2)}\right) & \text{if } Class_{cur\_MB} = 2 \\ \min\left(\max\left(\dfrac{COST^{init}_{cur\_MB}}{Avg\_COST^{init}_{Class(3)}} \cdot \dfrac{ab_{Class\,(3)}}{nm^{pre}_{Class\,(3)}}, 0\right), AL_{MB\,\,max\,\,Class(3)}\right) & \text{if } Class_{cur\_MB} = 3 \end{cases}$$

C_{cur_MB} is the computation allocated to the current MB, $COST^{init}_{cur\_MB}$ is the initial COST of the current MB as in Equation 4.2, $Avg\_COST^{init}_{Class(i)}$ is the average of the initial COST for all the already-coded MBs belonging to Class i in the current frame, "$ab_{Class(i)}$" is the computation budget available in the additional layer for class i before coding the current MB, and "$nm^{pre}_{Class(i)}$" is the estimated number of remaining uncoded MBs for class i before coding the current MB. $BLC_{MB\_Class(2)}$ is equal to $BL_{MB\_Class(2)}$ if either $ab_{Class(2)} > 0$ or $nm_{Class(2)} > 1$, and equal to $BL_{MB\_Class(3)}$ otherwise. It should be noted that $BLC_{MB\_Class(2)}$ is defined to follow the tight strategy where a larger ML–DL budget ($BL_{MB\_Class(2)}$) is used if the available budget is sufficient and a smaller ML–BL budget ($BL_{MB\_Class(3)}$) is used otherwise. $AL_{MB\_max\_Class(2)}$ and $AL_{MB\_max\_Class(3)}$ are the same as in Equation 4.5 and are set in order to avoid too many useless SPs allocated to the current MB. In the experiments of this chapter, we set $AL_{MB\_max\_Class(i)} + BL_{MB\_Class(i)} = 250$ for a search range of ±32 pixels. It should be noted that since we cannot get the exact number of remaining MBs for each class before coding the whole frame, $nm^{pre}_{Class(i)}$ is estimated by the parameters of the previous frame. "$ab_{Class(i)}$" and "$nm^{pre}_{Class(i)}$" are set as $AL_{Class(i)}$ and $NM^{pre}_{Class(i)}$, respectively, at the beginning of each frame and are updated before coding the current MB as in

$$
\begin{cases}
ab_{Class(i)} = ab_{Class(i)} - \left(CA_{per\_MB} - BL_{per\_MB}\right) & \text{if } Class_{per\_MB} = i \\
nm^{per}_{Class(i)} = nm^{per}_{Class(i)} - 1 & \text{if } Class_{per\_MB} = i
\end{cases}
\tag{4.7}
$$

where

the definition of $NM^{pre}_{Class(i)}$ is the same as in Equation 4.5

CA_{pre_MB} and BL_{pre_MB} represent the actual computation consumed and the basic layer computation allocated for the MB right before the current MB, respectively

From Equations 4.5 through 4.7, we can see that the CLA and MLA steps are based on classification using our CIM method, where Class 1 MBs are always allocated a fixed small number of SPs, and Class 2 and Class 3 MBs are first separated into independent class budgets and then allocated based on their *init_COST* value within each class budget. Thus, the proposed CCME algorithm can combine the class information and COST information for a more precise computation allocation.

4.3.3 Step-Level Computation Allocation

The SLA process will allocate the computation budget for an MB into each ME step. Since the SHS method is used to perform ME in this chapter, we will describe our SLA step based on the SHS algorithm. However, our SLA method can easily be applied to other ME algorithms (Li et al., 1994; Lin et al., 2008; Po and Ma, 1996; Zhou et al., 2004; Zhu and Ma, 2000).

The SLA process can be described as

$$
\begin{cases}
C_{Small\_Local\_Search} = C_{Step\_min} \\
C_{Cross\_Search} = NS_{Cross\_Search} \cdot CS_{Cross\_Search} \\
C_{Multi\_Hex\_Search} = NS_{Multi\_Hex\_Search} \cdot CS_{Multi\_Hex\_Search} \\
C_{Small\_Hex\_Search} = \begin{cases} \text{Let it go} & \text{if } (NS_{Cross\_Search} + NS_{Multi\_Hex\_Search}) > 1 \\ 0 & \text{if } (NS_{Cross\_Search} + NS_{Multi\_Hex\_Search}) \leq 1 \end{cases} \\
C_{Small\_Diamond\_Search} = \begin{cases} \text{Let it go} & \text{if } NS_{Cross\_Search} > 1 \\ 0 & \text{if } NS_{Cross\_Search} \leq 1 \end{cases}
\end{cases}
\tag{4.8}
$$

where $C_{Small\_Local\_Search}$, C_{Cross_Search}, $C_{Multi\_Hex\_Search}$, $C_{Small\_Hex\_Search}$,   and $C_{Small\_Diamond\_Search}$ are the computations allocated to each ME step of the SHS algorithm. C_{Step_min} is the minimum guaranteed computation for the *Small Local Search Step*. In the case of the SHS method, C_{Step_min} is set to be 4. CS_{Cross_Search} and $CS_{Multi\_Hex\_Search}$ are the number of SPs in each substep of the *Cross Search Step* and the *Multi Hexagon Search Step*, respectively. For the SHS method, CS_{Cross_Search} and $CS_{Multi\_Hex\_Search}$ are equal to 4 and 16, respectively (Yi et al., 2005). "Let it go" in Equation 4.8 means performing the regular motion search step. NS_{Cross_Search} and $NS_{Multi\_Hex\_Search}$ are the number of substeps in the *Cross Search Step* and the *Multi Hexagon Search Step*, respectively. They are calculated as

$$
\begin{cases}
NS_{Cross\_Search} = \left\lfloor \dfrac{RT_{Cross\_Search} \cdot (C_{cur\_MB} - C_{Step\_min})}{CS_{Cross\_Search}} \right\rfloor \\
NS_{Multi\_Hex\_Search} = \left\lfloor \dfrac{RT_{Multi\_Hex\_Search} \cdot (C_{cur\_MB} - C_{Step\_min})}{CS_{Multi\_Hex\_Search}} \right\rfloor
\end{cases}
\tag{4.9}
$$

where C_{cur_MB} is the computation budget for the whole MB as in Equation 4.6. RT_{Cross_Search} and $RT_{Multi\_Hex\_Search}$ are the predefined ratios by which the MB's budget C_{cur_MB} is allocated to the *Cross Search Step* and the *Multi Hexagon Search Step*. In the case of SHS method, we set RT_{Cross_Search} to be 0.32 and $RT_{Multi\_Hex\_Search}$ to be 0.64. This means that 32% of the MB's budget will be allocated to the *Cross Search Step* and 64% of the MB's budget will be allocated to the *Cross Search Step*. We use the floor function ($\lfloor \cdot \rfloor$) in order to make sure that the integer substeps of SPs are allocated.

From Equation 4.8, we can see that the SLA process will first allocate the minimum guaranteed computation to *the Small Local Search Step*. Then most of the available computation budget will be allocated to the *Cross Search Step* (32%) and the *Multi Hexagon Search Step* (64%). If there is still enough computation left after these two steps, the regular *Small Hexagon Search* and *Small Diamond Search* will be performed to refine the final MV. If there is not enough budget for the current MB, some motion search steps such as the *Small Hexagon Search* and *Small Diamond Search* will be skipped. In the extreme case, for example, if the MB's budget has only six SPs, then all the steps after the *Small Local Search* will be skipped, and the SLA process will end up with only performing a *Small Local Search*. It should be noted that since the SLA is proceeded before the ME process, the computation will be allocated to the *Cross Search* and the *Multi Hexagon Search Steps*, no matter whether these steps are skipped in the later ME process (i.e., skipped by th_2 in Figure 4.3).

4.4 Experimental Results

We implemented our proposed CCME algorithm on the H.264/MPEG-4 AVC reference software JM10.2 version (HHI, 2011). Motion search was based on SHS (Yi et al., 2005), where th_1 and th_2 in Figure 4.3 is set to be

1000 and 5000, respectively. For each of the sequences, 100 frames were coded, and the picture coding structure was IPPP.... It should be noted that the first *P* frame was coded by the original SHS method (Yi et al., 2005) to obtain initial information for each class. In the experiments, only the 16 × 16 partition was used with one reference frame coding for the *P* frames. The *QP* was set to be 28, and the search range was ±32 pixels.

4.4.1 Experimental Results for the CCME Algorithm

In this section, we show experimental results for our proposed CCME algorithm. We fix the target computation (or SP) budget for each frame. The results are shown in Table 4.3 and Figure 4.9.

Table 4.3 shows *PSNR*, *Bit Rate*, *the average number of search points actually used per frame (Actual SP)*, and *the average number of search points per MB (Actual SP/MB)* for different sequences. The *Budget* columns in the table represent the target SP budget for performing ME, where 100% in the *Scale* column represents the original SHS (Yi et al., 2005). Since we fix the target SP budget for each frame, the values in the *Scale* column are measured in terms of *the number of SPs per frame* (e.g., 40% in the *Scale* column means the target SP budget *for each frame* is 40% of the average-SP-per-frame value of the original SHS [Yi et al., 2005]). Similarly, the values in the *Budget SP* column represent the corresponding number of SPs per frame for the budget scale levels indicated by the *Scale* column. Figure 4.9 shows the number of SPs used for each frame as well as the target SP budgets for each frame under 60% budget levels for *Football_Cif*. Similar results can be found for other sequences.

TABLE 4.3

Experimental Results for the Tight Strategy When Fixing the Target Budget for Each Frame

Sequence	Budget		Actual SP	PSNR (dB)	Bit Rate (kbps)	Actual SP/MB
	Scale	Budget SP				
Football_Cif	100%	22,042	22,042	35.96	1661.62	55
	60%	13,225	10,692	35.96	1678.38	27
	40%	8,816	8,615	35.96	1682.57	21
Mobile_Cif	100%	9,871	9,871	33.69	2150.60	24
	60%	5,922	5,785	33.69	2152.56	15
	40%	3,948	3,825	33.68	2165.31	10

Source: From Lin et al., A computation control motion estimate method for complexity-scalable video coding, *IEEE Trans. Circuits Syst. Video Technol.,* 20(11), 1533, Table 3, With permission.

Note: The *Budget SP* and the *Actual SP* columns are measured in terms of the number of SPs per frame.

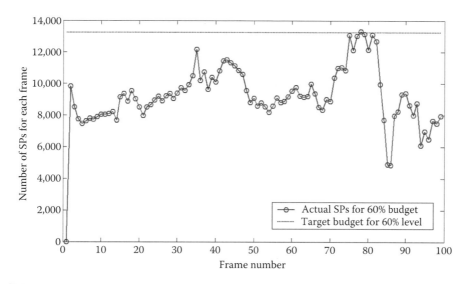

FIGURE 4.9

The number of SPs used for each frame vs. the target frame-level budgets for the tight strategy for *Football_Cif*. (From Lin et al., A computation control motion estimate method for complexity-scalable video coding, *IEEE Trans. Circuits Syst. Video Technol.*, 20(11), 1533, Fig. 7. With permission.)

Comparing the *Actual SP* column with the *Budget SP* column in Table 4.3, we can see that *the number of SPs actually used* is always smaller than the *target SP budget* for all target budget levels. This demonstrates that our CCME algorithm can efficiently perform computation allocation to meet the requirements of different target computation budgets. From Table 4.3, we can also see that our CCME algorithm has good performance even when the available budget is low (40% for Football and Mobile). This demonstrates the allocation efficiency of our algorithm. Furthermore, from Figure 4.9, we can see that since the CCME algorithm is based on the *tight strategy*, which does not allow computation borrowing from other frames, the number of SPs used in each frame is always smaller than the target frame-level budget. Thus, the average SPs per frame for the *tight strategy* is always guaranteed to be smaller than the target budget.

4.4.2 Comparison with Other Methods

In the previous sections, we have shown experimental results for our proposed CCME algorithm. In this section, we will compare our CCME methods with other methods.

Similar to the previous section, we fixed the target computation budget for each frame to prevent the effect of FLA. The following three methods are compared. It should be noted that all these three methods use our SLA method for a fair comparison:

TABLE 4.4

Performance Comparison for CCME Algorithms

	Budget (%)	Proposed			COST Only			(0,0) SAD		
		PSNR	BR	SPs	PSNR	BR	SPs	PSNR	BR	SPs
Bus_Cif	100	34.31	1424	35	34.31	1424	35	34.31	1424	35
	60	34.31	1459	20	34.29	1484	19	34.29	1482	20
	40	34.29	1524	13	34.25	1628	12	34.27	1642	13
Mobile_Cif	100	33.69	2151	24	33.69	2151	24	33.69	2151	24
	50	33.68	2153	12	33.69	2187	12	33.69	2196	11
	30	33.68	2167	7	33.66	2276	7	33.66	2283	7
Stefan_Cif	100	35.12	1354	22	35.12	1354	22	35.12	1354	22
	50	35.11	1369	11	35.09	1404	10	35.09	1394	11
	35	35.10	1376	7	34.98	1703	7	35.05	1642	7
Dancer_Cif	100	39.09	658	16	39.09	658	16	39.09	658	16
	60	39.10	701	9	39.12	746	9	39.11	732	8
	50	39.10	717	8	39.11	768	7	39.12	756	7
Foreman_Cif	100	36.21	515	16	36.21	515	16	36.21	515	16
	70	36.21	520	11	36.21	519	10	36.22	520	10
	50	36.22	522	8	36.21	522	7	36.22	523	8
Football_Cif	100	35.96	1662	55	35.96	1662	55	35.96	1662	55
	60	35.96	1678	27	35.96	1681	29	35.97	1689	28
	40	35.96	1682	21	35.95	1719	21	35.96	1711	21

Source: From Lin et al., A computation control motion estimate method for complexity-scalable video coding, *IEEE Trans. Circuits Syst. Video Technol.*, 20(11), 1533, Table 4, With permission.

1. Perform the proposed CCME algorithm with the *tight strategy* (*Proposed* in Table 4.4).

2. Do not classify the MBs into classes and allocate computation only based on their *Init_COST* (Chen et al., 2006; Huang et al., 2005) (*COST only* in Table 4.4).

3. First search the (0,0) points of all the MBs in the frame, and then allocate SPs based on (0,0) SAD. This method is the variation of the strategy for many multi-pass methods (Tai et al., 2003; Yang et al., 2005) ((0,0) *SAD* in Table 4.4).

Table 4.4 compares PSNR (in dB), bit rate (BR, in kbps), and the average number of search points per MB (SPs). The definition of the *Budget Scale* column of the table is the same as in Table 4.3. Figure 4.10 shows the *BR Increase* vs. *Budget Level* for these methods where the *BR Increase* is defined by the ratio between the current bit rate and its corresponding 100% Level bit rate.

From Table 4.4 and Figure 4.10, we can see that our proposed CCME method can allocate SPs more efficiently than the other methods at different

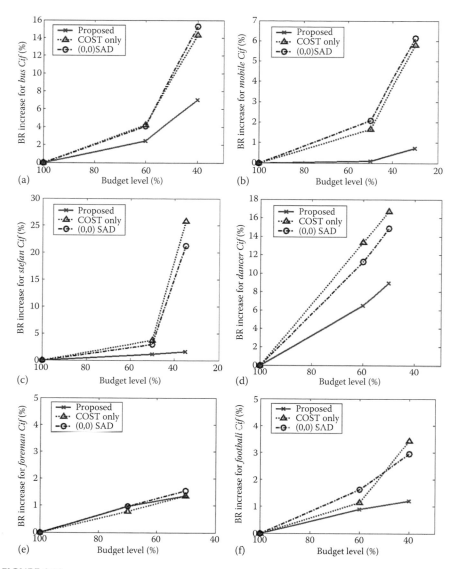

FIGURE 4.10
Performance comparison for different CCME algorithms. (a) *Bus_Cif*, (b) *Mobile_Cif*, (c) *Stefan_Cif*, (d) *Dancer_Cif*, (e) *Foreman_Cif*, (f) *Football_Cif*. (From Lin et al., A computation control motion estimate method for complexity-scalable video coding, *IEEE Trans. Circuits Syst. Video Technol.*, 20(11), 1533, Fig. 8. With permission.)

computation budget levels. This demonstrates that our proposed method, which combines the class and the COST information of the MB, can provide a more accurate way to allocate SPs.

For a further analysis of the result, we can compare the bit-rate performance of the Mobile sequence (i.e., Figure 4.10b) with its MB classification result (i.e., Figures 4.5b through d). When the budget level is low, our proposed algorithm can efficiently extract and allocate more SPs to the more important Class 2 MBs (Figure 4.5c) while reducing the unnecessary SPs from Class 3 (Figure 4.5d). This keeps the performance of our method as high as possible. Furthermore, since the number of extracted Class 2 MBs is low (Figure 4.5c), our proposed algorithm can still keep high performance at very low budget levels (e.g., 5% budget level in Figure 4.10b). Compared to our method, the performances of the other methods will significantly decrease when the budget level becomes low.

However, the results in Table 4.4 and Figure 4.10 also show that for some sequences (e.g., Foreman and Football), the advantages of our CCME algorithm are not so obvious from the other methods. This is because

1. For some sequences such as *Football*, the portion of Class 2 MBs is large. In this case, the advantages of our CCME method from MB classification become less obvious. In extreme cases, if all MBs are classified into Class 2, our proposed CCME algorithm will be the same as the *COST only* algorithm.

2. For some sequences such as *Foreman*, the performance will not decrease much even when very few points are searched for each MB (e.g., our experiments show that the performance for *Foreman_Cif* will not decrease much even if we only search six points for each MB). In this case, different computation allocation strategies will not make much difference.

Table 4.5 shows the results for sequences with different resolutions (*Mobile_Qcif* and *Mobile_SD*) or using different *QPs* (Bus with *QP* = 23 or 33). Table 4.5 shows the efficiency of our algorithm under different resolutions

TABLE 4.5

Experimental Results for Sequences with Different Resolutions or Using Different QPs

		Proposed			COST Only			(0,0) SAD		
	Budget	PSNR	BR	SPs	PSNR	BR	SPs	PSNR	BR	SPs
Bus_Cif	100%	38.28	2639	33	38.28	2639	33	38.28	2639	33
QP = 23	50%	38.26	2762	14	38.23	2912	13	38.24	2896	14
Bus_Cif	100%	30.47	722	40	30.47	722	40	30.47	722	40
QP = 33	50%	30.46	789	16	30.41	902	15	30.41	879	15
Mobile Qcif	100%	32.90	545	16	32.90	545	16	32.90	545	16
QP = 28	50%	32.90	545	7	32.90	546	7	32.90	545	7
Mobile SD	100%	34.07	7766	24	34.07	7766	24	34.07	7766	24
QP = 28	30%	34.07	7776	7	34.06	8076	7	34.05	8124	7

Source: From Lin et al., A computation control motion estimate method for complexity-scalable video coding, *IEEE Trans. Circuits Syst. Video Technol.*, 20(11), 1533, Table 5. With permission.

and different *QP*s. Furthermore, we can also see from Table 4.5 that the performance of our algorithm is very close to the other methods for *Mobile_Qcif*. The reason is similar to the case of *Foreman_Cif* (i.e., a local search for each MB can still get good performance and thus different computation allocation strategies will not make much difference).

From these discussions, we can summarize the advantages of our proposed CCME algorithm as follows:

1. The proposed algorithm uses a *more suitable way to measure MB importance* by differentiating MBs into different classes. When the available budget is small, the proposed method can save unnecessary SPs from Class 3 MBs so that more SPs can be allocated to the more important Class 2 MBs, which keeps the performance as high as possible. When the available target budget is large, the method will have more spare SPs for Class 3 MBs, which can overcome the possible performance decrease from MB misclassification and further improve the coding performance.

2. The proposed algorithm can reduce the impact of not having a global view of the whole frame for one-pass methods by (a) setting the basic and the additional layers, (b) using previous frame information as the global view estimation, (c) guaranteeing Class 2 MBs a higher minimum SPs, and (d) using three independent class budgets so that an unsuitable allocation in one class will not affect other classes.

Furthermore, we also believe that the framework of our CCME algorithm is general and can easily be extended. Some possible extensions of our algorithm can be described as follows:

1. As mentioned, other FLA or SLA methods (Chen et al., 2006; He et al., 2005; Huang et al., 2005; Kim et al., 2006; Tai et al., 2003; Yang et al., 2005) can easily be implemented into our CCME algorithm. For example, in some time-varying motion sequences, an FLA algorithm may be very useful to allocate more computation to those high-motion frames and further improve the performance.

2. In this chapter, we only perform experiments on the 16×16 partition size and the IPPP... picture type. Our algorithm can easily be extended to ME with multiple partition sizes and multiple reference frames such as in H.264/AVC (Wiegand et al., 2003) as well as other picture types.

3. In this chapter, we define three MB classes and perform CCME based on these three classes. Our method can also be extended by defining more MB classes and developing different CLA and MLA steps for different classes.

4.5 Extending Our MB Classification Method for Other Video-Processing Applications

In this section, we discuss using our proposed MB classification method for other video processing applications. From Equation 4.4 and Figure 4.5, we can further draw the following observations: From Figures 4.5b and f, we can see that most Class 1 MBs include backgrounds or flat areas that can find good matches in the previous frames. From Figure 4.5c and g, we can see that our classification method can effectively detect *irregular areas* and classify them into Class 2 (e.g., the *edge between the calendar and the background* as well as *the bottom circling ball* in (c), and *the running bus* as well as *the down-right logo* in (g)). From Figures 4.5d and h, we can see that most complex-texture areas are classified as Class 3, such as the complex background and calendar in (d) as well as the flower area in (h).

Therefore, based on the proposed MB class information in Equation 4.4, we can also develop various algorithms for different applications. Note that since our proposed method is directly defined based on the information readily available from the ME process or from the compressed video bit-stream, it is with low computational complexity and is applicable to various video applications, especially for those with low-delay and low-cost require-ments. In the following, we will propose algorithms for the three example applications: shot change detection, motion discontinuity (MD) detection, and outlier rejection for GME.

4.5.1 Shot Change Detection

In this chapter, we define a "shot" as a segment of continuous video frames captured by one camera action (i.e., a continuous operation of one camera), and a "shot change" as the boundary of two shots. Figure 4.11 shows an example of an abrupt shot change.

From the previous discussions, we can outline the ideas of applying our approach to shot change detection as follows: Since shot changes (including abrupt, gradual, fade-in or fade-out) (Lin et al., 2010) always happen between

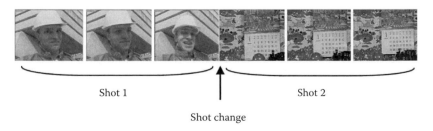

Shot 1   Shot change   Shot 2

FIGURE 4.11
An example of an abrupt shot change.

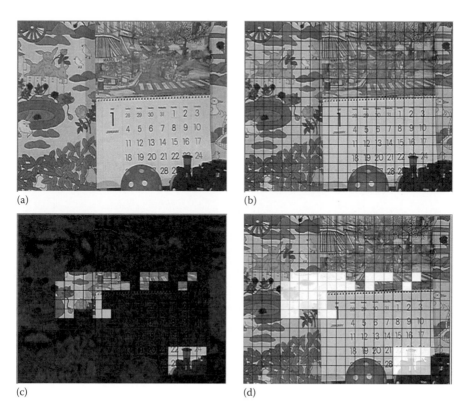

FIGURE 4.12
The MB distributions at the abrupt shot change frame from *Bus_Cif* to *Mobile_Cif*: (a) Original frame, (b) Class 1, (c) Class 2, (d) Class 3.

two uncorrelated video shots, the content correlation between frames at shot changes will be low. Therefore, we can use the information of Class 1 as the primary feature to detect shot changes. Furthermore, since the motion pattern will also change at shot changes, the information of Class 2 and Class 3 can be used as additional features for shot change detection.

Figure 4.12 shows an example of the effectiveness in using our class information for shot change detection. More results will be shown in the experimental results. Figures 4.12b through d show the MB distributions of three classes at the abrupt shot change from *Bus_Cif* to *Mobile_Cif*. We can see that the information of Class 1 can effectively indicate the low content correlation between frames at the shot change (i.e., no MB is classified as Class 1 in Figure 4.12b). Furthermore, a large number of MBs are classified as Class 2. This indicates the rapid motion pattern change at the shot change.

Based on these discussions, we can propose a Class-Based Shot Change detection (CB-Shot) algorithm. It is described as in Equation 4.10:

$$Fg_{shot}(t) = \begin{cases} 1 & \text{if} \quad N_{Class\_1}(t) \leq T_1 \ \text{and} \ N_{Intra\_MB}(t) - N_{IR}(t) \geq T_4 \\[2mm] & \text{or if} \begin{cases} N_{Class\_1}(t) \leq T_2 \ \text{and} \ N_{Intra\_MB}(t) - N_{IR}(t) \geq T_4 \ \text{and} \\[1mm] \left| N_{Class\_2}(t) - N_{Class\_2}(t-1) \right| + \left| N_{Class\_3}(t) - N_{Class\_3}(t-1) \right| \geq T_3 \end{cases} \\[4mm] 0 & \text{else} \end{cases}$$

$$(4.10)$$

where

     t is the frame number

     $Fg_{shot}(t)$ is a flag indicating whether a shot change happens at the current frame t or not

$Fg_{shot}(t)$ will be equal to 1 if there is a shot change and will be equal to 0 else. $N_{Intra\_MB}(t)$ is the number of intra-coded MBs at frame t; $N_{IR}(t)$ is the number of intra-refresh MBs in the current frame (i.e., forced intra-coding MBs [HHI, 2011]); $N_{Class\_1}(t)$, $N_{Class\_2}(t)$, and $N_{Class\_3}(t)$ are the total number of Class 1, Class 2, and Class 3 MBs in the current frame t, respectively. T_1, T_2, T_3, and T_4 are the thresholds for deciding the shot change. In this chapter, T_1 through T_4 are calculated by Equation 4.11:

$$T_1 = \frac{(N_{MB}(t) - N_{IR}(t))}{40}$$

$$T_2 = \frac{(N_{MB}(t) - N_{IR}(t))}{30}$$

$$(4.11)$$

$$T_3 = \frac{(N_{MB}(t) - N_{IR}(t))}{4}$$

$$T_4 = T_1$$

where $N_{MB}(t)$ is the total number of MBs of all classes in the current frame.

It should be noted that in Equation 4.10, the Class 1 information is the main feature for detecting shot changes (i.e., $N_{Class\_1}(t) \leq T_1$ and $N_{Class\_1}(t) \leq T_2$ in Equation 4.10). The intuitive of using the Class 1 information as the major feature is that it is a good indicator of the content correlation between frames. The Class 2 and Class 3 information is used to help detect frames at the beginning of some gradual shot changes where a large change in motion pattern has been detected but the number of Class 1 MBs has not yet decreased to a small number. The intra-coded MB information can help discard the possible false alarm shot changes due to the MB misclassification. From Equations 4.10 and 4.11, we can also see that when intra-refresh functionality is enabled (i.e., when $N_{IR}(t) > 0$), our algorithm can be extended by simply excluding these intra-refreshed MBs and only performing shot change detection based on the remaining MBs.

Furthermore, note that Equation 4.10 is only one implementation of using our class information for shot change detection. We can easily extend Equation 4.10 by using more sophisticated methods such as cross-validation (Lin et al., 2008) to decide the threshold values in an automatic way. Furthermore, some machine learning models can also be used to decide the shot detection rules and take the place of the rules in Equation 4.10. For example, we can train a support vector machine (SVM) or a Hidden Markov Model (HMM) based on our class information for detecting shot changes (Liu et al., 2004; SVMLight, 2011). By this way, we can avoid the tediousness of manually tuning multiple thresholds simultaneously. This point will be further discussed in the experimental results.

4.5.2 Motion Discontinuity Detection

We define motion discontinuity as the boundary between two smooth camera motions (SCMs), where SCM is a segment of continuous video frames captured by one single motion of the camera (such as zooming, panning, or tilting) (Lin et al., 2008; Shu and Chau, 2005). For example, in Figure 4.13 the first several frames are captured when the camera has no or little motion. Therefore, they form the first SCM (SCM1). The second several frames form another SCM (SCM2) because they are captured by a single camera motion of rapid rightward. Then, an MD can be defined between these two SCMs. It should be noted that the major difference between shots and SCMs is that a shot is normally composed of multiple SCMs.

Basically, motion discontinuity can be viewed as motion unsmoothness or the change of motion patterns. The detection of motion discontinuity can be very useful in video content analysis or video coding performance improvement (Boyce, 2004; Lee and Lee, 2001). Since our class information, especially Class 2 information, can efficiently reflect the irregular motion patterns, it can be easily used for MD detection.

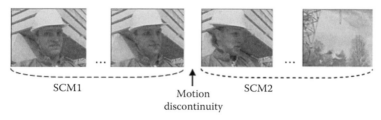

SCM1   Motion   SCM2
              discontinuity

FIGURE 4.13
An example of motion discontinuity.

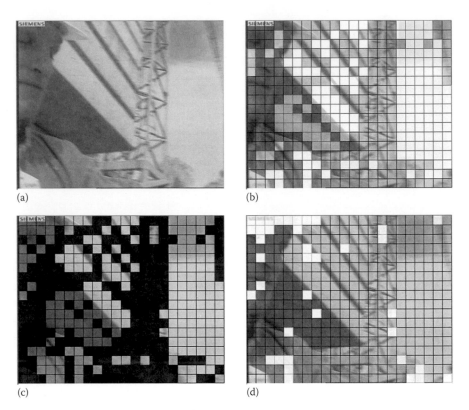

(a)   (b)

(c)   (d)

FIGURE 4.14
The MB distributions at a motion discontinuity frame in *Foreman_Cif*: (a) Original frame, (b) Class 1, (c) Class 2, (d) Class 3.

The ideas of applying our MB class information into MD detection can be outlined as follows: Since MD happens between different camera motions, the motion smoothness will be disrupted at the places of MDs. Therefore, we can use the Class 2 information as the primary feature to detect MDs. Furthermore, since frames at MDs belong to the same camera action (i.e., the same shot), their content correlation will not decrease. Therefore, the information of Class 1 can be used to differentiate shot changes from MDs.

Figure 4.14 shows an example of the effectiveness in using our class information in MD detection. Figure 4.14b through d shows the MB distributions of a MD frame in *Foreman_Cif* when the camera starts to move rightward rapidly. The large number of Class 2 MBs indicates the motion unsmoothness due to the MD. Furthermore, the big number of Class 1 MBs indicates the high content correlation between frames, which implies that it is not a shot change.

Therefore, we can propose a class-based MD detection (CB-MD) algorithm. It is described as

$$Fg_{MD}(t) = \begin{cases} 1 & \text{if } N_{Class\_1}(t) \geq th_1^{MD} \quad \text{and} \quad \sum_{i=0}^{k} I\left(N_{Class\_2}(t-i) \geq th_3^{MD}\right) = k+1 \\ 0 & \text{else} \end{cases}$$

(4.12)

where $I(f)$ is an indicator. I will equal to 1 if f is true, and 0 if f is false. Equation 4.12 means that an MD will be detected only if the number of Class 2 MBs is larger than a threshold for $k+1$ consecutive frames. This is based on the assumption that an obvious camera motion change will affect several frames rather than one. By including the information of several frames, the false alarm rate can be reduced. Furthermore, similar to shot change detection, the decision rules in Equation 4.12 can also be extended to avoid the manual setting of thresholds.

4.5.3 Outlier Rejection for Global Motion Estimation

Global motion estimation is another useful application of our class information. Since a video frame may often contain various objects with different motion patterns and directions, motion segmentation is needed to filter out these motion regions before estimating the global motion parameters of the background. Since our class information can efficiently describe the motion patterns of different MBs, it is very useful in filtering out the irregular motion areas (outliers). For example, we can simply filter out Class 2 or Class 2 + Class 3 MBs and perform GME based on the remaining MBs.

Based on the MB class information, the proposed GME algorithm can be described as follows:

Step 1: Use our class information to get a segmentation of the irregular motion MBs, as in

$$Foreground = \begin{cases} \text{all Class 2 and Class 3 MBs} & \text{if } (N_{Class\_2}(t) + N_{Class\_3}(t)) < Th_F \\ \text{Class 2 MBs} & \text{else} \end{cases}$$

(4.13)

where
  $N_{Class\_2}(t)$ and $N_{Class\_3}(t)$ are the number of Class 2 and Class 3 MBs in t
  Th_F is a threshold

Step 2: Estimate the global motion parameters based on the remaining background MVs. In this chapter, we use the six-parameter model as the global motion model, as described in

$$
\begin{pmatrix} x' \\ y' \\ 1 \end{pmatrix} = GMV^{6p}(x,y;a,b,c,d,e,f) = S(a,b,c,d,e,f) \times \begin{pmatrix} x \\ y \\ 1 \end{pmatrix} \quad (4.14)
$$

where

$$
S(a,b,c,d,e,f) = \begin{pmatrix} a & b & c \\ d & e & f \\ 0 & 0 & 1 \end{pmatrix} \text{ is the six-parameter model}
$$

(x,y) and (x',y') represent the pixel's original and global motion-compensated location, respectively

There are many ways to estimate S. In this chapter, we use the Least-Square method (Soldatov et al., 2006), which searches parameters in S that minimizes a given cost function (mean-square error), as in

$$
S(a,b,c,d,e,f) = \underset{S(a,b,c,d,e,f)}{\arg\min} \left(\sum_{x,y} \left\| V(x,y) - GMV^{6p}(x,y;a,b,c,d,e,f) \right\|^2 \right) \quad (4.15)
$$

where $V(x,y) = (V^x \; V^y \; 1)^T$ and (V^x,V^y) are the MV-terminate coordinates for pixel (x,y).

Figure 4.15 shows some results of using our class information for motion region segmentation. From Figures 4.15a and b, we can see that our class information can efficiently locate the foreground motion regions. From Figure 4.15c, we can also see that our algorithm focuses more on detecting the "motion regions" instead of the foreground object. In Figure 4.15c, since only the person's left hand is moving while the other parts of the person keep unchanged, only those blocks corresponding to the left hand are identified as motion regions. However, it is noted that our class information can also be extended to detect real foreground objects by combining with texture information such as DC and AC coefficients (Porikli et al., 2010).

4.5.4 Experimental Results

In this section, we perform experiments for the proposed methods in Sections 4.5.1 through 4.5.3. The algorithms are implemented on the H.264/MPEG-4 AVC reference software JM10.2 version (HHI, 2011). The picture resolutions for the sequences are CIF and SIF. For each of the sequences, the picture coding structure was IPPP.… In the experiments, only the 16×16 partition was used with one reference frame coding for the P frames. The QP was set to be 28, the search range was ±32 pixels, and the frame rate was 30 frames/s. The motion estimation is based on our proposed Class-based Fast Termination

FIGURE 4.15
Examples of using our class information for motion region segmentation for global motion estimation ((A) original frames; (B) segmented frames): (a) Original frame for *Dancer_cif*, (b) Segmented frame for *Dancer_cif* in (a), (c) Original frame for *Stefan_cif*, (d) Segmented frame for *Stefan_cif* in (c), (e) Original frame for *Silent_cif*, (f) Segmented frame for *Silent_cif* in (e).

method (Lin et al., 2009). Note that our MB classification method is general regardless of the ME algorithms used. It can easily be extended to other ME algorithms (Li et al., 1994; Po and Ma, 1996). Furthermore, we disable the intra-refresh functionality (HHI, 2011) in the experiments in this chapter in order to focus on our class information. However, from our experiments, the shot detection results will not differ by much when intra-refresh is enabled.

4.5.4.1 Experimental Results for Shot Change Detection

We first perform experiments for shot change detection. Four shot change detection algorithms are compared:

1. *Detect shot changes based on the number of Intra-MBs* (Eom and Choe, 2007; Zhang and Kittler, 1999) (*Intra-based* in Table 4.6). A shot change will be detected if the number of Intra-MBs in the current frame is larger than a threshold.
2. *Detect shot changes based on motion smoothness* (Akutsu et al., 1992; Shu and Chau, 2005) (*MV-Smooth-based* in Table 4.6). The motion smoothness can be calculated by the Square of Motion Change (Shu and Chau, 2005), as in

$$SMC(t) = \sum_{i \in current\_frame} \left(\left(MV_x^i(t) - MV_x^i(t-1) \right)^2 + \left(MV_y^i(t) - MV_y^i(t-1) \right)^2 \right)$$

(4.16)

where
$SMC(t)$ is the value of the *Square of Motion Change(SMC)* at frame t
$MV_x^i(t)$ and $MV_y^i(t)$ are the x and y components of the motion vector for Macroblock i of frame t, respectively

From Equation 4.16, we can see that SMC is just the "sum of squared motion vector difference" between colocated MBs of neighboring

TABLE 4.6

Performance Comparison of Different Algorithms in Detecting the Shot Changes in the Extended TRACVID Dataset

	Miss (%)	False Alarm (%)	TEFR
Intra-based	25.24	4.27	13.50
MV-Smooth-based	43.72	17.36	22.75
Intra + MV-Smooth	24.71	3.49	12.58
Proposed-Class 1 only	8.34	3.81	5.51
Proposed-All Class + Intra	6.13	2.91	3.23

frames. Based on Equation 4.16, a shot change can be detected if $SMC(t)$ is larger than a threshold at frame t.

3. *Detect shot changes based on the combined information of Intra-MB and motion smoothness.* (Shu and Chau, 2005) (*Intra + MV-Smooth* in Table 4.6). In this method, the Intra-MB information is included into the *SMC*, as in

$$SMC_{Intra\_included}(t) = \sum_{i \in current\_frame} MC(i) \qquad (4.17)$$

where $SMC_{Intra\_included}(t)$ is the *SMC with Intra-MB information included*. $MC(i)$ is defined as in

$$MC(i) = \begin{cases} \left(MV_x^i(t) - MV_x^i(t-1)\right)^2 + \left(MV_y^i(t) - MV_y^i(t-1)\right)^2 & \text{if } i \text{ is inter-coded} \\ L & \text{if } i \text{ is intra-coded} \end{cases}$$

$$(4.18)$$

where
i is the MB number
L is a large fixed number

In the experiment of this chapter, we set L to be 500. From Equations 4.17 and 4.18, we can see that the *Intra + MV-Smooth* method is similar to the *MV-Smooth-based* method except that when MB i is intra-coded, a large value L will be used instead of the *squared motion vector difference*. It should be noted that when the number of intra-MBs is low, the *Intra + MV-Smooth* method will be close to the *MV-Smooth-based* method. If the number of intra-MBs is high, the *Intra + MV-Smooth* method will be close to the *Intra-based* method.

4. *The proposed Class-based shot change detection algorithm*, which uses the Class 1 information as the major feature for detection, as in Equation 4.10 (*Proposed All Class + Intra* in Table 4.6).

It should be noted that we choose methods (I) through (III) as the reference algorithms to compare with our methods because they are all computationally efficient methods (with the average operation time less than 5 ms). Thus, they are suitable for the application of shot change detection for video coding.

Figure 4.16 shows the curves of features that are used in the aforementioned algorithms. Since all the algorithms perform well in detecting abrupt shot changes, we only show the curves of a gradual shot change in Figure 4.16.

Figures 4.16a through e are the feature curves of a gradual shot change sequence as in Figure 4.17a. In this sequence, the first five frames are *Bus_Cif,*

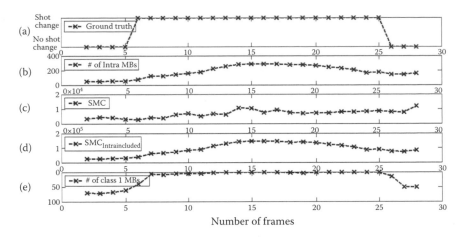

FIGURE 4.16
Feature curves of a gradual shot change sequence. (a) The ground-truth shot change frames, (b) the curve of the number of Intra-MBs in each frame, (c) the curve of $SMC(t)$, (d) the curve of $SMC_{Intra\_included}(t)$, and (e) the curve of the number of Class 1 MBs in each frame.

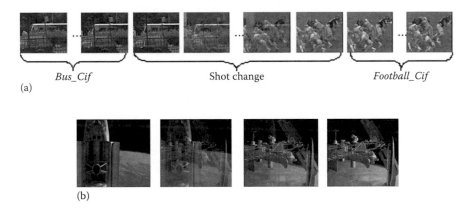

FIGURE 4.17
Example sequences in the extended TRACVID dataset. (a) An example sequence that we created. (b) The example sequence from TRECVID dataset (NIST, 2001).

the last five frames are *Football_Cif,* and the middle 20 frames are the period of the gradual shot change. Figure 4.16a is the ground truth for the shot change sequence; Figure 4.16b shows the curve of *the number of Intra-MBs* in each frame; Figure 4.16c shows the curve of $SMC(t)$; Figure 4.16d shows the curve of $SMC_{Intra\_included}(t)$; and Figure 4.16e shows the curve of *the number of Class 1 MBs* in each frame. It should be noted that we reverse the *y-axis* of Figure 4.16e so that the curve has the same concave shape as others.

Figure 4.16 shows the effectiveness of using our class information for shot change detection. From Figure 4.16e, we can see that *the number of Class 1*

MBs suddenly decreases to 0 when a shot change happens and then quickly increases to a large number right after the shot change period. Therefore, our proposed algorithms can effectively detect the gradual shot changes based on the Class 1 information. Compared to our class information, the method based on the *Intra-MB number*, *SMC(t)*, and $SMC_{Intra\_included}(t)$ has low effectiveness in detecting the gradual shot changes. We can see from Figures 4.16b through d that *the Intra-MB number*, *SMC(t)*, and $SMC_{Intra\_included}(t)$ have similar values for frames inside and outside the shot change period. This makes them very difficult to differentiate the gradual shot change frames. Figure 4.16c shows that *SMC(t)* is the least effective. This implies that only using motion smoothness information cannot work well in detecting shot changes. Our experiments show that the effectiveness of *SMC(t)* will be further reduced when both of the sub-sequences before and after the shot change have similar patterns or low motions. In these cases, the motion unsmoothness will not be so obvious at the shot change.

Table 4.6 compares the *Miss* rate, the *False Alarm* rate, and the total error frame rate (TEFR) (Lin et al., 2008) for different algorithms in detecting the shot changes in an extended TRECVID dataset. The extended TRECVID dataset has totally 60 sequences, which include both the sequences from the public TRECVID dataset (NIST, 2001) and the sequences that we create. There are totally 16 abrupt shot change sequences and 62 gradual shot change sequences with different types (gradual transfer, fade-in, and fade-out) and with different lengths of shot-changing period (e.g., 10 frames, 20 frames, and 30 frames). The example sequences of the dataset are shown in Figure 4.17. The *Miss* rate is defined by N_{FA}^k/N_-^k, where N_{FA}^k is the total number of misdetected shot change frames in sequence k and N_-^k is the total number of shot change frames in sequence k. The *False Alarm* rate is defined by N_{FA}^k/N_-^k, where N_{FA}^k is the total number of *false alarmed* frames in sequence k and N_-^k is the total number of *non-shot change* frames in sequence k. We calculate the *Miss* rate and the *False Alarm* rate for each sequence and average the rates. The *TEFR* rate is defined by N_{t_miss}/N_{t_f}, where N_{t_miss} is the total number of misdetected shot change frames for all sequences and N_{t_f} is the total number of frames in the dataset. The *TEFR* rate reflects the overall performance of the algorithms in detecting all sequences.

In order to have a fair comparison, we also list the results of only using Class 1 information for detection (i.e., detect a shot change frame if $N_{class\_1}(t) < T_1$, *Proposed-Class 1 only* in Table 4.6). In the experiments of Table 4.6, the thresholds for detecting shot changes in Method (1) (*Intra-based*), Method (2) (*MV-Smooth-based*), and Method (3) (*Intra + MV_Smooth*) are set to be 200, 2,000, and 105,000, respectively. These thresholds are selected based on the experimental statistics.

From Table 4.6, we can see that the performances of our proposed algorithms (*Proposed-Class 1 only* and *Proposed-All Class + Intra*) are better than the other methods.

Furthermore, several other observations can be drawn from Table 4.6 as follows:

1. Basically, our Class 1 information, the Intra-MB information (Eom and Choe, 2007; Zhang and Kittler, 1999), and the residue information (Arman et al., 1994) can all be viewed as the features to measure the content correlation between frames. However, from Table 4.6, we can see that the performance of our *Proposed-Class 1 only* method is obviously better than the *Intra-based* method. This is because the Class 1 information includes both the residue information and the motion information. Only those MBs with both regular motion patterns (i.e., *MV* close to *PMV* or (0,0) *MV*) and low-matching-cost values are classified as Class 1. We believe that these MBs can reflect more efficiently the nature of the content correlation between frames. In our experiment, we found that there are large portions of MBs in the gradual shot change frames where neither intra- nor inter-prediction can perform well. The inter/intra mode selections for these MBs are quite random, which affect the performance of the *Intra-based* method. Compared to the *Intra-based* method, our algorithm can work well by simply classifying these MBs outside Class 1 and discarding them from the shot change detection process.

2. The performance of the *Proposed-All Class + Intra* method can further improve the performance from the *Proposed-Class 1 only* method. This implies that including Class 2 and Class 3 can help detect those frames that cannot be easily differentiated by only using the Class 1 information at the boundary of the shot change period. Furthermore, the reduced FA rate of the *Proposed-All Class + Intra* method also implies that including the intra-coded MB information can help discard false alarm frames due to MB misclassification.

4.5.4.2 Experimental Results for Motion Discontinuity Detection

In this section, we perform experiments for MD detection. The following four methods are compared. Methods (1) through (3) are the same as in the previous section:

1. Detect MD based on the number of Intra-MBs (*Intra-based*).
2. Detect MD based on motion smoothness (*MV-Smooth-based*).
3. Detect MD based on the combined information of Intra-MB and motion smoothness (*Intra + MV-Smooth*).
4. Our proposed MD detection algorithm as in Equation 4.12 (*Proposed*).

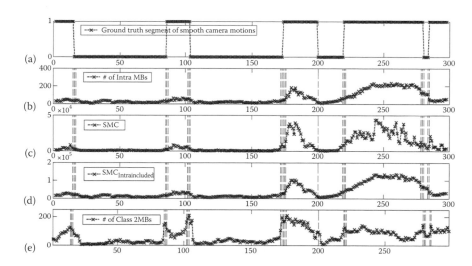

FIGURE 4.18
Feature curves for the MD detection in *Stefan_Sif.* (a) The ground-truth shot change frames, (b) the curve of the number of Intra-MBs in each frame, (c) the curve of $SMC(t)$, (d) the curve of $SMC_{Intra\_included}(t)$, and (e) the curve of the number of Class 2 MBs in each frame.

Figure 4.18 shows the curves of features that are used in the aforementioned algorithms for *Stefan_Sif* sequence. Figure 4.18a shows the ground-truth segment of SCMs. In Figure 4.18a, the segments valued 0 represent SCMs with low or no camera motion and the segments with value 1 represent SCMs with high or active camera motion. For example, the segment between frame 177 and 199 represents an SCM where there is a rapid rightward of the camera; and the segment between frame 286 and 300 represents an SCM of a quick zoom-in of the camera. The frames between SCMs are the MD frames that we want to detect. The ground-truth MD frames are labeled as the vertical dashed lines in Figures 4.18b through e. It should be noted that most MDs in Figure 4.18 include several frames instead of only one. Figure 4.18b through e show the curves of *the number of Intra-MBs*, $SMC(t)$, $SMC_{Intra\_included}(t)$, and *the number of Class 2 MBs*, respectively.

Several observations can be drawn from Figures 4.18b through e as follows:

1. Our Class 2 information is more effective in detecting the MDs. For example, in Figure 4.18e, we can see that our Class 2 information has strong response when the first three MDs happen. Comparatively, the other features in Figures 4.18b through d have low or no response. This implies that Methods (I) through (III) will easily miss these MDs.

2. Our Class 2 information has quicker and sharper response to MDs. For example, the value of our Class 2 information increases quickly at the places of the fourth (around frame 175) and sixth (around frame 220) MDs, while the other features respond much slower or more gradual.

3. Figure 4.18 also demonstrates that our Class 2 information is a better measure of the motion unsmoothness. Actually, the largest camera motion in *Stefan_Sif* takes place in the segment between frame 222 and frame 278. However, we can see from Figure 4.18e that the values of the Class 2 information are not the largest in this period. This is because although the camera motion is large, the motion pattern is pretty smooth during the period. Therefore, a large number of MBs will have regular and predictable motions and will not be classified as Class 2. In most cases, our Class 2 information will have the largest responses when the motion pattern changes or the motion smoothness disrupts. Compared to our Class 2 information, other features are more sensitive to the "motion strength" rather than the "motion unsmoothness." Furthermore, although SMC can also be viewed as a measure of the motion smoothness, we can see from Figure 4.18 that our Class 2 information is obviously a better measure for motion unsmoothness.

Figure 4.19b shows the MD detection result of the *proposed* method based on the Class 2 information in Figure 4.18e, where k, th_1^{shot}, and th_3^{shot} in Equation 4.12 are set to be 4, 50, and 100, respectively. From Figure 4.19, we can see that:

1. The *proposed* method can detect most MDs except the one at frame 200. The *frame*-200 MD is missed because we use a large window size of five frames (i.e., $k=4$ in Equation 4.12). This MD can also be detected if we select a smaller window size.

2. Since the *proposed* method detects MDs based on the information of several frames, some delay may be introduced. We can see that the MDs detected in Figure 4.19b have a delay of a couple of frames from the ground truth in Figure 4.19a.

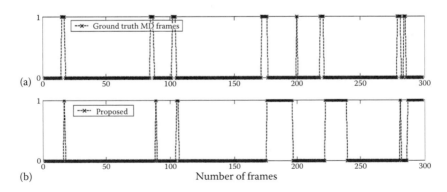

FIGURE 4.19
The detection result of the *proposed* algorithm in *Stefan_Sif*. (a) The ground-truth MD frames and (b) the MD detection results by our proposed method.

3. There are also some false alarms such as the period between frames 180 and 190. This is because the camera motions in these periods are too rapid. In these cases, the motion prediction accuracy will be decreased and some irregular global motions will be included. These factors will prevent the number of Class 2 MBs from decreasing after the *MD* finishes. In these cases, some postprocessing steps may be needed to discard these false alarms.

4.5.4.3 Experimental Results for Global Motion Estimation

We compare the following four GME algorithms. For all of the methods, we use the same six-parameter model for estimating the global motions, as in Equation 4.14:

1. Do not discard the foreground MBs and directly use the Least-Square method (Soldatov et al., 2006) to estimate the global model parameters (LS-6).
2. Use the MPEG-4 VM GME method (Sikora, 1997) (MPEG-4).
3. Use the method of Soldatov et al. (2006) for GME. In the work of Soldatov et al. (2006), an MV histogram is constructed for parameter estimation to speed up the GME process (MSU).
4. Use the method of Porikli et al. (2010) to segment and discard foreground MBs and perform GME on the background MBs (P-Seg).
5. Use our MB class information to segment and discard foreground MBs and perform GME on the background MBs, as described in Section 4.3.3 (Proposed).

Table 4.7 compares the mean square error (MSE) of the global motion-compensated results of the five algorithms. Normally, a small MSE value can be expected if the global motion parameter is precisely estimated. Table 4.8 compares the average MSE and the average operation time for different methods. Furthermore, Figure 4.20 also shows the subjective global motion-compensated results for the five methods.

TABLE 4.7

Comparison of Global-Motion-Compensated MSE Results for Different GME Methods

	LS-6	MPEG-4	MSU	P-Seg	Proposed
Bus	27.73	22.67	22.85	23.42	22.32
Stefan	22.71	20.99	19.09	19.36	19.52
Flower tree	24.92	20.66	21.51	20.83	19.72

TABLE 4.8

Comparison of Average MSE and Average Operation Time for
Different GME Methods

	LS-6	MPEG-4	MSU	P-Seg	Proposed
Average MSE	25.12	21.44	21.15	21.20	20.52
Average operation time (ms)	17	376	56	37	25

Note: The operation time for the object segmentation part for P-Seg is
taken from Porikli et al. (2010).

Some observations can be drawn from Tables 4.7 and 4.8, and Figure 4.20
as follows:

1. Since the LS-6 method does not differentiate foreground and background, it cannot estimate the global motion of the background precisely. We can see from Table 4.7 that the LS-6 method has larger
   MSE values. Furthermore, Figure 4.20a also shows that there are
   obvious background textures in the compensated frame.

2. Compared to the LS-6 method, the other four methods will segment and discard the foreground MBs before estimating the
   global motion for the background. We can see that our proposed
   method can achieve performance similar to the MEPG-4 and MSU
   methods.

3. Since the MEPG-4 algorithm uses a three-layer method to find the
   outlier (foreground) pixels, its computation complexity is high.
   Although the MSU and the P-Seg algorithms reduce the complexity
   by constructing histograms or performing volume growth for estimating the foreground area, they still require several steps of extra
   computations for estimating the global parameters. Compared with
   these two methods, our proposed method segments the foreground
   based on the readily available class information, and the extra computation complexity is obviously minimum. Note that this operation
   time reduction will become very obvious and important when the
   GME algorithms are integrated with the video compression module
   for real-time applications.

4. Although P-Seg can create good object segmentation results, its
   GME performance is not as good as our method. This is because
   our proposed algorithm focuses on detecting and filtering the
   "irregular motion" blocks while P-Seg focuses more on segment-
   ing a complete object. By using our algorithm, blocks that do not
   belong to the foreground but have irregular motions will also be
   filtered from the GME process. This further improves the GME
   performance.

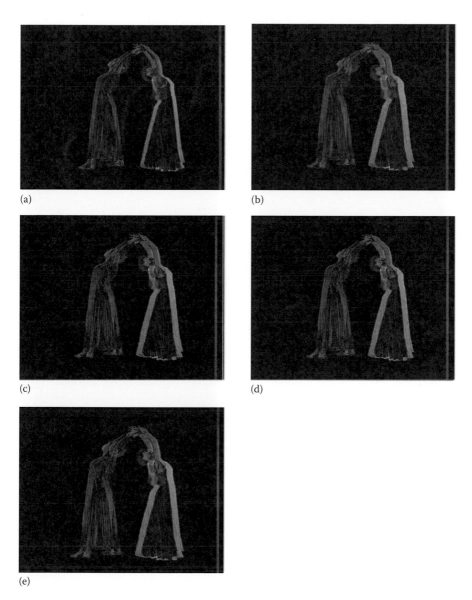

(a)

(b)

(c)

(d)

(e)

FIGURE 4.20
Subjective global motion-compensated results of the four methods for *Dancer_cif.* (a) LS-6, (b) MPEG-4, (c) MSU, (d) P-Seg, (e) Proposed.

4.6 Summary

In this chapter, we propose a more accurate MB Importance Measure method by classifying MBs into different classes. A new one-pass CCME is then proposed based on the new measure method. The four computation allocation steps of FLA, CLA, MLA, and SLA in the proposed CCME algorithm are introduced in the chapter. Furthermore, based on the proposed MB classification method, we further propose several algorithms for various video processing applications, including shot change detection, MD detection, and GME. Experimental results demonstrate the effectiveness of our proposed MB classification method as well as the corresponding video processing algorithms.

Acknowledgments

This chapter is supported in part by the following grants: Chinese National 973 grants (2010CB731401 and 2010CB731406), National Science Foundation of China Grants (61001146, 61103124, and 61101147).

References

Akutsu, A., Tonomura, Y., Hashimoto, H., and Ohba, Y. 1992. Video indexing using motion vectors. *Proc. SPIE Visual Commun. Image Process.*, 1818: 1522–1530.

Arman, F., Hsu, A., and Chiu, M. Y. 1994. Image processing on encoded video sequences. *Multimedia Syst.*, 1: 211–219.

Boyce, J. M. 2004. Weighted prediction in the H.264/MPEG AVC video coding standard. *IEEE Int. Symp. Circuits Syst.*, 3: 789–792.

Burleson, W., Jain, P., and Venkatraman, S. 2001. Dynamically parameterized architectures for power-aware video coding: Motion estimation and DCT. *IEEE Workshop on Digital and Computational Video*, Tampa, FL, pp. 4–12.

Chen, Y.-H., Chen, T.-C., Tsai, C.-Y., Tsai, S.-F., and Chen, L.-G. 2009. Algorithm and architecture design of power-oriented H.264/AVC baseline profile encoder for portable devices. *IEEE Trans. Circuits Syst. Video Technol.*, 19(8): 1118–1128.

Chen, C., Huang, Y., Lee, C., and Chen, L. 2006. One-pass computation-aware motion estimation with adaptive search strategy. *IEEE Trans. Multimedia*, 8(4): 698–706.

Eom, M. and Choe, Y. 2007. Scene change detection on H.264/AVC compressed video using intra mode distribution histogram based on intra prediction mode, *Proceedings of the Conference on Applications of Electrical Engineering*, Istanbul, Turkey, pp. 140–144.

He, Z., Liang, Y., Chen, L., Ahmad, I., and Wu, D. 2005. Power-rate-distortion analysis for wireless video communication under energy constraints. *IEEE Trans. Circuits Syst. Video Technol.*, 15(5): 645–658.

HHI, 2011. H.264/AVCreference software, JM 10.2 [Online]. Available at: <http://iphome.hhi.de/suehring/tml/download/old_jm/> [Accessed December 30, 2011].

Huang, Y., Lee, C., Chen, C., and Chen, L. 2005. One-pass computation-aware motion estimation with adaptive search strategy. *IEEE Int. Symp. Circuits Syst.*, 6: 5469–5472.

Kim, C., Xin, J., and Vetro, A. 2006. Hierarchical complexity control of motion estimation for H.264/AVC. *SPIE Conference on Visual Communications and Image Processing*, San Jose, CA, Vol. 6077, pp. 109–120.

Lee, M.-S. and Lee, S.-W. 2001. Automatic video parsing using shot boundary detection and camera operation analysis. *International Conference on Pattern Recognition*, Brisbane, Australia, Vol. 2, pp. 1481–1483.

Li, R., Zeng, B., and Liou, M. L. 1994. A new three-step search algorithm for block motion estimation. *IEEE Trans. Circuits Syst. Video Technol.*, 4(4): 438–442.

Lin, W., Baylon, D., Panusopone, K., and Sun, M.-T. 2008. Fast sub-pixel motion estimation and mode decision for H.264. *IEEE International Symposium on Circuits and Systems*, Seattle, WA, Vol. 6, pp. 3482–3485.

Lin, W., Panusopone, K., Baylon, D., and Sun, M.-T. 2009. A new class-based early termination method for fast motion estimation in video coding. *IEEE International Symposium on Circuits and Systems*, Taipei, Taiwan, pp. 625–628.

Lin, W., Panusopone, K., Baylon, D., and Sun, M.-T. 2010. A computation control motion estimate method for complexity scalable video coding, *IEEE Trans. Circuits Syst. Video Technol.*, 20(11): 1533–1543.

Lin, W., Sun, M.-T., Li, H., and Hu, H. 2010. A new shot change detection method using information from motion estimation. *Pacific-Rim Conference on Multimedia*, Shanghai, China, Vol. 6298, pp. 264–275.

Lin, W., Sun, M.-T., Poovendran, R., and Zhang, Z. 2008. Activity recognition using a combination of category components and local models for video surveillance. *IEEE Trans. Circuits Syst. Video Technol.*, 18(8): 1128–1139.

Liu, Y., Wang, W., Gao, W., and Zeng, W. 2004. A novel compressed domain shot segmentation algorithm on H.264/AVC. *International Conference on Image Processing*, Singapore, pp. 2235–2238.

NIST, 2001. TREC Video Retrieval Evaluation [Online]. Available at: <http://www-nlpir.nist.gov/projects/trecvid/> [Accessed December 30, 2011].

Po, L.-M. and Ma, W. C. 1996. A novel four-step search algorithm for fast block motion estimation. *IEEE Trans. Circuits Syst. Video Technol.*, 6(3): 313–317.

Porikli, F., Bashir, F., and Sun, H. 2010. Compressed domain video object segmentation. *IEEE Trans. Circuits Syst. Video Technol.*, 20(1): 2–14.

Shu, S. and Chau, L. P. 2005. A new scene change feature for video transcoding. *IEEE International Symposium on Circuits and Systems*, Kobe, Japan, Vol. 5, pp. 4582–4585.

Sikora, T. 1997. The MPEG-4 video standard verification model. *IEEE Trans. Circuits Syst. Video Technol.*, 7(1): 19–31.

Soldatov, S., Strelnikov, K., and Vatolin, D. 2006. Low complexity global motion estimation from block motion vectors. *Spring Conference on Computer Graphics*, Bratislava, Slovakia, pp. 1–8.

SVMLight, 2011. Light support vectormachine software [Online]. Available at: <http://svmlight.joachims.org/> [Accessed December 30, 2011].

Tai, P., Huang, S., Liu, C., and Wang, J. 2003. Computational aware scheme for software-based block motion estimation. *IEEE Trans. Circuits Syst. Video Technol.*, 13(9): 901–913.

Vanam, R., Riskin, E. A., Hemami, S. S., and Ladner, R. E. 2007. Distortion-complexity optimization of the H.264/MPEG-4 AVC encoder using the GBFOS algorithm. *Data Compression Conference*, Snowbird, UT, pp. 303–312.

Vanam, R., Riskin, E. A., and Ladner, R. E. 2009. H.264/MPEG-4 AVC encoder parameter selection algorithms for complexity distortion tradeoff. *Data Compression Conference*, Snowbird, UT, pp. 372–381.

Weigand, T., Schwarz, H., Joch, A., Kossentini, F., and Sullivan, G. 2003. Rate-constrained coder control and comparison of video coding standards. *IEEE Trans. Circuits Syst. Video Technol.*, 13(7): 688–703.

Wiegand, T., Sullivan, G.-J., Bjontegaard, G., and Luthra, A. 2003. Overview of the H.264/AVC video coding standard. *IEEE Trans. Circuit Syst. Video Technol.*, 13(7): 560–576.

Yang, Z., Cai, H., and Li, J. 2005. A framework for fine-granular computational-complexity scalable motion estimation. *IEEE International Symposium on Circuits and Systems*, Vol. 6, pp. 5473–5476.

Yi, X. and Ling, N. 2005. Scalable complexity-distortion model for fast motion estimation. *SPIE Conference on Visual Communications and Image Processing*, Beijing, China, Vol. 5960, pp. 1343–1353.

Yi, X., Zhang, J., Ling, N. and Shang, W. 2005. Improved and simplified fast motion estimation for JM. *The 16th MPEG/VCEG Joint Video Team Meeting*, Poznan, Poland, pp. 1–8.

Zhang, J. and He, Y. 2003. Performance and complexity joint optimization for H.264 video coding. *IEEE International Symposium on Circuits and Systems*, Vol. 8, pp. 888–891.

Zhang, K. and Kittler, J. 1999. Using scene-change detection and multiple-thread background memory for efficient video coding. *Electron. Lett.*, 35(4): 290–291.

Zhou, Z., Sun, M.-T., and Hsu, Y.-F. 2004. Fast variable block-size motion estimation algorithms based on merge and split procedures for H.264/MPEG-4 AVC. *IEEE International Symposium on Circuits and Systems*, Vancouver, British Columbia, Canada, pp. 725–728.

Zhu, S. and Ma, K.-K. 2000. A new diamond search algorithm for fast block matching motion estimation. *IEEE Trans. Image Process.*, 9(2): 287–290.

5

Transmission Rate Adaptation in Multimedia WLAN: A Dynamic Games Approach*

Jane Wei Huang, Hassan Mansour, and Vikram Krishnamurthy

CONTENTS

---

* This chapter is based on the following publication.: J. W. Huang, H. Mansour, and V. Krishnamurthy, A dynamical games approach to transmission rate adaptation in multimedia WLAN, *IEEE Transactions on Signal Processing*, 58(7):3635–3646, July 2010.

This chapter considers the scheduling, rate adaptation, and buffer management in a multiuser wireless local area network (WLAN) where each user transmits scalable video payload. Based on opportunistic scheduling, users access the available medium (channel) in a decentralized manner. The rate adaptation problem of the WLAN multimedia networks is then formulated as a general-sum switching control dynamic Markovian game by modeling the video states and block-fading channel* qualities of each user as a finite-state Markovian chain. A value iteration algorithm is proposed to compute the Nash equilibrium policy of such a game, and the convergence of the algorithm is also proved. We also give assumptions on the system so that the Nash equilibrium transmission policy of each user is a randomization of two pure policies with each policy nondecreasing on the buffer state occupancy. Based on this structural result, we use policy-gradient algorithm to compute the Nash equilibrium policy.

5.1 Introduction

Advances in video coding technology and standardization [1,2] along with the rapid developments and improvements of network infrastructures, storage capacity, and computing power are enabling an increasing number of video applications. Application areas today range from multimedia messaging, video telephony, and video conference over mobile TV, wireless and wired Internet video streaming, standard and high-definition (HD) TV broadcasting to DVD, blue-ray disc, and HD DVD optical storage media. For these applications, a variety of video transmission and storage systems may be employed.

Modern video transmission and storage systems using the Internet and mobile networks are typically based on real-time transport protocol (RTP)/ Internet protocol (IP) for real-time services (conversational and streaming)

---

* Block-fading channel assumes that the channel states remain constant within a time slot and that they vary from one time slot to another.

and on computer file formats like mp4 or 3 gp. Most RTP/IP access networks are typically characterized by a wide range of connection qualities and receiving devices. The connection quality would be adjusted if the network requires time-varying data throughput. The variety of devices with different capabilities ranging from cell phones with small screens and restricted processing power to high-end personal computers with HD displays results from the continuous evolution of these endpoints.

Scalable video coding (SVC) is a highly attractive solution to the problems posted by the characteristics of modern video transmission system. The time-varying nature of wireless channels and video content motivate the need for scalable media, which can adapt the video bit rate without significantly sacrificing the decoded video quality. The SVC project provides spatial, temporal, and signal-to-noise ratio (SNR) scalability, which ensures a graceful degradation in video quality when faced with channel fluctuations [3].

Efficient SVC provides a number of benefits in terms of applications [4,5]. In the scenario of a video transmission service with heterogeneous clients, multiple bitstreams of the same source content differing in coded picture size, frame size, and bit rate should be provided simultaneously. With the application of a properly configured SVC scheme, the source content has to be encoded only once, for the highest required resolution and bit rate, resulting in a scalable bitstream from which representations with lower resolution and/or quality can be obtained by discarding selected data.

Real-time video streaming applications belong to a group of fast spreading services that are quickly reaching the mobile arena. Such applications have increased the need to supply large volumes of coded video data to mobile and heterogeneous users [6]. However, capacity restrictions, heterogeneous device and network capabilities, and error-prone transmissions are just some of the problems resulting from the characteristics of modern video communication systems, to which SVC offers an attractive solution [7]. In this chapter, we present a rate adaptation scheme for real-time encoding and streaming of multiple scalable video streams in a WLAN system. The aim is to adapt the video layers of each user to the fluctuations in channel quality and the transmission control polices, such that the transmission buffer of each user does not saturate and the transmission delay constraint is not violated.

Rate allocation for multimedia transmission in wireless networks has been studied extensively. Previous works that are of interest include [8–10]. In [8], joint radio link buffer management and scheduling strategies are compared for wireless video streaming over high speed downlink packet access (HSDPA) networks. We use the results of [8] to choose an appropriate buffer management strategy for our proposed switching control stochastic dynamic game solution. In [9] and [10], video rate and distortion models are used to improve the transmitted video quality for low-latency multimedia traffic. Rate and distortion modeling has become a keystone in model-based video rate and transmission control algorithms. The models are used to predict the coded video packet size and the distortion of the corresponding decoded picture prior to the actual encoding process in order to be used in a constrained optimization framework. However, these models cannot be used for characterization of scalable video data.

5.1.1 Motivation for Stochastic Game Formulation

The concept of a stochastic game, first introduced by Lloyd Shapley in early 1950s, is a dynamic game played by one or more players. The elements of a stochastic game include system-state set, action sets, transition probabilities, and utility functions. It is an extension of the single player Markov decision processes (MDPs) to include the multiple players whose actions all impact the resulting payoffs and next state. A switching control game [11–13] is a special type of stochastic dynamic game where the transition probability in any given state depends on only one player. It is known that the Nash equilibrium for such a game can be computed by solving a sequence of MDPs.

In this chapter, we consider an Institute of Electrical and Electronics Engineers (IEEE) 802.11 [14] WLAN, which deploys the carrier sense multiple access with collision avoidance (CSMA/CA). We propose a modified CSMA/CA channel access mechanism (Section 5.3.3), which takes into account the dynamic behaviors of the video variation, as well as the channel quality and transmission delay of each user. Due to the fact that there is no central controller for resource allocation in a WLAN system, there is strong motivation to study the case where individual users seek to selfishly maximize their transmission rate while taking into account the access rule. This is akin to individuals minimizing their tax (throughput subject to latency constraint in our case) without violating the tax law (the modified CSMA/CA channel access rule in our case). The interactions among system users can be naturally formulated using game theory. Furthermore, by exploiting the correlated time-varying sample paths of the channels and buffers, the transmission rate adaptation problem can be formulated as a Markovian dynamic game. Section 5.2 describes the details of the WLAN system under study as well as the dynamic behavior of video sources and the scalable rate-distortion models used to represent this behavior.

The main results of this chapter are summarized as follows.

First, we consider a multiuser WLAN system where each user is equipped with a scalable video quality encoder delivering video bitstream that conforms with SVC [7], which is the scalable extension of H.264/AVC. We address the scheduling, rate control, and buffer management of such system, and formulate the problem as a constrained switching control stochastic dynamic game combined with rate-distortion modeling of the source. The video states and the block-fading channel qualities of each user are formulated as a finite-state Markovian chain, and each user aims to optimize its own utility under a transmission latency constraint when it is scheduled for transmission. We then formulate the rate adaptation in the WLAN multimedia system as a *constrained* switching control dynamic Markovian game [12,15].

Second, we propose a value iteration algorithm that obtains the Nash equilibrium transmission policy of the constrained general-sum switching control game in Section 5.4. The Nash equilibrium policy of such constrained game is a randomization of two deterministic policies. The value iteration algorithm involves a sequence of dynamic programming problems with Lagrange multipliers. In each iteration of the value iteration algorithm, a Lagrange multiplier is used to combine the latency cost with the transmission reward for each user; the Lagrange multiplier is then updated based on the delay constraint. The algorithm converges to the Nash equilibrium policy of the constrained general-sum switching control game.

We then proceed to the main structural result of this chapter. It is shown that under reasonable assumptions, the Nash equilibrium policy is monotone-nondecreasing on the buffer state occupancy. This structural result on the transmission policy enables us to search for the Nash equilibrium policy in the policy space via a policy-gradient algorithm. The structural result is especially useful in dealing with multiuser systems where each user has a large buffer size. Finally, numerical results of the proposed switching control Markovian dynamic game policy in transmitting scalable video are presented.

5.2 System Description and the Video Rate-Distortion Model

This section introduces the time-slotted WLAN system model (Figure 5.1), which consists of multiple users' attempts to transmit their data over available channels (medium) [14]. The time synchronization among users can be done by downlink pilots obtaining relevant time information. Each user is equipped with a buffer, a decentralized scheduler, and a rate adaptor to transmit a scalable video payload, which consists of a base layer and multiple enhancement layers. We use game theory to formulate the decentralized behaviors of all the users accessing the common spectrum resource. It will be shown in Section 5.4 that the state information of other users is needed in order to compute the optimal transmission policy of the current user. This can be realized by allowing all the users to broadcast their state information (buffer size, channel) during the guard interval at the beginning of each time slot.

5.2.1 System Description

We consider a K user WLAN system where only one user can access the channel at each time slot according to a modified CSMA/CA mechanism (Section 5.3). The rate-control problem of each user can be formulated as a constrained dynamic Markovian game by modeling the correlated block-fading channel as a Markov chain. The "users" that are referred to in this

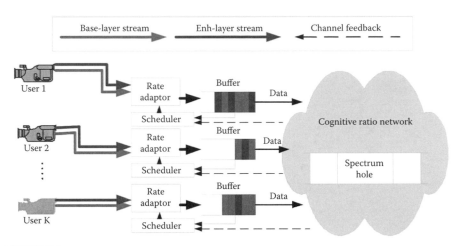

FIGURE 5.1
(**See color insert.**) A WLAN system where each user is equipped with a size B buffer, a decentralized scheduler, and a rate adaptor for transmission control. The users transmit a scalable video payload in which enhancement layers provide quality refinements over the base layer bitstream.

chapter are video sources who are uploading video streams to the network, and mobile end users are on the other side of the WLAN system downloading the video streams. Under the predefined decentralized access rule (Section 5.3.3), the problem presented is a special type of game, namely, a switching control Markovian dynamic game.

We assume that K users (indexed by $k = 1, \ldots, K$) are equipped with scalable video encoders delivering video bitstreams that conform with SVC [7]. In SVC, quality scalability is achieved using medium-grained or coarse-grained scalability (MGS/CGS), where scalable enhancement packets deliver quality refinements to a preceding layer representation by requantizing the residual signal using a smaller quantization step size and encoding only the quantization refinements in the enhancement layer packets [7]. Moreover, MGS/CGS enhancement layer packets are coded using the *key picture concept*. For each picture a flag is transmitted, which signals whether the base quality reconstruction or the enhancement layer reconstruction of the reference pictures is employed for motion-compensated prediction. In order to limit the memory requirements, a second syntax element signals where the base quality representation of a picture is additionally reconstructed and stored in drop priority based (DPB). In order to limit the decoding overhead for such key pictures, SVC specifies that motion parameters must not change between the base and enhancement layer representation of key picture, and thus also for key pictures, the decoding can be done with a single motion-compensation loop. This approach achieves a trade-off between the coding efficiency of enhancement layers and the drift at the decoder [7].

Assume that each user K encodes its video stream at fixed base and enhancement layer quantization parameter (QP) values $q_{l,k}$, where $l=0$ corresponds to the base layer, and $l \geq 1$ corresponds to every subsequent enhancement layer. This results in a fluctuation of the video bit rate and distortion as the scene complexity and level of motion vary. The video variation can be captured using the mean absolute difference (MAD) of the prediction residual, since it quantifies the mismatch between the original uncoded picture and the intra/inter prediction. Therefore, we resort to video rate and distortion models to predict the video distortion of the corresponding decoded picture prior to the actual encoding process. In [16], new rate and distortion models are proposed that capture the variation of MGS/CGS scalable coded video content as a function of two encoding parameters, namely, the QP and the MAD of the residual signal. We will summarize the results in the following subsection.

5.2.2 Scalable Rate-Distortion Modeling

In this section, we present new rate and distortion estimation models for individual frame representation of MGS/CGS scalable coded video content. Two real-time encoding parameters are commonly used in rate-distortion estimation models, namely, the QP and the MAD of the residual signal.

Dropping the user subscript k, let q_l, $q_{l'} \in \{0,1,...51\}$ [1] be the QP values of two distinct video layers l and l', respectively. We have found a simple relationship that estimates the residual MAD $\tilde{m}_l$ of a specific frame given the residual MAD of the same frame at an initial $q_{l'}$ as shown in the following:

$$\tilde{m}_l = \tilde{m}_{l'} \cdot 2^{\varsigma(q_l - q_{l'})}, \tag{5.1}$$

where ς is a model parameter typically valued around 0.07 for most sequences [1,16]. In MGS/CGS scalability, enhancement layer packets contain refinements on the quantization of residual texture information [7]. Therefore, we use the expression in (5.1) to estimate the perceived MAD of each of the MGS/CGS enhancement layers.

5.2.2.1 Rate Model

Here we present the rate model for base and CGS enhancement layer SVC coded video frames. The MGS/CGS [10] base and enhancement layer bit rate in SVC can be expressed as follows:

$$r(l, \tilde{m}_{l'}) = c_l (\tilde{m}_l)^u 2^{-q_l/6} = c_l \left(\tilde{m}_{l'} \cdot 2^{\varsigma(q_l - q_{l'})} \right)^u 2^{-q_l/6}, \tag{5.2}$$

where
c_l is a model parameter
u is a power factor that depends on the frame type, such that $u=1$ for inter-coded frames and $u \approx 5/6$ for intra-coded frames [1,16]

5.2.2.2 Distortion Model

The distortion model estimates the decoded picture luminance peak signal-to-noise ratio (Y-PSNR). Let q_l be the QP value of layer l, and let $\tilde{m}_l$ be the prediction MAD estimate. The expression of the video PSNR achieved by decoding all layers up to layer l is as follows:

$$\delta(l, \tilde{m}_{l'}) = v_1 \log_{10}\left((\tilde{m}_l)^u + 1\right) \cdot q_l + v_2$$

$$= v_1 \log_{10}\left(\left(\tilde{m}_{l'} 2^{a(q_l - q_{l'})}\right)^u + 1\right) q_l + v_2, \tag{5.3}$$

where

u is the same power factor described in (5.2)

v_1 and v_2 are sequence-dependent model parameters typically valued at 0.52 and 47, respectively [16]

The parameters v_1 and v_2 can be refined for each sequence during encoding. Figure 5.2 illustrates the performance of the proposed distortion model compared to actual encoding of two reference video sequences: Foreman and Football. Figures 5.2a and b measure the Y-PSNR estimation results, while Figures 5.2c and d measure the rate estimation results. The video streams used in the simulation have one base layer and two CGS enhancement layers. The accuracy of this rate-distortion model (5.3) can be easily seen in Figure 5.2.

Note that the parameters ς, v_1, v_2, and c_l are calculated only once and remain constant for the entire video sequence.

5.3 Uplink Rate Adaptation Problem Formulation

We assume every user has an uplink buffer that holds the incoming coded video frames. When a user is scheduled for transmission, the buffer output rate is chosen depending on the channel quality and buffer occupancy. Therefore, the rate adaptation problem is reformulated as a buffer control problem for every user. In this section, we give a Markovian dynamic game description of this problem.

We assume that the time slot coincides with the video frame duration (30–40 ms [7]). Each user has a fixed packet arrival rate, and the incoming packets are SVC bitstreams with both base and enhancement layers. We denote the incoming number of video frames of user k as $f_{in,k}$ ($k = 1, 2, \dots, K$).

Let b_k^n represent the buffer occupancy state of user k at time n and $b_k^n \in \{0, 1, \dots B\}$, where B is the maximum number of coded video frames that can be stored in the buffer. The composition of buffer states of all the K users is $\mathbf{b}^n = \{b_1^n, \dots, b_K^n\}$ and $\mathbf{b}^n \in \mathcal{B}$, where $\mathcal{B}$ is used to denote the space of system

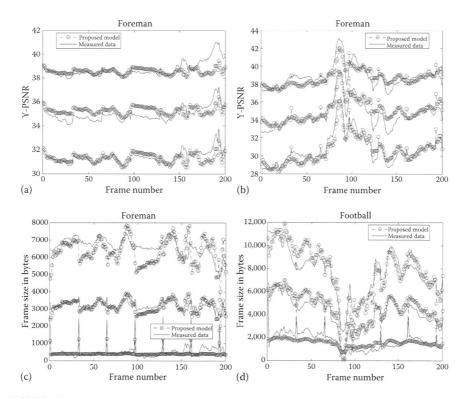

FIGURE 5.2
Illustration of the modeled PSNR estimates (a and b), and the modeled rate estimates (c and d) for the base layer and two CGS enhancement layers of the sequences Foreman and Football.

buffer states. Furthermore, we use $\mathbf{b}^n_{-k}$ to denote the buffer state composition of all the users excluding user k.

The channel quality of user k at time n is denoted as h^n_k. The channel is characterized using circularly symmetric complex Gaussian random variables, which depend only on the previous time slot. By quantizing the channel quality metric, we can denote the resulting discrete channel state space by $h^n_k \in \{1, 2, \ldots\}$. Let $\mathbf{h}^n = \{h^n_1, \ldots, h^n_K\}$ be the composition of channel states of all the K users. Assuming that $\mathbf{h}^n \in \mathcal{H}$, $n = 1, 2, \ldots, N$ is block-fading, with block length equal to each time period, the channel states constitute a Markov process with transition probability from time n to $(n+1)$ given by $\mathbb{P}(\mathbf{h}^{(n+1)} \mid \mathbf{h}^n)$.

Let m^n_k be the MAD of user k at time n. In [9], the MAD between two consecutive video frames is modeled as a stationary first-order Gauss Markov process. We quantize the range of m^n_k to achieve a video variability state space denoted by $m^n_k \in \mathcal{M} = \{0, 1, 2 \ldots, M\}$. The video states constitute a Markov process with transition probabilities given by $\mathbb{P}(m^{(n+1)}_k \mid m^n_k)$. The system video state at time n is a composition of the video states of all the users and $\mathbf{m}^n = \{m^n_1, m^n_2, \ldots, m^n_K\}$.

The system state at time n is composed of the channel state, video state, and buffer state, which is denoted as $\mathbf{s}^n = \{\mathbf{h}^n, \mathbf{m}^n, \mathbf{b}^n\}$. Let S denote the finite system state space, which comprises channel state space $\mathcal{H}$, video state space $\mathcal{M}$, and user buffer state space $\mathcal{B}$. That is, $S = \mathcal{H} \times \mathcal{M} \times \mathcal{B}$. Here $\times$ denotes a Cartesian product. S_k is used to indicate the states where user k is scheduled for transmission. $S_1, S_2, \ldots, S_K$ are disjoint subsets of S with the property of $S = S_1 \cup S_2 \cup \ldots \cup S_K$.

5.3.1 Actions, Transmission Reward, and Holding Cost

The action a_k^n by user k at time slot n is defined to be the number of coded video frames taken from the buffer and transmitted. The rate-control algorithm considered in this chapter adjusts the output frame rate a_k^n according to the channel quality and buffer occupancy. Figure 5.3 illustrates the buffer control mechanism adopted in this chapter.

5.3.1.1 Actions

Let a_k^n denote the action of the kth user at the nth time slot. We express a_k^n as the number of coded video frames included in the outgoing traffic payload. If $\mathbf{s}^n \in S_k$, then $a_k^n \in \mathcal{A} = \{a_{\min}, \ldots, a_{\max}\}$, otherwise, $a_k^n = 0$. $a_{\min}$ and $a_{\max}$ denote the minimum and maximum video frames that user k can output at time n, respectively.

We assume that the number of video layers $l_{a_k^n}$ at the output window at time slot n is dictated by a channel-dependent rule. Therefore, we express the output number of layers of the video frames at time slot n as follows:

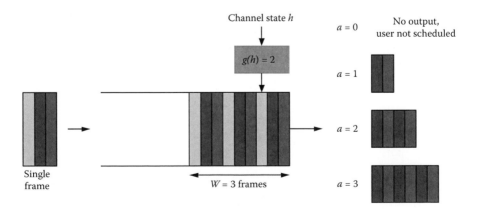

FIGURE 5.3
(**See color insert.**) Example of the buffer control mechanism assumed in this chapter where $f_{in,k} = 1$ ($k = 1, 2, \ldots, K$). At every time slot, a new coded video frame enters the buffer. The buffer output depends on the scheduling algorithm involved, the buffer occupancy, and the channel quality. If a user is scheduled for transmission, then the action taken will extract a specific number l of video frame layers from up to N frames stored in the buffer.

$$l_{a_k^n} = g(h_k^n), \tag{5.4}$$

where $g(.)$ is the channel-dependent rule, and it is an increasing function on the channel state. The remaining video layers $l > l_{a_k^n}$ are dropped. This assumption is to reduce the action space, and it allows us to derive the structural results as illustrated later in Section 5.4.2. Consequently, the action space constitutes the number of video frames to be transmitted and is given by $A = \{1, 2, \ldots, W\}$.

Let L_k be the number of video layers admitted into the buffer of user k; each input packet of user k has L_k layers. The video layer index is denoted as $l \in \mathcal{L} = \{0, 1, \ldots L_k - 1\}$, where $l = 0$ corresponds to the base layer, $l \geq 1$ to every subsequent enhancement layer.

The buffer output packet a_k^n of user k at time n is determined by the modified CSMA/CA decentralized channel access algorithm. If user k is scheduled for transmission at time slot n, $a_k^n > 0$, otherwise, $a_k^n = 0$. However, the system selectively transmits some packet layers according to the channel quality. For example, the system transmits all of the L_k layers of each packet if the channel quality h_k^n is good, while it only transmits the base layers of each packet if the channel quality h_k^n is poor.

5.3.1.2 Transmission Reward

Let $c_k(\mathbf{s}^n, a_1^n, \ldots, a_K^n)$ denote the transmission reward of user k at time n. Specifically, $c_k(\mathbf{s}^n, a_1^n, \ldots, a_K^n)$ is chosen to be the expected video PSNR resulting from transmitting $l_{a_k^n}$ video layers from a_k^n video frames stored in the buffer of user k. Let $\delta(l, m_k^n)$ be the video PSNR achieved by decoding all packets up to layer l of user k at time n achieved with video state m_k^n, where m_k^n is the base layer MAD. Let $p_l(h_k^n)$ be the packet error rate (PER) encountered by user k at the lth layer during the transmission. The transmission reward can be expressed as [16]

$$c_k(\mathbf{s}^n, a_1^n, \ldots, a_K^n)$$

$$= a_k^n \cdot \left((1 - p_0(h_k^n)) \cdot \delta(0, m_k^n) + \sum_{l=1}^{l_{a_k^n} - 1} \prod_{j=0}^{l} (1 - p_j(h_k^n)) \cdot \left[\delta(l, m_k^n) - \delta(l-1, m_k^n) \right] \right). \tag{5.5}$$

The bit errors are assumed to be independent and identically distributed (i.i.d.), and $\delta(l, m_k^n)$ is given in (5.3). Note here that the video distortion measure $\delta(l, m_k^n)$ can either be considered as the picture mean-squared error (MSE) or the PSNR. In the case of PSNR, the objective will be to *maximize*

$c_k(\mathbf{s}^n, a_k^n)$, otherwise it is minimized. In this chapter, we consider the PSNR as the video quality metric.

It can be seen from (5.5) that at time slot n if the kth user is scheduled for transmission, the performance of user k depends solely on its own channel state h_k^n and not on the actions. Thus, for $\mathbf{s}^n \in S_k$, the rewards of all the users in the system are

$$c_k(\mathbf{s}^n, a_1^n, \ldots, a_K^n) = c_k(\mathbf{s}^n, a_k^n) \geq 0 \tag{5.6}$$

$$c_i(\mathbf{s}^n, a_1^n, \ldots, a_K^n) = 0, \quad (i \neq k). \tag{5.7}$$

For notational convenience, in the following sections, we will drop the subscript k by defining

$$c(\mathbf{s}^n, a_k^n) := c_k(\mathbf{s}^n, a_k^n).$$

5.3.1.3 Holding Cost

Each user has an instantaneous quality of service (QoS) constraint denoted by $d_k(\mathbf{s}^n, a_1^n, \ldots, a_K^n)$, where $k = 1, \ldots, K$. If the QoS is chosen to be the delay (latency), then $d_k(\mathbf{s}^n, a_1^n, \ldots, a_K^n)$ is a function of the buffer state b_k^n of the current user k. The instantaneous holding costs will be subsequently included in an infinite horizon latency constraint.

We express the holding cost of user k with $\mathbf{s}^n \in S_k$ as follows:

$$d_\kappa(\mathbf{s}^n, a_\kappa^n) = \frac{1}{\kappa} \cdot ([b_\kappa^n - a_\kappa^n + f_{in,\kappa}]^+)^\tau, \quad \tau \geq 1, \tag{5.8}$$

where κ is the average output frame rate, which is determined by the system. τ is a constant factor, which is specified to be $\tau \geq 1$; this is due to the fact that the data overflow probability becomes higher with more data in the buffer. The latency cost at time slot n is evaluated by the buffer state after taking action a_k^n.

Since the transmission latency is independent of the actions of all the remaining users, it can be simplified as

$$d_k(\mathbf{s}^n, a_1^n, \ldots, a_K^n) = d_k(\mathbf{s}^n, a_k^n) \geq 0, \quad \mathbf{s}^n \in S_k. \tag{5.9}$$

For the remaining $K - 1$ users who are not scheduled for transmission, their holding costs can be expressed as

$$d_{i, i \neq k}(\mathbf{s}^n, a_k^n) = \frac{1}{\kappa} \cdot \min(b_i^n + f_{in,i}, B)^\tau. \tag{5.10}$$

5.3.2 Markovian Game Formulation

The intuition behind the Markovian game formulation for the rate adaptation problem in a WLAN system is as follows: A WLAN system does not have a central authority for the resource allocation, and the system access rule is typically a decentralized opportunistic scheme. Due to the selfish nature of the system users, each user aims to maximize its own payoff. The interaction among users is characterized as the competition for the common system resources, which can best be formulated using game theory. By modeling the transmission channel as correlated Markov process, the rate adaptation problem can further be formulated as a Markovian game.

At time instant n, assume user k is scheduled for transmission according to the system access rule specified in Section 5.3.3. We define $c(\mathbf{s}^n, a_k^n) \geq 0$ to be the instantaneous reward of user k when the system is in state $\mathbf{s}^n$. We use Φ_1, Φ_2, ..., Φ_K to represent the set of all the deterministic policies of each user, respectively. The infinite horizon expected total discounted reward of any user i, given its transmission policy π_i ($\pi_i \in \Phi_i$), can be written as

$$C_i(\pi_i) = \mathbb{E}_{\mathbf{s}^n}\left[\sum_{n=1}^{\infty} \beta^{n-1} \cdot c_i(\mathbf{s}^n, a_k^n) \,\Big|\, \pi_i \right] \qquad (5.11)$$

where $0 \leq \beta < 1$ is the discount factor, the state $\mathbf{s}^n \in S_k$, and the expectation is taken over action a_k^n as well as system state $\mathbf{s}^n$ evolution for $n = 1, 2, \ldots$. The holding cost of user i at that time is $d_i(\mathbf{s}^n, a_k^n)$ and the infinite horizon expected total discounted latency constraint can be written as

$$D_i(\pi_i) = \mathbb{E}_{\mathbf{s}^n}\left[\sum_{n=1}^{\infty} \beta^{n-1} \cdot d_i(\mathbf{s}^n, a_k^n) \,\Big|\, \pi_i \right] \leq \tilde{D}_i, \qquad (5.12)$$

with $\tilde{D}_i$ being a system parameter depending on the system requirement on user i.

Optimization problem: Given the policies of the other users, the optimal transmission policy of user i, π_i^*, is chosen so as to maximize the overall expected discounted reward subject to its constraint; this can be mathematically written as

$$\pi_i^* = \{\pi_i : \max_{\pi_i \in \Phi_i} C_i(\pi_i) \ \ s.t. \ D_i(\pi_i) \leq \tilde{D}_i, i = 1, \ldots, K\}. \qquad (5.13)$$

Note that the choice of the optimal rate transmission policy of ith user is a function of the policies of other users. Each user aims to optimize its own discounted reward (5.11) under the latency constraint (5.12).

5.3.3 System Access Rule

This chapter adopts a WLAN system model (IEEE 802.11 [14]) with a modified CSMA/CA mechanism, which will be implemented by a decentralized channel access rule. The decentralized channel access algorithm can be constructed as follows: At the beginning of a time slot, each user k attempts to access the channel after a certain time delay t_k^n. The time delay of user k can be specified via an opportunistic scheduling algorithm [17], such as

$$t_k^n = \frac{\gamma_k}{b_k^n h_k^n}. \tag{5.14}$$

Here, γ_k is a user-specified QoS parameter and $\gamma_k \in \{\gamma_p, \gamma_s\}$. As soon as a user successfully accesses the channel, the remaining users detect the channel occupancy and stop their attempts to access. Let k^*n denote the index of the first user that successfully accesses the channel. If there are multiple users with the same minimum waiting time, k^*n is randomly chosen from these users with equal probability.

5.3.4 Transition Probabilities and Switching Control Game Formulation

With the aforementioned setup, the system feature satisfies the assumptions of a special type of dynamic game called a switching control game [11–13], where the transition probabilities depend on only one player in each state. It is known that the Nash equilibrium for such a game can be computed by solving a sequence of MDPs [12]. The decentralized transmission control problem in a Markovian block-fading channel WLAN system can now be formulated as a switching control game. In such a game [12], the transition probabilities depend only on the action of the kth user when the state $s \in S_k$; this enables us to solve such type of game by a finite sequence of MDPs.

According to the property of the switching control game, when the kth user is scheduled for transmission, the transition probability between the current composite state $s = \{h, m, b\}$ and the next state $s' = \{h', m', b'\}$ depends only on the action of the kth user a_k. The transition probability function of our problem can now be mathematically expressed by the following equation:

$$\begin{aligned}
&\mathbb{P}(s' \mid s, a_1, a_2, \ldots, a_K) \\
&= \mathbb{P}(s' \mid s, a_k) \\
&= \prod_{i=1}^{K} \mathbb{P}(h_i' \mid h_i) \cdot \prod_{i=1}^{K} \mathbb{P}(m_i' \mid m_i) \cdot \prod_{i=1, i \neq k}^{K} \mathbb{P}(b_i' \mid b_i) \cdot \mathbb{P}(b_k' \mid b_k, a_k)
\end{aligned} \tag{5.15}$$

As each user is equipped with a size B buffer, the buffer occupancy of user k evolves according to Lindley's equation [18,19]:

$$b'_k = \min([b_k - a_k + f_{in,k}]^+, B). \tag{5.16}$$

The evolution of the buffer state of user $i = 1, 2, \ldots, K$ when $i \neq k$ follows the following rule:

$$b'_i = \min(b_i + f_{in,i}, B).$$

For user k, the buffer state transition probability depends on the input and output date rates. This can be expressed as follows:

$$\mathbb{P}(b'_k \mid b_k, a_k) = I_{\{b'_k = \min([b_k - a_k + f_{in,k}]^+, B)\}}, \tag{5.17}$$

where $I_{\{\cdot\}}$ is the indicator function.

For those users that are not scheduled for transmission, the buffer state transition probability depends only on the distribution of incoming traffic, which can be written as

$$\mathbb{P}(b'_i \mid b_i) = I_{\{b'_i = \min(b_i + f_{in,i}, B)\}}. \tag{5.18}$$

Equations 5.11, 5.12, and 5.15 define the constrained switching control Markovian game that we consider. Our goal is to solve such games, that is, we seek to compute a Nash equilibrium* policy π^*_i, $i = 1, \ldots, K$ (which is not necessarily unique). However, if both the reward (5.11) and constraint (5.12) are zero-sum among all the users, then all the Nash equilibria have a unique value vector and are globally optimal [12].

The Markovian switching control game can be solved by constructing a sequence of MDP as described in Algorithm 5.1. We refer to [12, Chapter 3.2] and [11] for the proof. So, the constrained switching controlled Markovian game ((5.11) and (5.12)) can be solved by iteratively updating the transmission policy π^{*n}_i of user i, $i = 1, \ldots, K$ with the policy of remaining users fixed. Here, n denotes the iteration index as mentioned in Algorithm 5.1. At each step, ith user aims to maximize the overall expected discounted reward, subject to its constraint as specified in (5.13).

5.4 Nash Equilibrium Solutions to the Markovian Dynamic Game

This section first presents a value iteration algorithm to compute the Nash equilibrium solution to the formulated general-sum dynamic Markovian switching control game and proves the convergence of such algorithm.

---

* A Nash equilibrium [12] is a set of policies, one for each player, such that no player has an incentive to unilaterally change their actions. Players are in equilibrium if a change in policies by any one of them would lead that player to earn less than that if it remained with its current policy.

Then, we introduce four assumptions on the transmission reward, holding cost, and state transition probability functions, which lead us to the structural result on the Nash equilibrium policy. This structural result enables us to search for the Nash equilibrium policy in the policy space using a policy-gradient algorithm, and it has reduced computational cost.

5.4.1 Value Iteration Algorithm

In [12], a value iteration algorithm was designed to calculate the Nash equilibrium for an unconstrained general-sum dynamic Markovian switching control game. Therefore, we convert the problem in (5.13) to an unconstrained one using Lagrangian relaxation and use the value iteration algorithm specified in Algorithm 5.1 to find a Nash equilibrium solution.

Algorithm 5.1: Value iteration algorithm

Step 1:
Set $m = 0$; Initialize l
Initialize $\{\mathbf{V}_1^0, \mathbf{V}_2^0, \ldots, \mathbf{V}_K^0\}$, $\{\lambda_1^0, \lambda_2^0, \ldots, \lambda_K^0\}$
Step 2: Inner loop: Set $n = 0$
Step 3: Inner loop: Update transmission policies
for $k = 1$: K **do**
for each $\mathbf{s} \in \mathcal{S}_k$,

$$\pi_k^n(\mathbf{s}) = \arg\min_{\pi_k^n(\mathbf{s}) \in \mathcal{A}} \left\{ -c(\mathbf{s}, a_k) + \lambda_k^m \cdot d_k(\mathbf{s}, a_k) \right.$$

$$\left. + \beta \sum_{s'=1}^{|\mathcal{S}|} \mathbb{P}(\mathbf{s}' \mid \mathbf{s}, a_k) v_k^n(\mathbf{s}') \right\};$$

$$v_k^{n+1}(\mathbf{s}) = -c(\mathbf{s}, \pi_k^n(\mathbf{s})) + \lambda_k^m \cdot d_k(\mathbf{s}, \pi_k^n(\mathbf{s})) + \beta \sum_{s'=1}^{|\mathcal{S}|} \mathbb{P}(\mathbf{s}' \mid \mathbf{s}, \pi_k^n(\mathbf{s})) v_k^n(\mathbf{s}');$$

$$v_{i=1:K, i \neq k}^{n+1}(\mathbf{s}) = \lambda_i^m \cdot d_i(\mathbf{s}, \pi_k^n(\mathbf{s})) + \beta \sum_{s'=1}^{|\mathcal{S}|} \mathbb{P}(\mathbf{s}' \mid \mathbf{s}, \pi_k^n(\mathbf{s})) v_i^n(\mathbf{s}');$$

end for
Step 4: If $\| \mathbf{V}_k^{n+1} - \mathbf{V}_k^n \| \leq \varepsilon$, $k = 1, \ldots, K$, set $n = n + 1$, and return to Step 3; otherwise, go to Step 5
Step 5: Update Lagrange multipliers
for $k = 1$: K **do**

$$\lambda_k^{m+1} = \lambda_k^m + \frac{1}{l} \left[D_k(\pi_1^n, \pi_2^n, \ldots, \pi_K^n) - \tilde{D}_k \right]$$

end for
Step 6: The algorithm stops when λ_k^m, $k = 1, 2, \ldots, K$, converge; otherwise, set $m = m + 1$ and return to Step 2

The algorithm can be summarized as follows. We use $\mathbf{V}^n_{k=1,2,\ldots,K}$ and $\lambda^m_{k=1,2,\ldots,K}$ to represent the value vector and Lagrange multiplier, respectively, of a user k ($k=1, \ldots, K$) at the nth and mth time slot. The algorithm mainly consists of two parts: the outer loop and the inner loop. The outer loop updates the Lagrange multipliers of each user and the inner loop optimizes the transmission policy of each user under fixed Lagrange multipliers. The outer loop index and inner loop index are m and n, respectively. Note from Algorithm 5.1 that the interaction between users is through the update of value vectors since $v^{n+1}_{i=1:K, i \neq k}(s)$ is a function of $\pi^n_k(s)$.

In Step 1, we set the outer loop index m to be 0 and initialize the step size l, the value vectors $\mathbf{V}^0_{k=1,2,\ldots,K}$, and Lagrange multipliers $\lambda^0_{k=1,2,\ldots,K}$. Step 3 is the inner loop, where at each step we solve kth user-controlled game and obtain the new optimal strategy for that user with the strategy of the remaining players fixed. Note here that the objective of the system is to maximize the transmission reward and minimize the holding cost as shown in (5.13). Since $-c(\mathbf{s}, a_k)$ is used for each step, the optimal transmission policy $\pi^n_k(s)$ is obtained by doing the minimization. Variable ε in Step 4 is a small number chosen to ensure the convergence of $\mathbf{V}^n_k$. Step 5 updates the Lagrange multipliers based on the discounted delay value of each user given the transmission policies $\{\pi^n_1, \pi^n_1, \ldots, \pi^n_K\}$. $\dfrac{1}{l}$ is the step size that satisfies the conditions for convergence of the policy-gradient algorithm. This sequence of Lagrange multipliers $\{\lambda^m_1, \ldots, \lambda^m_K\}$ with $m = 0, 1, 2, \ldots$ converges in probability to $\{\lambda^*_1, \ldots, \lambda^*_K\}$, which satisfies the constrained problem defined in (5.13) [20,21]. The algorithm terminates when $\lambda^m_{k=1,2,\ldots,K}$ converge; otherwise, go to Step 2.

Since this is a constrained optimization problem, the optimal transmission policy is a randomization of two deterministic polices [18,19]. Use $\lambda^*_{k=1,2,\ldots,K}$ to represent the Lagrange multipliers obtained with Algorithm 5.1. The randomization policy of each user can be written as

$$\pi^*_k(\mathbf{s}) = q_k \pi^*_k(\mathbf{s}, \lambda_{k,1}) + (1 - q_k) \pi^*_k(\mathbf{s}, \lambda_{k,2}), \tag{5.19}$$

where $0 \leq q_k \leq 1$ is the randomization factor, and $\pi^*_k(\mathbf{s}, \lambda_{k,1})$ and $\pi^*_k(\mathbf{s}, \lambda_{k,2})$ are the unconstrained optimal policies with Lagrange multipliers $\lambda_{k,1}$ and $\lambda_{k,2}$. Specifically, $\lambda_{k,1} = \lambda^*_k - \Delta$ and $\lambda_{k,2} = \lambda^*_k + \Delta$ for a perturbation parameter Δ. The randomization factor of the kth user q_k is calculated by

$$q_k = \frac{\tilde{D}_k - D_k(\lambda_{1,2}, \ldots, \lambda_{K,2})}{D_k(\lambda_{1,1}, \ldots, \lambda_{K,1}) - D_k(\lambda_{1,2}, \ldots, \lambda_{K,2})}. \tag{5.20}$$

The convergence proof of the inner loop of Algorithm 5.1 is shown in Appendix 5.A.1, and this value iteration algorithm obtains a Nash equilibrium solution to the constrained switching control Markovian game with general-sum reward and general-sum constraint [12].

Remark: The primary purpose of the value iteration algorithm is to prove the structural results on the Nash equilibrium policy. The value iteration algorithm (Algorithm 5.1) is not easy to implement in a practical system, because at each iteration of the algorithm, a user k is required to know the channel states of all the other users and the system state transition probability matrix.

5.4.2 Structural Result on Randomized Threshold Policy

In this section, we characterize the structure of the Nash equilibrium achieved by Algorithm 5.1. First, we list four assumptions. Based on these four assumptions, Theorem 5.1 is introduced:

- *A1:* The set of policies that satisfy constraint (5.12) is nonempty, to ensure the delay constraint of the system is valid.

  A1 is the feasible assumption on the system constraint, and it is assumed to be satisfied.

- *A2:* Transmission reward $c(\mathbf{s}, a_k)$ is a supermodular* function of b_k, a_k for any channel quality h_k and MAD m_k of the current user. $c(\mathbf{s}, a_k)$ is also an integer-concave function of $b_k - a_k$ for any h_k and m_k.

  It can be seen from Section 5.3.1 that the transmission reward $c(\mathbf{s}, a_k)$ is linear in a_k (5.5) and independent of the buffer state b_k. Thus, the assumption A2 holds.

- *A3:* Holding cost $d_k(\mathbf{s}, a_k)$ is a submodular function of b_k, a_k for any channel quality h_k and MAD m_k of the current user. $d_k(\mathbf{s}, a_k)$ is also integer-convex in $b_k - a_k$ for any h_k and m_k.

  The holding cost of user k with $\mathbf{s} \in S_k$ is as follows:

  $$d_k(\mathbf{s}, a_k) = \kappa \cdot (b_k - a_k + f_{in,k})^\tau. \tag{5.21}$$

  The submodularity property of the holding cost function $d_k(\mathbf{s}, a_k)$ can be easily verified. It is also straightforward to show $d_k(\mathbf{s}, a_k)$ is integer-convex in $b_k - a_k$.

- *A4:* $\mathbb{P}(b_k' | b_k, a_k)$ is of second order, stochastically increasing on $(b_k - a_k)$ for any b_k and a_k.

  The $\mathbb{P}(b_k' | b_k, a_k)$ expression given by (5.17) shows it is first order, stochastically increasing in $(b_k - a_k)$. First-order stochastic dominance implies second-order stochastic dominance [18,19]; thus assumption A4 holds.

---

* If a function $f : \mathcal{A} \times \mathcal{B} \times \mathcal{C} \to \mathcal{R}$ is supermodular in (a, b) for any fixed $c \in C$, then for all $a' \geq a$ and $b' \geq b, f(a', b'; c) - f(a, b'; c) \geq f(a', b; c) - f(a, b; c)$ holds. f is submodular if $-f$ is supermodular.

The following theorem is one of our main results. It shows that the Nash equilibrium of a constrained dynamic switching control game is a randomization of two pure monotone policies.

Theorem 5.1 *For each user k and a given channel state, the Nash equilibrium policy $\pi_k^*(\mathbf{s})$ is a randomization of two pure policies $\pi_k^1(\mathbf{s})$ and $\pi_k^2(\mathbf{s})$. Each of these two pure policies is a nondecreasing function with respect to buffer occupancy state b_k.* □

The detailed proof of Theorem 5.1 is given in Appendix5.A.2.

5.4.3 Learning Nash Equilibrium Policy via Policy-Gradient Algorithm

The value iteration in Algorithm 5.1 is applicable to a general type of constrained switching control game. In the case that the system parameters satisfy assumptions A1–A4 as stated earlier, we can exploit the structural result (Theorem 5.1) to search for the Nash equilibrium policy in the policy space. This results in a significant reduction on the complexity of computing the Nash equilibrium policies.

The main idea of searching the Nash equilibrium policy in the policy space is as follows. If there are three actions available with action set $a_k = \{1, 2, 3\}$, the Nash equilibrium policy $\pi_k^*(\mathbf{s})$ is a randomized mixture:

$$\pi_k^*(s) = \begin{cases} 1 & 0 \leq b_k < b_1(\mathbf{s}) \\ p_1 & b_1(\mathbf{s}) \leq b_k < b_2(\mathbf{s}) \\ 2 & b_2(\mathbf{s}) \leq b_k < b_3(\mathbf{s}). \\ p_2 & b_3(\mathbf{s}) \leq b_k < b_4(\mathbf{s}) \\ 3 & b_4(\mathbf{s}) \leq b_k \end{cases} \tag{5.22}$$

Here, $\{p_1, p_2\} \in [0,1]$ is the randomization factor; $b_1(\mathbf{s})$, $b_2(\mathbf{s})$, $b_3(\mathbf{s})$, and $b_4(\mathbf{s})$ are the buffer thresholds. The search for each Nash equilibrium policy problem is now converted to the estimation of these six parameters. Note here that the policy search applies to a system with any number of actions; here we consider a three-action system as an example.

The simultaneous perturbation stochastic approximation (SPSA) method [20] is adopted to estimate the parameters. This method is especially efficient in high-dimensional problems in terms of providing a good solution for a relatively small number of measurements for the objective function. The essential feature of SPSA is the underlying gradient approximation, which requires only two objective function measurements per iteration regardless of the dimension of the optimization problem. These two measurements are made by simultaneously varying, in a properly random fashion, all the variables in the problem. The detailed algorithm is described in Algorithm 5.2.

Algorithm 5.2: Policy-gradient algorithm

1: **Initialization**: $\theta^{(0)}$, λ^0; $n = 0$; $\rho = 4$

2: Initialize constant perturbation step size β and gradient step size α

3: **Main Iteration**

4: **for** $k = 1: K$ **do**

5: $dim_k = 6 \times |h_k| \times |h_i| \times |b_i|$

6: Generate $\Delta^n = [\Delta_1^n, \Delta_2^n, \ldots, \Delta_{dim_k}^n]^T$ are Bernoulli random variables with $p = \dfrac{1}{2}$.

7: $\theta_{k+}^n = \theta_k^n + \beta \times \Delta^n$

8: $\theta_{k-}^n = \theta_k^n - \beta \times \Delta^n$

9: $\Delta C_k^n = \dfrac{c_k(s^n, \theta_{k+}^n) - c_k(s^n, \theta_{k-}^n)}{2\beta}[(\Delta_1^n)^{-1}, (\Delta_2^n)^{-1}, \ldots, (\Delta_{m_k}^n)^{-1}]^T$

10: $\Delta D_k^n = \dfrac{d_k(s^n, \theta_{k+}^n) - d_k(s^n, \theta_{k-}^n)}{2\beta}[(\Delta_1^n)^{-1}, (\Delta_2^n)^{-1}, \ldots, (\Delta_{m_k}^n)^{-1}]^T$

11: $\theta_k^{(n+1)} = \theta_k^n - \alpha \times \left(\Delta C_k^n + \Delta D_k^n \cdot \max\left[0, \lambda_k^n + \rho \cdot \left(D(s^n, \theta_k^n) - \tilde{D}_k \right) \right] \right)$

12: $\lambda_k^{(n+1)} = \max\left[(1 - \dfrac{\alpha}{\rho} \cdot \lambda_k^n), \lambda_k^n + \alpha \cdot \left(D(s^n, \theta_k^n) - \tilde{D}_k \right) \right]$

13: **end for**

14: The parameters of other users remain unchanged

15: $n = n + 1$

16: The iteration terminates when the values of the parameters θ^n converge; else return back to Step 3.

The first part of the algorithm initializes system variables. θ^n represents the union of all the parameters we search for at the nth time slot, and $\theta^n = \{\theta_1^n, \theta_2^n, \ldots, \theta_K^n\}$. θ_k^n indicates parameter vector of the kth user. The Lagrange multipliers of all the K users are defined to be $\lambda^n = \{\lambda_1^n, \lambda_2^n, \ldots, \lambda_K^n\}$. β and α denote the constant perturbation step size and constant gradient step size, respectively. When $\alpha \to 0$, $\dfrac{\alpha}{\beta^2} \to 0$, and $n \to \infty$, the policy-gradient algorithm converges weakly [22, Theorem 8.3.1]. In the main part of the algorithm, SPSA algorithm is applied to iteratively update system parameters. When the kth user is scheduled to transmit at time slot n, parameters θ_k^n and the Lagrange multiplier λ_k^n can be updated after introducing a random perturbation vector Δ^n. Meanwhile, the parameters of the other users remain unchanged. The algorithm terminates when θ^n converge.

Remark: At each iteration of the policy-gradient algorithm, user k is only required to know its own transmission reward and holding cost. The transmission reward and holding cost of user k are functions of its own channel state and buffer state (Section 5.4.2). Thus, the policy-gradient algorithm is distributed and implementable in a practical system.

5.5 Numerical Examples

In this section, we illustrate the performance improvement of the proposed dynamic switching control game policy over myopic policies. We also present numerical examples of the Nash equilibrium transmission policy. For our comparisons, we used the JSVM-9-8 reference software [23] to encode two reference video sequences in Common Intermediate Format (CIF) resolution: Foreman and Football. Each sequence is composed of 100 frames encoded in SVC, with an H.264/AVC compatible base layer and two CGS enhancement layers. The coded video streams are composed of I and P frames only, with only the first frame encoded as an I frame, a frame rate of 30 fps, and minimum QoS guarantee corresponding to the average base-layer QP value of $q_0 = 38$. The first and second CGS enhancement layers have average QP values of 32 and 26, respectively. The video MAD m_k is quantized into three states, $m_k \in \mathcal{M} = \{0,1,2\}$, such that

$$
m_k = \begin{cases}
0, & \text{if } MAD_k < \mu_k - \dfrac{\sigma_k}{2}, \\[2mm]
1, & \text{if } \mu_k - \dfrac{\sigma_k}{2} \leq MAD_k \geq \mu_k + \dfrac{\sigma_k}{2}, \\[2mm]
2, & \text{if } MAD_k > \mu_k + \dfrac{\sigma_k}{2}
\end{cases}
$$

where μ_k and σ_k are the mean and standard deviation of the MAD of user k.

Tables 5.1 and 5.2 list the bit rate in bits per second and average distortion in terms of Y-PSNR of each video layer of the two incoming video sequences.

In all the cases considered here, the channel quality measurements h_k of user k are quantized into two different states, $h_k \in \{1, 2\}$. In the models used, each user has a size 10 buffer with an incoming rate equivalent to increasing the buffer size by 1 during a transmission time slot ($f_{in,1} = f_{in,2} = \cdots = f_{in,K} = 1$). One buffer unit is equivalent to one video frame. The transmission time slot duration is set equal to 10 ms, thus allowing a maximum transmission delay of 100 ms.

TABLE 5.1

Average Incoming Rate and Distortion (Y-PSNR)
Characteristics of Video User Foreman

| Sequence | Foreman | |
Video Layer	Bit Rate $\bar{r}_{l,1}$ (kbps)	Y-PSNR (dB)
Base $l=0$	40.9	30.75
First CGS $l=1$	103.36	34.59
Second CGS $l=2$	242.64	38.87

TABLE 5.2

Average Incoming Rate and Distortion (Y-PSNR)
Characteristics of Video User Football

| Sequence | Football | |
Video Layer	Bit Rate $\bar{r}_{l,2}$ (kbps)	Y-PSNR (dB)
Base $l=0$	55.82	28.17
First CGS $l=1$	139.03	32.47
Second CGS $l=2$	252.97	37.29

This system configuration ensures that the transmission rewards, holding costs, and buffer transition probability matrices satisfy assumptions A2 and A3 specified in Section 5.4.2. The channel is assumed to be Markovian, and the transition probability matrices are generated randomly.

In the system models used, each user has three different action choices when it is scheduled for transmission. The three different actions are $a_k = 1, 2, 3$, and correspond to the user buffer output of one, two, and three frames, respectively. The action $a_k = 0$ corresponds to the *No-Transmit* state. Therefore, when the Nash equilibrium policy is 0, the current user is not scheduled for transmission, and $a_k = 0$. However, the outgoing traffic is still one frame, $f_{in,k} = 1$.

5.5.1 General-Sum Constrained Game: Randomized Monotone Transmission Policy

Figure 5.4 considers the same two-user system of the previous example with a 25 ms transmission delay constraint. As it is a constrained switching controlled Markovian game, the Nash equilibrium policy is a randomization of two pure policies. The figure shows that the optimal transmission policies are no longer deterministic but are a randomization of two pure policies, and that each Nash equilibrium policy is monotone-increasing on the buffer state. The results of Figure 5.4 are obtained by applying the value iteration algorithm. The first subfigure of Figure 5.4 is the optimal policy of user 1

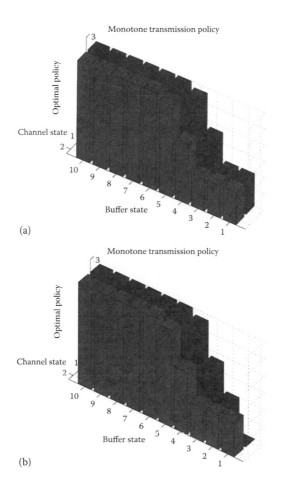

(a)

(b)

FIGURE 5.4

The Nash equilibrium transmission policy obtained via value iteration algorithm for users 1 and 2. The first subfigure is obtained when $h_2=1$ and $b_2=1$, and the second subfigure is obtained when $h_1=1$ and $b_1=1$, respectively. The transmission delay constraint is specified to be 25 ms. Each Nash equilibrium transmission policy is a randomization of two pure monotone policies.

when $h_2=1$ and $b_2=1$, while the second subfigure of Figure 5.4 shows the optimal policy of user 2 when $h_1=1$ and $b_1=1$.

5.5.2 Result by Policy-Gradient Algorithm

The policy-gradient algorithm (Algorithm 5.2) we propose uses the structural result on the optimal transmit policy, and each policy can be determined by three parameters, namely, lower threshold $b_l(\mathbf{s})$, upper threshold $b_h(\mathbf{s})$, and randomization factor p, as described in (5.22) (assume there are two actions). The simulation results of the policy of user 1 with $h_2=1$, $b_2=1$ are shown in

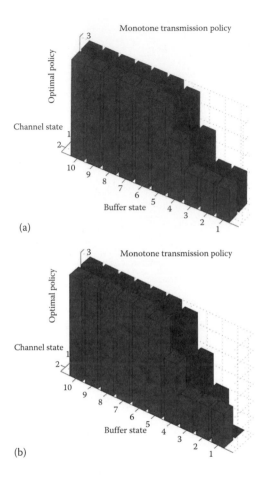

(a)

(b)

FIGURE 5.5
The Nash equilibrium transmission control policy obtained via policy-gradient algorithm
for user 1 and 2. The result is obtained when (a) $h_2=1$ and $b_2=1$ and (b) $h_1=1$ and $b_1=1$. The
transmission delay constraint is specified to be 25 ms. The transmission policy is monotone-
nondecreasing on its own buffer state.

Figure 5.5. In the system setup, each user has a buffer size of 10 and three
actions, and the system transmission delay constraint is 25 ms. By comparing Figure 5.5 to the first subfigure of Figure 5.4, we can see that the results
obtained by the value iteration algorithm and policy-gradient algorithm are
very close.

5.5.3 Transmission Buffer Management

We adopt a priority-based packet drop similar to the DPB buffer management
policy proposed in [8]. With this policy, low-priority packets that have resided
longest in the transmission buffer are dropped when the buffering delay

approaches the delay constraint. In [8], the DPB policy was found to be second best to the *drop dependency based* (DDB), which requires video dependency information that cannot be collected in real time. The DPB policy is attractive since it requires minimal priority information about the video payload, which is readily available in the video network abstraction layer unit (NALU) header of a MGS/CGS-coded video stream. We implement this policy by shifting the user action a_k to a lower state that satisfies the buffer delay constraint.

In order to demonstrate the effectiveness of the proposed dynamic switching control game algorithm, we compare its performance to that of a myopic policy that selects at each time slot the user action that maximizes the video PSNR while satisfying the transmission buffer delay constraint. The time slot duration equals the video frame capturing duration, which is around 30 ms. The myopic policy does not consider the effect of current actions on future states. We assume that the delay constraint corresponds to the packet play-out deadline; therefore, all buffered packets that exceed the delay constraint are assumed to be lost and are discarded from the user buffer. Figures 5.6 and 5.7 show the results of simulating the transmission of two users Foreman and Football for both the proposed switching control game policy and the myopic policy. The figures show that under the same channel conditions, the proposed policy has better Y-PSNR performance and transmission buffer utilization for both users.

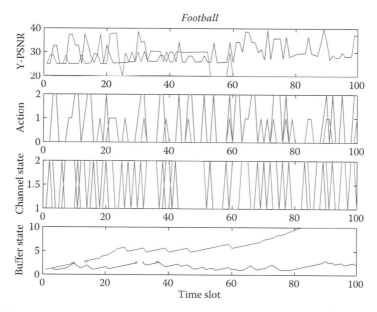

FIGURE 5.6

(**See color insert.**) Result of the transmission of the *Football* sequence comparing the performance in terms of video PSNR and buffer utilization between the proposed switching control game policy (blue lines) and the myopic policy (red lines) with 80 ms delay constraint. The result shows that the proposed switching control game policy performs better than the myopic policy.

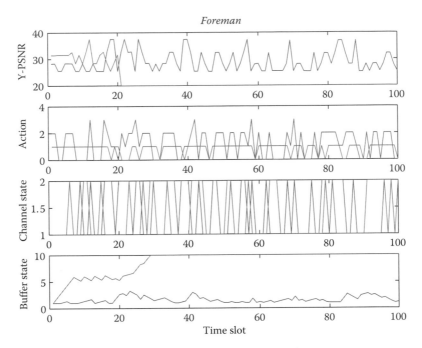

FIGURE 5.7
(See color insert.) Result of the transmission of the *Foreman* sequence comparing the performance in terms of video PSNR and buffer utilization between the proposed switching control
game policy (blue lines) and the myopic policy (red lines) with 80 ms delay constraint. The
result shows that the proposed switching control game policy performs better than the myopic
policy.

5.6 Conclusions

This chapter considers a time-slotted multiuser WLAN system where each
user delivers SVC video bitstream. The modified CSMA/CA mechanism
of such system is implemented through an opportunistic scheduling system access rule. By modeling video states and block-fading channels as a
finite-state Markov chain, the rate adaptation problem among users can
be formulated as a constrained general-sum switching control Markovian
game. The Nash equilibrium transmission policy of such a game can be
computed through a value iteration algorithm. Given four assumptions on
the transmission reward, holding cost, and transition probability functions
(Section 5.4.2), the Nash equilibrium policy is a randomized mixture of two
pure policies with each policy monotone-nondecreasing on the buffer state
occupancy. This structural result enables us to search for the Nash equilibrium policy in the policy space via a policy-gradient algorithm.

Glossary

CGS	Coarse-grained scalability
CIF	Common Intermediate Format
CSMA/CA	Carrier sense multiple access with collision avoidance
DPB	Drop priority based
HD	High definition
HSDPA	High speed downlink packet access
IEEE	Institute of Electrical and Electronics Engineers
IP	Internet protocol
MAD	Mean absolute difference
MDP	Markov decision process
MGS	Medium-grained scalability
MSE	Mean-square-error
NALU	Network Abstraction Layer Unit
PER	Packet error rate
PSNR	Peak signal-to-noise ratio
QoS	Quality of service
QP	Quantization parameter
RTP	Real-time transport protocol
SNR	Signal-to-noise ratio
SPSA	Simultaneous perturbation stochastic approximation
SVC	Scalable video coding
WLAN	Wireless local area network
Y PSNR	Luminance peak signal-to-noise ratio

Symbols

k	User index
K	Total number of users
l	Layer index
n	Time index
r	Bit rate
u	Power factor
B	Maximum number of coded video frames stored in the buffer
W	Number of actions
a_k^n	Action of user k at time n
b_k^n	Buffer occupancy state of user k at time n
c_k	Transmission reward of user k
d_k	Holding cost of user k
h_k^n	Channel state of user k at time n
m_k^n	MAD at the lth layer

t_k^n	Channel access time delay of user k at time n
p_l	PER at the lth layer
$q_{l,k}$	Quantization parameter at the lth layer of the kth user
m_l	MAD at the lth layer
$f_{in,k}$	Incoming number of video frames of user k
$v_k^n(\mathbf{s})$	Value of user k at time n at state $\mathbf{s}$
L_k	Number of video layers admitted into the buffer of user k
C_k	Discounted transmission reward of user k
D_k	Discounted holding cost of user k
$\tilde{D}_k$	Discounted holding cost constraint
V_k^n	Value
$\mathbf{b}^n$	Composition of buffer states of all the K users at time n
$\mathbf{h}^n$	Composition of channel states of all the K users at time n
$\mathbf{m}^n$	Composition of MAD states of all the K users at time n
$\mathbf{s}^n$	System state at time n
$\mathbf{V}_k^n$	Value vector of user k at time n
δ	Video PSNR
β	Discount factor
γ_k	QoS parameter of user k
$\boldsymbol{\theta}^n$	Union of all the parameters at nth time slot
π_k	Transmission policy of user k
π_k^*	Optimal transmission policy of user k
λ_k^m	Lagrange multiplier of user k at time m
Φ_k	Set of all the deterministic policies of user k
$\boldsymbol{\Delta}^n$	Bernoulli random variables
$\mathcal{P}$	Transition matrix

Appendix 5.A.1 Proof of Convergence of Algorithm 5.1

1. By following the definition of $\pi_k^n(\mathbf{s}, \mathbf{s} \in \mathcal{S}_k)$ from Algorithm 5.1, we can deduce that

$$\mathbf{V}_k^{(n+1)} \leq \mathbf{V}_k^n. \tag{5.23}$$

2. When $\pi_k^n(\mathbf{s}, \mathbf{s} \in \mathcal{S}_k) = \pi_k^{(n+\Delta)}(\mathbf{s}, \mathbf{s} \in \mathcal{S}_k)$, we have $\mathbf{V}_k^n = \mathbf{V}_k^{(n+\Delta)}$. In view of (5.23), if $\mathbf{V}_k^n \neq \mathbf{V}_k^{(n+\Delta)}$, then $\pi_k^n(\mathbf{s}, \mathbf{s} \in \mathcal{S}_k) \neq \pi_k^{(n+\Delta)}(\mathbf{s}, \mathbf{s} \in \mathcal{S}_k)$ for any $\Delta = 1, 2, 3, \ldots$.

3. The payoff function of the matrix game with $\mathbf{s} \in \mathcal{S}_l$ is of the form $\left[-c(\mathbf{s}, a_k) + \lambda_k^m \cdot d_k(\mathbf{s}, a_k) \right] + \beta \sum_{\mathbf{s}'=1}^{|S|} \mathbb{P}(\mathbf{s}' \mid \mathbf{s}, a_k) v_k^n(\mathbf{s}')$. Please note that the

first term of this payoff function $\left[-c(\mathbf{s}, a_k) + \lambda_k^m \cdot d_k(\mathbf{s}, a_k)\right]$ is independent of $\mathbf{V}_k^n$. By using the result from [12], Lemma 6.3.2, $\pi_k^n(\mathbf{s}, \mathbf{s} \in \mathcal{S}_k)$ equals to some extreme optimal action in a submatrix game with payoff function $\left[-c(\mathbf{s}, a_k) + \lambda_k^m \cdot d_k(\mathbf{s}, a_k)\right]$. As there are only finite number of extreme optimal action candidates for $\pi_k^n(\mathbf{s}, \mathbf{s} \in \mathcal{S}_k)$, there exist n and $\Delta \geq 1$, such that $\pi_k^n(\mathbf{s}, \mathbf{s} \in \mathcal{S}_k) = \pi_k^{(n+\Delta)}(\mathbf{s}, \mathbf{s} \in \mathcal{S}_k)$, which in turn implies $\mathbf{V}_k^n = \mathbf{V}_k^{(n+\Delta)}$.

4. If $\mathbf{V}_k^n = \mathbf{V}_k^{(n-1)}$, $\pi_k^n(\mathbf{s}, \mathbf{s} \in \mathcal{S}_k)$ is the optimal action policy for user k ($k \in \{1,...,K\}$) of the game with a fixed Lagrange multiplier λ_k^m.

Based on these observations 1–4, we conclude that Algorithm 5.1 will converge in a finite number of iterations with fixed Lagrange multipliers $\lambda_{k=1,2,...,K}^m$. $\square$

Appendix 5.A.2 Proof of Theorem 5.1

We choose delay constraint parameters $\tilde{D}_{k,k=1,2,...,K}$ in the optimization problem (5.13) carefully to ensure the existence of an optimal policy. The optimal policy can be obtained when the objective function (5.11) is maximized and the equality of the delay constraint (5.12) is held.

According to Algorithm 5.1, $\forall s \in S$, the optimal policy and value matrix of kth user are updated by the following steps:

$$\pi_k^n(\mathbf{s}) = \arg\min_{\pi_k^n(\mathbf{s})} \left\{ -c(\mathbf{s}, a_k) + \lambda_k^m \cdot d_k(\mathbf{s}, a_k) \right.$$

$$\left. + \beta \sum_{s'=1}^{|S|} \mathbb{P}(\mathbf{s}' \mid \mathbf{s}, a_k) v_k^n(\mathbf{s}') \right\}; \tag{5.24}$$

$$v_k^{(n+1)}(\mathbf{s}) = -c(\mathbf{s}, \pi_k^n(\mathbf{s})) + \lambda_k^m \cdot d_k(\mathbf{s}, \pi_k^n(\mathbf{s}))$$

$$+ \beta \sum_{s'=1}^{|S|} \mathbb{P}(\mathbf{s}' \mid \mathbf{s}, \pi_k^n(\mathbf{s})) v_k^n(\mathbf{s}'). \tag{5.25}$$

A sufficient condition for the optimal transmission action policy $\pi_k^n(\mathbf{s})$ to be monotone-nondecreasing in the buffer state b_k is that the right-hand side of (5.24) is a submodular function of (b_k, a_k). According to assumptions A2 and A3, $\left[-c(\mathbf{s}, a_k) + \lambda_k^m \cdot d_k(\mathbf{s}, a_k)\right]$ is a submodular function of (b_k, a_k); thus, we only need to demonstrate that is also submodular in the pair (b_k, a_k).

Recall the overall state space S is the union of all the K subspaces, $S = S_1 \cup S_2 \ldots \cup S_K$. Thus, we can write $\sum_{s'=1}^{|S|} \mathbb{P}(s' \mid s, a_k) v_k^n(s')$ as a summation of K terms according to the partition of the state space, where the ith term $(i = 1, 2, \ldots, K)$ is denoted by $Q_i(s, a_k)$. By using the property of the state transition probability (5.15) and Assumption A4, we have the following result:

$$Q_i(s, a_k) = \sum_{s', s \in S_i} \mathbb{P}(s' \mid s, a_k) v_k^n(s')$$

$$= \prod_{l=1}^{K} \mathbb{P}(h_l' \mid h_l) \prod_{l=1}^{K} \mathbb{P}(m_l' \mid m_l) \prod_{l=1, l \neq k}^{K} \mathbb{P}(b_l' \mid b_l) \mathbb{P}(b_k' \mid b_k + a_k) v_k^n(s'). \quad (5.26)$$

In order to show the submodularity of $\sum_{s'=1}^{|S|} \mathbb{P}(s' \mid s, a_k) v_k^n(s')$, we only need to prove the submodularity property of $Q_i(s, a_k)$ for $i = 1, 2, \ldots, K$.

The proof of Theorem 5.1 needs the result from Lemma 5.1, the proof of which will be given after the proof of the Theorem in Appendix 5.A.3.

Lemma 5.1 *Under the conditions of Theorem 1, $Q_i(s, a_k)$ is of the form $Q_i(s, a_k) = \bar{Q}_i(b_k - a_k, \{\mathbf{h}, \mathbf{m}, \mathbf{b}_{-k}\})$. Function $\bar{Q}_i(x, y)$ is integer-convex function of x for any given $\mathbf{y} = \{\mathbf{h}, \mathbf{m}, \mathbf{b}_{-k}\}$.*

According to Lemma 5.1, $\bar{Q}_i(x, y)$ is integer-convex in x, which can be written as

$$\bar{Q}_i(x' + \Delta, \mathbf{y}) - \bar{Q}_i(x', \mathbf{y})$$
$$\leq \bar{Q}_i(x + \Delta, \mathbf{y}) - \bar{Q}_i(x, \mathbf{y}) \quad (5.27)$$

for $x' \geq x$ and $\Delta \geq 0$. Substituting in Equation 5.27 with $x' = b_k - a_k$, $x = b_k - a_k'$, and $\Delta = b_k' - b_k$ for $b_k' \geq b_k$ and $a_k' \geq a_k$, we obtain

$$Q_i(\{b_k', a_k\}, \mathbf{y}) - Q_i(\{b_k, a_k\}, \mathbf{y})$$
$$\leq Q_i(\{b_k', a_k'\}, \mathbf{y}) - Q_i(\{b_k, a_k'\}, \mathbf{y}), \quad (5.28)$$

which is the definition of submodularity of $Q_i(s, a_k)$ in b_k and a_k. Furthermore, the linear combination of submodular functions still remains the submodular property. Thus, $\sum_{s'=1}^{|S|} \mathbb{P}(s' \mid s, a_k) v_k^n(s')$ is submodular in the pair (b_k, a_k).

Based on these results, we can prove the submodularity of the right-hand side of Equation 5.24. This proves that the optimal transition action policy

$\pi_k^*(\mathbf{s})^n$ is monotone-nondecreasing on the buffer state b_k under fixed positive Lagrange constant.

According to [24], the optimal randomized policy with a general constraint is a mixed policy comprising of two pure policies that can be computed under two different Lagrange multipliers. As discussed earlier, we have already shown that each of these two pure policies is nondecreasing on the buffer state occupancy. Thus, the mixed policy also owns the nondecreasing property on the buffer state. This concludes the proof. □

Appendix 5.A.3 Proof of Lemma 5.1

Let us first assume that $v_k^n(\mathbf{s}')$ is an integer-convex function of b_k'. Then, according to Proposition 2 of [25] and A4 in Section 5.4.2 along with (5.26), it can be concluded that $Q_i(\mathbf{s},a_k)$ is an integer-convex function of $b_k - a_k$.

Thus, we only need to prove that $v_k^n(\mathbf{s}')$ is an integer-convex function of b_k', which can be proved via backward induction. The value vector can be updated by the value iteration algorithm as follows:

$$v_k^{(n)}(\mathbf{s}) = -c(\mathbf{s},a_k) + \lambda_k^m \cdot d_k(\mathbf{s},a_k) + \beta \sum_{s'=1}^{|S|} \mathbb{P}(\mathbf{s}' \mid \mathbf{s},a_k) v_k^{n-1}(\mathbf{s}'). \quad (5.29)$$

If $v_k^{n-1}(\mathbf{s}')$ is integer-convex in b_k', as well as the A2 and A3 from Section 5.4.2, it can be proved that $v_k^n(\mathbf{s})$ is integer-convex in b_k by using the key Lemma 61 from [26]. The convergence of the value iteration algorithm is not affected by the initial values. Choosing $v_k^0(\mathbf{s}')$ to be an integer-convex function of b_k', $v_k^0(\mathbf{s})$ is integer-convex in b_k followed by induction. This concludes the proof. □

References

1. ITU-T Rec. H.264 and ISO/IEC 14496-10 (MPEG-4 AVC), ITU-T and ISO/IEC JTC 1. *Advanced Video Coding for Generic Audiovisual Services*, June 2005.
2. ISO/IEC 14492-2 (MPEG-4 vISUAL), ISO/IEC JTC 1. *Coding of Audio-Visual Objects—Part 2: Visual*, May 2004.
3. J. Reichel, H. Schwarz, and M. Wien. Joint scalable video model JSVM-9. Technical Report N 8751, ISO/IEC JTC 1/SC 29/WG 11, Marrakech, Morocco, January 2007.
4. T. Schierl and T. Wiegand. Mobile video transmission using scalable video coding. *IEEE Transactions on Circuits and Systems for Video Technology*, 17(9):1204–1217, September 2007.

5. M. Wien, R. Cazoulat, A. Graffunder, A. Hutter, and P. Amon. Real-time system for adaptive video streaming based on SVC. *IEEE Transactions on Circuits and Systems for Video Technology*, 17(9):1227–1237, September 2007.
6. M. Luby, T. Gasiba, T. Stockhammer, and M. Watson. Reliable multimedia download delivery in cellular broadcast networks. *IEEE Transactions on Broadcasting*, 53(1):235–246, March 2007.
7. H. Schwarz, D. Marpe, and T. Wiegand. Overview of the scalable video coding extension of the H.264/AVC standard. *IEEE Transactions on Circuits and Systems for Video Technology*, 17(9):1103–1120, 2007.
8. G. Liebl, H. Jenkac, T. Stockhammer, and C. Buchner. Radio link buffer management and scheduling for wireless video streaming. *Springer Telecommunication Systems*, 30(1–3):255–277, 2005.
9. X. Zhu and B. Girod. Analysis of multi-user congestion control for video streaming over wireless networks. In *Proceedings of IEEE International Conference on Multimedia and Expo (ICME)*, Toronto, Ontario, Canada, July 2006.
10. D. K. Kwon, M. Y. Shen, and C. C. J. Kuo. Rate control for H.264 video with enhanced rate and distortion models. *IEEE Transactions on Circuits and Systems for Video Technology*, 17(5):517–529, 2007.
11. S. R. Mohan and T. E. S. Raghavan. An algorithm for discounted switching control stochastic games. *OR Spektrum*, 55(10):5069–5083, October 2007.
12. J. A. Filar and K. Vrieze. *Competitive Markov Decision Processes*. Springer-Verlag, New York, 1997.
13. O. J. Vrieze, S. H. Tijs, T. E. S. Raghavan, and J. A. Filar. A finite algorithm for the switching control stochastic game. *OR Spektrum*, 5(1):15–24, March 1983.
14. IEEE Std 802.11-2007. IEEE standard for local and metropolitan area networks, Part 11: Wireless LAN medium access control (MAC) and physical layer (PHY) specifications, 2007.
15. E. Altman. *Constrained Markov Decision Processes*. Chapman & Hall, London, U.K., 1999.
16. H. Mansour, P. Nasiopoulos, and V. Krishnamurthy. Real-time joint rate and protection allocation for multi-user scalable video streaming. In *Proceedings of IEEE Personal, Indoor, and Mobile Radio Communications (PIMRC)*, Cannes, France, pp. 1–5, September 2008.
17. A. Farrokh and V. Krishnamurthy. Opportunistic scheduling for streaming users in HSDPA multimedia systems. *IEEE Transactions on Multimedia Systems*, 8(4):844–855, August 2006.
18. D. V. Djonin and V. Krishnamurthy. MIMO transmission control in fading channels—A constrained Markov decision process formulation with monotone randomized policies. *IEEE Transactions on Signal Processing*, 55(10):5069–5083, October 2007.
19. D. V. Djonin and V. Krishnamurthy. Q-Learning algorithms for constrained Markov decision processes with randomized monotone policies: Application to MIMO transmission control. *IEEE Transactions on Signal Processing*, 55(5):2170–2181, May 2007.
20. J. C. Spall. *Introduction to Stochastic Search and Optimization: Estimation, Simulation and Control*. Wiley-Interscience, Hoboken, NJ, 2003.
21. V. Krishnamurthy and G. G. Yin. Recursive algorithms for estimation of hidden Markov models and autoregressive models with Markov regime. *IEEE Transactions on Information Theory*, 48(2):458–476, February 2002.

22. H. J. Kushner and G. Yin. *Stochastic Approximation and Recursive Algorithms and Applications*. Springer-Verlag, New York, 2003.

23. ISO/IEC JTC 1/SC 29/WG 11 N8964. JSVM-10 software, 2007.

24. F. J. Beutler and K. W. Ross. Optimal policies for controlled Markov chains with a constraint. *Journal of Mathematical Analysis and Applications*, 112:236–252, November 1985.

25. J. E. Smith and K. F. McCardle. Structural properties of stochastic dynamic programs. *Operations Research*, 50(5):796–809, September–October 2002.

26. E. Alman, B. Gaujal, and A. Hordijk. *Discrete-Event Control of Stochastic Networks: Multimodularity and Regularity*. Springer, Berlin, Germany, 2003.

6

Energy and Bandwidth Optimization in Mobile Video Streaming Systems

Mohamed Hefeeda, Cheng-Hsin Hsu, and Joseph Peters

CONTENTS

6.1 Introduction

In recent years, mobile devices have become more powerful in terms of computing power, memory, screen size, and screen quality. This provides an improved experience for viewing TV and multimedia content and has resulted in increasing demand for multimedia services for mobile devices. However, video streaming to mobile devices still has many challenges that need to be addressed. One challenge for mobile video is the limited wireless bandwidth in wide-area wireless networks. The wireless bandwidth is not only limited, but also quite expensive. Thus, for commercially viable mobile video services, network operators should maximize the utilization of their license-based wireless spectrum bands.

Another challenge is that mobile devices can be equipped only with small batteries that have limited lifetimes. Thus, conservation of the energy of mobile devices during streaming sessions is needed to prolong the battery lifetimes and enable users to watch videos for longer periods. *Time slicing* is a common technique in which the base station sends each video stream in bursts at a much higher bit rate than the encoding rate of the video. The base station computes the next burst time and includes it in the header field of the current burst. Time slicing enables a mobile device to receive a burst of traffic and then put its network interface to sleep to save energy until the next burst arrives.

In practice, the aggregate bit rate of the video streams can be greater than the available bandwidth. Controlling the aggregate bit rate is possible, to some extent, using joint video coders consisting of joint rate allocators, decoders, and variable bit rate (VBR) coders to encode video streams and dynamically allocate bandwidth among them so that the network is not overloaded. But in practice, many broadcast networks are not equipped with these components due to the added expense and complexity, and it is more common for videos to be encoded using stand-alone coders. In these cases, the aggregate bit rate of the video streams could instantaneously exceed the network bandwidth, and the receivers may experience playout glitches. Careful management of the levels of the buffers at the receivers is essential to the provision of good service quality.

In this chapter, we will consider the problem of multicasting multiple video streams from a wireless base station to many mobile receivers over a common wireless channel. This problem arises in wide-area wireless networks that offer multimedia content using multicast and broadcast services, such as WiMAX (worldwide interoperability for microwave access) [27,58], 3G/4G cellular networks that enable the multimedia broadcast multicast services (MBMS) [42], DVB-H (digital video broadcasting—handheld) [20,22,34,38], MediaFLO (media forward link only) [11], and ATSC M/H (advanced television systems committee—mobile/handheld) [5]. Our objective is to maximize bandwidth utilization, energy saving, and perceived quality of the transmitted videos.

We start in Section 6.2 by providing some background on the different coding techniques for video streams, and we describe various models for streaming videos over wireless networks. The section also presents the general architecture of mobile video streaming networks and introduces the energy conservation problem in such networks. In Section 6.3, we state, formulate, and prove the hardness of the energy optimization problem in mobile video streaming networks. The section then presents an algorithm that solves this problem under a certain assumption that usually holds in practice. This algorithm is analytically analyzed and empirically evaluated in a mobile video streaming test bed. We extend the energy optimization problem to consider maximizing the utilization of the wireless bandwidth in Section 6.4, where we employ statistical multiplexing of VBR video streams. An algorithm is then presented for the extended problem and analyzed both theoretically and empirically in the mobile video streaming test bed. We conclude the chapter in Section 6.5.

6.2 Background

We begin this section with some background about video coding standards and models for streaming video over wireless networks. We then give an overview of mobile video streaming services, which is followed by a description of a common energy conservation technique.

6.2.1 Video Coding Standards

Video coders can be roughly categorized into constant bit rate (CBR) and VBR coders. CBR coders adjust the coding parameters to maintain a fixed frame size, and thus a constant bit rate, throughout a video. Algorithms that analyze video complexity and determine coding parameters for individual frames are called rate control algorithms. CBR-encoded streams have fixed bandwidth requirements, which largely simplify the problem of bandwidth allocation in packet-switched networks. Encoding a video in CBR requires either an oversubscription of the network bandwidth in best-effort networks, or a bandwidth reservation in reservation-based networks. While streaming videos encoded in CBR is less complicated, doing so leads to degraded user experience and lower bandwidth efficiency. This is because each video consists of scenes with diverse complexities, and encoding all scenes at the same bit rate results in quality fluctuations, which are annoying to users. Moreover, to maintain a minimum target quality, a video must be encoded at a bit rate high enough for the most complex scene to achieve that target quality [36]. This in turn leads to wasted bandwidth, because less complex scenes are unnecessarily encoded at high quality with little or no impact on human perception.

To address these two issues, VBR coders can be used to provide constant video quality and avoid wasting bandwidth. This is achieved by dynamically distributing the available bandwidth among frames of the same video so that more complex scenes and more critical frames get more bits. That is, VBR coders enable bit budget redistribution along the time axis of a given video and achieve better coding and bandwidth efficiency. VBR coders that encode videos without considering buffer and network status are called unconstrained VBR (UVBR) coders. Since there are no constraints on the individual frame sizes, UVBR coders encode videos only from the perspective of coding efficiency and achieve the best possible video quality. UVBR streams, however, can have high bit rate variability, and the variability is increasingly significant in modern video coders such as H.264/AVC [57]. Smoothly streaming UVBR streams encoded by modern coders can be very challenging because the encoded streams may require instantaneous bit rates much higher than the bandwidth of the underlying networks.

Constrained VBR (CVBR) coders take target values of one or more streaming parameters as input and create a VBR stream that can be smoothly streamed in streaming environments with these parameter values. Common streaming parameters include channel bandwidth, smoothing buffer size, and initial buffering delay [13]. To ensure smooth playout, CVBR coders implement rate control algorithms, which monitor the complexities of scenes and frames, determine the target size of each frame, and constrain the frame sizes while encoding a video. The rate control algorithms are similar to those used in CBR coders, but CVBR coders allow bit rate variability as long as it is within the constraints imposed by the target streaming parameters.

To prepare encoded streams for packet-switched networks, service providers may choose different video coders for various reasons. Service providers that would trade bandwidth efficiency for deployment simplicity may choose CBR coders and oversubscribe the network bandwidth. Service providers who care about bandwidth utilization may prefer VBR coders for the higher coding efficiency. When distribution networks support bandwidth reservation, or when target streaming parameter values are known at encoding time, service providers may choose CVBR coders because they produce fewer playout glitches and thus deliver better streaming quality. Finally, service providers may choose UVBR coders if the streaming parameter values are unknown at the encoding time.

Statistical multiplexing refers to the capability to share the network bandwidth among multiple data streams that have VBR requirements. Statistical multiplexing allows the network to achieve higher bandwidth efficiency. Packet-switched networks achieve statistical multiplexing by dividing each data stream into small packets and routing the packets independently over potentially diverse paths to the destination. This allows routers to interleave packets of different data streams on network links in order to share the link bandwidth among data streams. More specifically, packets arriving at each router are first stored. The router then runs a packet scheduling algorithm to determine the next packet to be transmitted. In general, using VBR streams enables network operators to multiplex more videos over a bandwidth-limited network, because the bit rate of each video is proportional to its current scene complexity, and usually only a few VBR streams require high bit rates at any given time. Sharing the network bandwidth among multiple VBR streams achieves statistical multiplexing, and the increase in the number of video streams that can be streamed at a given target video quality by VBR coding is called statistical multiplexing gain.

6.2.2 Models for Wireless Video Streaming

Video streams can be delivered to mobile devices using unicast communications in cellular networks. That is, for each mobile device, the cellular base station allocates a unicast connection and transmits a video stream to that mobile device. Unicast communication is not bandwidth-efficient if multiple mobile devices tune to the same video stream, because the same video data are sent over the shared air medium several times. In contrast, multicast communications leverage the *broadcast* nature of wireless networks and significantly reduce the video data redundancy over the wireless spectrum. Therefore, multicast communications can better utilize network resources, especially in urban areas where there are many users interested in the same video streams. Today, the deployed cellular networks are unable to multicast video streams, because they can only multicast 90-character messages for alarm and warning purposes [51]. In fact, field measurements indicate that each UMTS HSDPA cell can only support four to six mobile video users at 256 kbps [29], which renders large-scale commercial video streaming services less viable.

Video streaming in unicast-only cellular networks is not considered in this chapter due to its low bandwidth efficiency. We focus on multicast-enabled networks that have the potential to serve video streams to a large number of mobile devices. In particular, the following cellular network standards are considered in this chapter: MBMS [42], WiMAX [27,58], DVB-H [20,22,34,38], MediaFLO [11], and ATSC M/H [5]. These multicast-enabled networks may employ legacy cellular networks to enable user interaction with the backend servers but not to transmit videos. We briefly introduce these wireless networks in the following.

The wireless telephony companies' consortium 3GPP has proposed MBMS [2] and evolved MBMS (eMBMS) [3] to deliver video streams over 3G and LTE cellular networks, respectively. MBMS/eMBMS specifications include features like electronic service guide (ESG), transparent handover, and bounded channel switching delays. More importantly, MBMS/eMBMS allow many mobile devices, within the range of a base station, to receive the same copy of a video stream for higher bandwidth efficiency. While MBMS/eMBMS can utilize the existing cellular infrastructure for pervasive connectivity, it has to compete with other mobile data and voice traffic for available bandwidth. To the best of our knowledge, there is no commercially deployed MBMS/eMBMS service at the time of writing.

IEEE 802.16 WiMAX [27,58] is a standard for metropolitan area wireless networks. WiMAX is envisioned as an alternative solution to traditional wire-based broadband access. For emerging economies like China and India, WiMAX is a cost-effective last-mile solution, and they are expected to be the major WiMAX markets. WiMAX has the capability of delivering high-speed services up to a range of 50 km. WiMAX uses orthogonal frequency division multiplexing (OFDM) and orthogonal frequency division multiple access (OFDMA) to improve the transmission range and increase bandwidth utilization. OFDM prevents interchannel interference among adjacent wireless channels, which allows WiMAX to achieve high network bandwidth. In the physical layer, WiMAX data are transmitted over multiple carriers in time division duplex (TDD) frames. As illustrated in Figure 6.1, each frame contains header information and upload/download maps followed by bursts of user data. Since video dissemination is expected to be a prevalent traffic pattern in future networks, the WiMAX standard defines multicast and broadcast service (MBS) in the link layer to facilitate multicast and broadcast [14]. Using MBS, a certain area in each TDD frame can be set aside for multicast- and broadcast-only data, as shown in Figure 6.1. The entire frame can also be designated as a download-only broadcast frame.

There are several example systems and standards for dedicated video multicast networks, including T-DMB [12], ISDB-T [55], DVB-H [22,34], and ATSC M/H [5]. T-DMB [12] is an extension of the DAB (digital audio broadcast) standard [18] that adds video streaming services to the high-quality

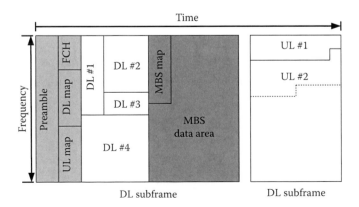

FIGURE 6.1
The frame structure in WiMAX.

audio services offered by DAB. The extension includes both source coding, such as using MPEG-4/AVC encoding, and channel coding, such as employing Reed–Solomon code. The development of T-DMB is supported by the South Korean government, and T-DMB is the first commercial mobile video streaming service. In addition to South Korea, several European countries may deploy T-DMB as they already have equipment and experience with DAB systems.

ISDB-T [55] is a digital video streaming standard defined in Japan, which is not only for fixed video receivers but also for mobile receivers. ISDB-T divides its band into 13 segments, with 12 of them being used for high-definition video streams and 1 for mobile video streams. ATSC M/H [5] is a mobile extension of the digital television (DTV) technologies developed by ATSC. ATSC M/H allows service providers to leverage the available spectrum, freed by all-digital TV broadcast, to provide video streaming services to mobile devices. MediaFLO is a video streaming network developed by Qualcomm and the FLO forum [25]. MediaFLO is designed from scratch for video streaming services to mobile devices. While the physical layer specification of MediaFLO has been published [11], to our knowledge, the design of its higher layers is not public.

DVB-H is an extension of the DVB-T (Digital video broadcast–terrestrial) standard [19], tailored for mobile devices. The DVB-H standard defines protocols below the network layer and uses IP as the interface with higher-layer protocols such as UDP and RTP. Standards such as IP datacast [34,38] complement DVB-H by defining higher-level protocols for a complete end-to-end solution. DVB-H [22,34] is an open international standard [20]. Figure 6.2 illustrates the protocol stack for video streaming over DVB-H networks.

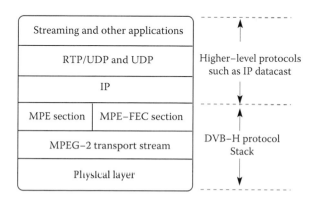

FIGURE 6.2

The protocol stack of video streaming services using the DVB-H standard. (From Hsu, C. and Hefeeda, M., Using simulcast and scalable video coding to efficiently control channel switching delay in mobile TV broadcast networks, *ACM Transactions on Multimedia Computing, Communications, and Applications*, 7(2), 8:1. Copyright 2011 ACM; Hefeeda, M. and Hsu, C., On burst transmission scheduling in mobile TV broadcasting networks, *IEEE/ACM Trans. Netw.*, 18(2), 612. Copyright 2010 IEEE. With permission.)

DVB-H uses a physical layer compatible with DVB-T, which employs OFDM modulation. DVB-H encapsulates IP packets using multi-protocol encapsulation (MPE) sections to form MPEG-2 transport streams (TSs). Thus, data from a specific video stream form a sequence of MPEs. MPEs are optionally FEC-protected before transmission over the wireless medium.

6.2.3 Architecture of Mobile Video Streaming Networks

A mobile video streaming network, illustrated in Figure 6.3, consists of three entities: content providers, network operators, and mobile users. Content providers are companies that create videos. Since content providers send the same video to multiple network operators that have diverse needs, the newly created videos are encoded as high-quality streams and sent to network operators. Each network operator may transcode the video stream according to its quality requirements and available bandwidth. Network operators are companies that manage base stations and provide services to mobile users. A network operator multiplexes several videos into an aggregated stream and transmits it over a wireless network with a fixed bandwidth. Because the wireless network is bandwidth-limited, the multiplexer must ensure that the bit rate of the aggregated stream does not exceed the network bandwidth.

One way to control the bit rate is to employ *joint video coders*, which encode multiple videos and dynamically allocate available network bandwidth among them, so that the aggregate bit rate of the encoded streams never exceeds the network bandwidth. As shown in Figure 6.3, a joint video coder consists of a joint rate allocator, several decoders, and several VBR coders. The joint rate allocator collects scene complexities from decoders/coders, distributes available bandwidth among video streams, and instructs the coders to avoid overloading the wireless network by controlling their encoding bit rates. There are several joint video coders proposed in the literature, e.g., [30,33,45,54]. Commercial joint coders, such as Ref. [39], are also available in the market. Streaming systems with joint video coders are called *closed-loop*

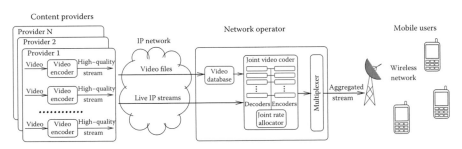

FIGURE 6.3
Main components of a closed-loop video streaming network. (From Hsu, C. and Hefeeda, M., Statistical multiplexing of variable bit rate video streamed to mobile devices, *ACM Transactions on Multimedia Computing, Communications, and Applications*, 7(2), 12:1. Copyright 2011 ACM. With permission.)

streaming networks. While deploying joint video coders can simplify the design of the multiplexer, doing so may not always be possible for several reasons, such as complex business agreements and higher deployment costs. Streaming networks with stand-alone video coders are called *open-loop* streaming networks. We note that video streaming networks that transmit CBR video streams are essentially closed-loop streaming networks, because the aggregate bit rate of all video streams is constant in these systems. In either closed- or open-loop networks, the multiplexer generates an aggregated stream, which in turn is transmitted over a wireless network to mobile users. Mobile users use mobile devices to receive video streams over the wireless network. Each user typically tunes to one video stream at a time.

6.2.4 Energy Conservation in Wireless Networks

Mobile devices are battery-powered and sensitive to high energy consumption. Therefore, various wireless multicast networks, including WiMAX [27,58], MBMS [2], DVB-H [16], and ATSC M/H [4] all try to minimize the energy consumption of mobile devices by periodically turning their wireless interfaces off. For example, in WiMAX networks, multiple video streams carrying the content and data streams containing the service information reach the base station through high-speed backhaul links. At a base station, the video streams are stored in individual channel queues. After the data and video are packetized into stream-specific MAC PDUs and security mechanisms are applied, these packets are handed over to the MBS manager, which is a software component at the base station. The MBS manager packs the MAC PDUs into the MBS blocks of the TDD frames, which are then transmitted over OFDMA. The MBS manager constructs a number of MBS areas at a time, consisting of a series of TDD frames, to efficiently utilize the MBS space. The data for a particular video stream are transmitted in a time-sliced manner to save energy. Time slicing, illustrated in Figure 6.4, refers to the technique of transmitting each video stream in bursts at a bit rate much higher than the video bit rate so as to allow mobile devices to turn off their wireless interfaces between bursts to save energy. More specifically, energy conservation is achieved by setting the mobile device to sleep mode when it is not receiving any TDD frames. In this chapter, we consider the burst scheduling problem that determines the optimal start times and durations of individual bursts for all video streams to minimize the energy consumption of mobile devices and maximize the bandwidth utilization of the wireless network. The start time and duration of each sleep period are computed by the MBS manager and stored in the first few TDD frames of each burst.

In DVB-H networks, IP packets are encapsulated using MPE sections, and the packets of a specific video stream form a sequence of MPEs. The MPE encapsulation is done in a software module called the IP encapsulator. The encapsulated data are then fed to an RF signal modulator to be transmitted over the air medium. The IP encapsulator realizes time slicing in DVB-H

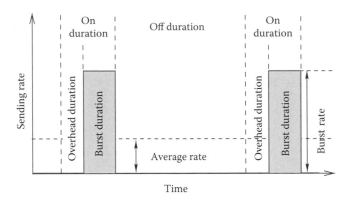

FIGURE 6.4
Time slicing in mobile video streaming networks to save energy. (From Hefeeda, M. and Hsu, C., On burst transmission scheduling in mobile TV broadcasting networks, *IEEE/ACM Trans. Netw.*, 18(2), 611. Copyright 2010 IEEE; From Hsu, C. and Hefeeda, M., Broadcasting video streams encoded with arbitrary bit rates in energy-constrained mobile TV networks, *IEEE/ACM Transactions on Networking*, 18(3), 681. Copyright 2010 IEEE. With permission.)

networks. In particular, once the burst start times and durations are determined, MPEs belonging to a given video stream are transmitted in bursts accordingly. In DVB-H networks, the time period between two adjacent bursts (the off period) is flexible in the sense that the time offset between the start time of the next burst and the start time of the current MPE section is sent as part of the MPE header. This enables DVB-H systems to adopt variable burst durations and off durations not only for different video streams but also for the same video stream at different times. The burst scheduling algorithms developed in this chapter are designed for implementation in the MBS manager of WiMAX base stations and the IP encapsulator of DVB-H base stations.

6.3 Energy Optimization in Mobile Video Streaming

Common mobile video streaming networks, such as WiMAX [27,58], MBMS [2], DVB-H [16], and ATSC M/H [4], allow/dictate the use of a time slicing scheme (see Figure 6.4) to increase the viewing time of mobile devices. Although time slicing leads to energy conservation, burst transmission schedules, which specify the burst start times and sizes, must be carefully composed to guarantee service quality and proper functioning of the system for several reasons. First, since mobile devices have limited receiving buffer capacity, arbitrary burst schedules can result in buffer overflow/ underflow instances that cause playout glitches and degrade viewing experience. Second, as several video streams share the same air medium, burst schedules must not have any burst conflicts, which occur when two or more bursts intersect with each other in time. Third, turning on wireless interfaces is not instantaneous as it takes some time to search for and lock onto RF signals. This imposes overhead on the energy consumption, and this overhead must be considered in constructing burst schedules. Fourth, burst schedules directly impact the channel switching delay, which is the average time after requesting a change of video stream that a user waits before starting to view the selected video stream. Long and variable channel switching delays are annoying users and could turn them away from the mobile video streaming service.

Current practices for computing burst schedules are rather ad hoc. For example, the heuristic method proposed in the DVB-H standard document [21, p. 66] provides schedules for only one video stream. This heuristic simply allocates a new burst after the data of its preceding burst are consumed by the player at the mobile device. This cannot be generalized to multiple video streams with *different* bit rates, because the computed schedule may have burst conflicts and may result in buffer overflow/underflow instances. Thus, mobile video streaming networks may resort to encoding all video streams at the same bit rate in order to use the heuristic burst scheduling technique in the standard. This is clearly inefficient and may yield large quality variations among video streams carrying different types of multimedia content, which is the common case in practice. For example, encoding a high-motion soccer game requires a much higher bit rate than encoding a low-motion talk show. If all video streams are encoded at the same high bit rate, some video streams may unnecessarily be allocated more bandwidth than they require and this extra bandwidth yields marginal or no visual quality improvement. Thus, the expensive wireless bandwidth of the wireless network could be wasted. On the other hand, if we encode all video streams at the same low or moderate bit rate, not all video streams will have good visual quality, which could be annoying to users of a commercial service. Therefore, a systematic approach to address the burst scheduling problem for mobile devices receiving video streams at different bit rates is desired.

In this section, we study the burst scheduling problem in mobile video streaming networks in which videos are encoded as CBR streams at different bit rates. In Section 6.4, we consider more general streaming networks in which videos are encoded as VBR streams. In this section, we first formulate the burst scheduling problem in mobile video streaming networks, and we show that this problem is NP-complete for video streams with arbitrary bit rates. We then propose a practical simplification of the general problem, which allows video streams to be classified into multiple classes with each class having a different bit rate. The bit rate r_c of class c can take any value of the form of $r_c = 2^i \times r_1$, where $i \in \{0,1,2,3, \ldots\}$, and the bit rate r_1 of the lowest class can be any arbitrary bit rate. For example, the bit rates 800, 400, 200, and 100 kbps could be used for four different classes to encode sports events, movies, low-motion episodes, and talk shows, respectively. This classification of video streams also enables network operators to offer differentiated services: higher bit rate classes can carry premium services for higher fees. This service differentiation is not possible with the current burst scheduling schemes. Using the simplification provided earlier, we develop an optimal (in terms of energy consumption) burst scheduling algorithm, which is quite efficient: its time complexity is $O(S \log S)$, where S is the total number of video streams. We implement our algorithm in a real mobile video streaming test bed and demonstrate its practicality and effectiveness in saving energy.

Energy conservation in multicast-enabled wireless networks has been considered in the literature. Yang et al. [59] estimate the effectiveness of the time slicing technique for given burst schedules. They show that time slicing enables mobile devices to turn their wireless interfaces off for a significant fraction of the time. The work by Yang et al. does not solve the burst scheduling problem; it only computes the achieved energy saving for given predetermined burst schedules. Balaguer et al. [6] propose to save energy by not receiving some parity bytes of error correction codes as soon as the received data are sufficient to successfully reconstruct the original video data. Zhang et al. [61] consider mobile devices with an auxiliary short-range wireless interface and construct a cooperative network among several devices over this short-range wireless network. Mobile devices share received video packets over the short-range network, so that each mobile device receives only a small fraction of IP packets directly from the base station. This allows receivers to reduce the frequency of turning on their wireless interfaces. Assuming that sending and receiving IP packets through the short-range network is more energy-efficient than receiving from the base station, this cooperative strategy can reduce energy consumption. The proposals in Refs. [6,61] are orthogonal and complementary to our work, as they reside in the mobile devices themselves and try to achieve additional energy saving on top of that achieved by time slicing. In contrast, our algorithm is to be implemented in the base station that is transmitting multiple video streams to mobile devices.

6.3.1 Problem Statement and Hardness

We consider a mobile video streaming network in which a base station concurrently transmits S video streams to clients with mobile devices over a wireless medium with bandwidth R kbps. Each video stream s, $1 < s < S$, has a bit rate r_s kbps, which is typically much less than R. The base station transmits each video stream in bursts at bit rate R kbps. After receiving a burst of data, the wireless interfaces are switched off until the time of the next burst, which is computed by the base station and included in the header fields of the current burst. The wireless interfaces of the mobile devices must be activated slightly before the burst time because it takes some time to wake up and synchronize the circuitry before it can start to receive data. This time is called the overhead duration and is denoted by T_o. With current technology, T_o is in the range of 50–250 ms [21,34]. The energy saving achieved by mobile devices receiving video stream s is denoted γ_s, and it is calculated as the ratio of time that the network interface is off to the total time [21,59]. We define the system-wide energy saving metric over all video streams as $\gamma = \left(\sum_{s=1}^{S} \gamma_s \right) / S$. Both the energy saving and the burst scheduling are typically performed on a recurring time window called a frame. The length of each frame is a system parameter and is denoted by p. With these definitions, we can state the burst scheduling problem.

Problem 6.1 (Burst scheduling in mobile video streaming networks). We are given S video streams of different bit rates to be simultaneously multicast to mobile devices. Each video stream is transmitted as bursts of data to save the energy of mobile devices. The problem is to find the optimal transmission schedule for bursts of all video streams to maximize the system-wide energy saving metric γ. The transmission schedule must specify the number of bursts for each video stream in a frame as well as the start and end times for each burst. The schedule cannot have burst collisions, which happen when two or more bursts have a nonempty intersection in time. In addition, given a link-layer receiver of buffer size B, the schedule must ensure that there are no receiver buffer violations for any video stream. A buffer violation occurs when the receiver has either no data in the buffer to pass on to the decoder for playout (buffer underflow) or has no space to store data during a burst transmission (buffer overflow).

Figure 6.5 illustrates a simple example of the burst scheduling problem. Notice that the bursts have different sizes, are disjoint in time, and are

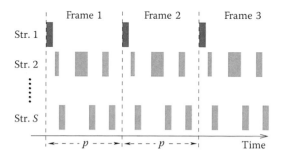

FIGURE 6.5
The burst scheduling problem in mobile video streaming networks. (From Hefeeda, M. and Hsu, C., Energy optimization in mobile TV broadcast networks, in *Proc. of IEEE International Conference on Innovations in Information Technology (Innovations'08)*, pp. 430–434, 2008. Copyright 2008 IEEE; Hefeeda, M. and Hsu, C., On burst transmission scheduling in mobile TV broadcasting networks, *IEEE/ACM Trans. Netw.*, 18(2), 613. Copyright 2010 IEEE; From Hsu, C. and Hefeeda, M., Broadcasting video streams encoded with arbitrary bit rates in energy-constrained mobile TV networks, *IEEE/ACM Transactions on Networking*, 18(3), 681. Copyright 2010 IEEE. With permission.)

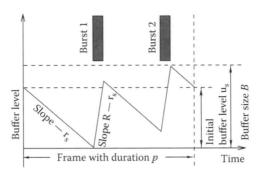

FIGURE 6.6
The dynamics of the receiver buffer in successive frames. (From Hsu, C. and Hefeeda, M., Time slicing in mobile TV broadcasting networks with arbitrary channel bit rates, in *Proc. of IEEE INFOCOM'09*, pp. 2231–2239, 2009. Copyright 2009 IEEE; Hefeeda, M. and Hsu, C., On burst transmission scheduling in mobile TV broadcasting networks, *IEEE/ACM Trans. Netw.*, 18(2), 614. Copyright 2010 IEEE; From Hsu, C. and Hefeeda, M., Broadcasting video streams encoded with arbitrary bit rates in energy-constrained mobile TV networks, *IEEE/ACM Transactions on Networking*, 18(3), 681. Copyright 2010 IEEE. With permission.)

repeated in successive frames. In addition, there can be multiple bursts for a video stream in each recurring frame to ensure that there are no buffer overflow/underflow instances. To illustrate the receiver buffer dynamics for a valid solution of the burst scheduling problem, the buffer level as a function of time is shown in Figure 6.6. This is shown for a receiver of a sample video stream s with two bursts in each frame. We make two observations about this figure. First, during a burst, the buffer level increases at a rate (slope of the line) of $R - r_s$, which is much larger than the consumption rate of $-r_s$ when there is no burst. Second, the frame starts with an initial buffer level (denoted by u_s) and ends at the same buffer level. Clearly, this is a requirement for any valid burst scheduling solution; otherwise there is no guarantee that the buffer will not suffer from overflow or underflow instances.

The following theorem shows that the burst scheduling problem is NP-complete.

Theorem 6.1 (Burst scheduling). The burst scheduling problem stated in Problem 6.1 is NP-complete.

Proof: We first show that the problem of maximizing energy saving (Problem 6.1) is the same as the problem of minimizing the total number of bursts in each frame for a given frame length p. To maximize energy saving γ, we have to minimize the wireless interface on time for receivers. Notice that the wireless interface on time can be divided into two parts: burst and overhead durations, as illustrated in Figure 6.4. The burst duration is the time during which mobile devices receive the video data. Since we consider steady burst schedules in which the number of received bits is equal to the number of consumed bits in each frame for all mobile devices, the burst duration must be constant across all feasible burst schedules. Notice also that each burst incurs a fixed overhead T_o. Therefore, minimizing the wireless interface on time is equivalent to minimizing the total number of bursts in each frame, because it minimizes the total overhead.

Next, we reduce the NP-complete problem of task sequencing with release times and deadlines [26, p. 236] to the problem of minimizing the total number of bursts in each frame. The task sequencing problem consists of T tasks, where each task t, $1 \leq t \leq T$, is released at time x_t with length y_t s and deadline z_t. The problem is to determine whether there is a single machine (non-preemptive) schedule that meets all constraints on release times and deadlines. For any task sequencing problem, we set up a burst scheduling problem as follows. Let $S = T$ and map every task to a video stream. Let $p = p^*$ be the optimum frame length, which will be derived in the next subsection. We choose an arbitrary burst bit rate R. We set the bit rate of each video stream s, $1 \leq s \leq S$, to $r_s = y_s R / p^*$ to balance the number of received bits and the number of consumed bits. We set the initial buffer level to $u_s = (z_s - y_s) r_s$, which guarantees that a burst with a length y_s will be scheduled (and finished) *before* the deadline z_s, and mobile devices will not run out of data for play-out (underflow). We set the receiver buffer size to $b_s = u_s + y_s R - (x_s + y_s) r_s - \epsilon$, where $\epsilon > 0$ is an arbitrary small number. Selecting such a b_s guarantees that a burst with length y_s will be scheduled *after* the release time x_s, and mobile devices will not run out of buffer space (overflow).

Clearly, we can set up the burst scheduling problem in polynomial time. Furthermore, solving the burst scheduling problem leads to the solution of the task sequencing problem because the minimum total number of bursts is equal to S if and only if there is a non-preemptive schedule that satisfies the constraints on release times and deadlines of the task sequencing problem. Thus, the burst scheduling problem is NP-hard. Finally, determining whether a given burst schedule is a valid solution for Problem 6.1 that meets the collision-free and buffer violation-free requirements takes polynomial time. Hence, the burst scheduling problem is NP-complete.

This theorem might seem counterintuitive at first glance because the burst scheduling problem looks somewhat similar to preemptive machine scheduling problems. However, there is a fundamental difference between our burst scheduling problem and most machine scheduling problems: most of the machine scheduling problems consider a *costless* preemption model [9]. In contrast, our burst scheduling problem adopts a *costly* preemption model because our problem aims to minimize the total number of bursts in a frame, which is essentially the number of preemptions. Therefore, the algorithms developed for various machine scheduling problems are not applicable to our burst scheduling problem (see Ref. [9] for a comprehensive list of machine scheduling problems). The costly preemption model has been considered only in a few papers [7,8,40,49]. Braun and Schmidt [8] and Motwani et al. [40] partly cope with preemption costs by adding constraints to limit

the number of preemptions. Bartal et al. [7] solve the problem of minimizing the weighted sum of the total task flow time and the preemption pen-
alty, where the weight is heuristically chosen. Schwiegeishohn [49] considers
the problem of minimizing weighted completion time and task makespan
under a given preemption cost. Unlike these problems, our burst scheduling
problem uses only the preemption cost as the objective function and does
not allow any late tasks, which render the algorithms proposed in [7,8,40,49]
inapplicable to our problem.

6.3.2 Burst Scheduling Algorithm

Next, we present an algorithm that optimally and efficiently solves the
NP-complete burst scheduling problem under a certain assumption that usually holds in practice. To simplify the presentation, we first give an overview of
the algorithm and an illustrative example. We then analyze the algorithm and
prove its correctness, optimality, and efficiency. Then, we analyze the trade-
off between the achieved energy saving and the channel switching delay.

6.3.2.1 Overview of the Algorithm

We propose an optimal algorithm for the burst scheduling problem for which
the bit rate of each video stream s, $1 \leq s \leq S$, is given by $r_s = 2^i \times r_1$ for any $i \in$
$\{0,1,2,3, \ldots\}$, and r_1 can be any arbitrary bit rate. As aforementioned, the video
streams are divided into classes with each class containing similar types of
multimedia content encoded at the same bit rate. Without loss of generality,
we assume that the bit rates of the S video streams are ordered such that
$r_1 \leq r_2 \leq \cdots \leq r_S$; otherwise, a relabeling based on the bit rates is applied. We
also assume that the bandwidth of the wireless medium satisfies $R = 2^k \times r_1$,
where k is a positive integer. We present in Figure 6.7 an optimal algorithm,
called power-of-two optimal (P2OPT), for solving the burst scheduling problem in this case.

The basic idea of our algorithm is as follows. The algorithm first computes
the optimal value for the frame length $p^* = B/r_1$. (We prove that p^* is optimal
in Theorem 6.3.) It then divides p^* into bursts of equal size $p^* r_1$ bits. Thus,
there are $(p^* R)/(p^* r_1) = R/r_1$ bursts in each frame. Then, each video stream is
allocated a number of bursts proportional to its bit rate. That is, video stream s,
$1 < s < S$, is allocated r_s/r_1 bursts and video stream 1 is allocated only one burst
in each frame. Moreover, bursts of video stream s are equally spaced within
the frame, with an inter-burst distance of $p^*/(r_s/r_1)$ s. This ensures that there
will be no underflow instances in the receiver buffer, because the consumption rate of the data in the buffer for video stream s is r_s bps and the burst size
is $p^* r_1$ bits. Since the optimal frame length is $p^* = B/r_1$, the size of each burst
is $p^* r_1 = B$, which is no larger than the receiver buffer size B. This ensures
that there are no buffer overflow instances. Finally, bursts of different video
streams are arranged such that they do not intersect in time, i.e., the resulting
schedule is conflict free.

Power-of-Two Optimal (P2OPT) Algorithm

---

1. **compute** optimal frame length $p^* = B/r_1$
2. allocate a leaf node l for every stream s, **let** $l.str = s$
3. push each leaf node s into a priority queue P with key r_s/r_1
4. **while true** {
5.   **let** $m_1 = \text{pop\_min}(P)$, $m_2 = \text{pop\_min}(P)$;
6.   **if** m_2 is **null**, or $m_1.key < m_2.key$ { // no sibling
7.   **if** m_2 is not **null**, push(P, m_2); // return m_2 back to P
8.   allocate a dummy node m_2, where $m_2.key = m_1.key$
9.   }
10.   **create** an internal node n with children $n.left$ & $n.right$
11.   **let** $n.left = m_1$, $n.right = m_2$;
12.   **let** $n.key = m_1.key + m_2.key$;
13.   push(P, n); // insert this internal node
14.   **if** $n.key \_ R/r_1$ **break**;
15. }
16. **if** $|P| > 1$ **return** $\varnothing$; // no feasible schedule
17. **let** $\mathbb{T} = \varnothing$; // start composing schedule $\mathbb{T}$
18. Traverse tree labelling each leaf node with an *offset* that is the bit pattern on the leaf to root path
19. **foreach** leaf node l {
20.   **for** $i = 0$ to $l.key - 1$ {
21.   start = offset + $i \times (R/r_1)/l.key$;
   // add a burst to stream $l.str$ at time start $\times p^* r_1/R$
22.   insertBurst($\mathbb{T}$, start $\times p^* r_1/R$, $l.str$);
23.   }
24. }
25. **return** $\mathbb{T}$;

FIGURE 6.7

The P2OPT burst scheduling algorithm. (From Hefeeda, M. and Hsu, C., Energy optimization in mobile TV broadcast networks, in *Proc. of IEEE International Conference on Innovations in Information Technology (Innovations'08)*, pp. 430–434, 2008. Copyright 2008 IEEE; Hefeeda, M. and Hsu, C., On burst transmission scheduling in mobile TV broadcasting networks, *IEEE/ACM Trans. Netw.*, 18(2), 615. Copyright 2010 IEEE. With permission.)

To achieve the earlier steps in a systematic way, the algorithm builds a binary tree bottom-up. Leaf nodes representing video streams are created first, with the leaf node of video stream s being assigned the value r_s/r_1. The algorithm uses this value as a key and inserts all leaf nodes into a priority queue. This priority queue is implemented as a binary heap to efficiently find the node with the smallest key. The algorithm then repeatedly merges the two nodes in the heap that have the smallest key values into a new internal node. This new internal node has a key value equal to the sum of the key values of its children. This is done by popping the two smallest values from the heap and then pushing the newly created node into the heap. The merging of nodes continues until the tree has a height of $\log(R/r_1)$. The last merged node becomes the root of the binary tree. Note that if the wireless medium is fully utilized by the video streams, i.e., $\sum_{s=1}^{S} r_s = R$, the computed bursts of the different video streams will completely fill the frame p. If the utilization is less than 100%, the wireless medium will be idle during some periods within the frame. The algorithm represents these idle periods as dummy nodes in the tree.

Once the binary tree has been created, the algorithm constructs the burst schedule. It allocates a number of bursts that is equal to its key value to each video stream. In order to ensure a conflict-free schedule, the algorithm computes the start time for each burst as follows. The algorithm traverses the tree assigning bit patterns to each leaf node representing a video stream. When a left branch is followed down from an internal node, it contributes a 0 bit and a right branch contributes a 1 bit. The bit pattern assigned to a leaf node is the bit pattern corresponding to the leaf to root path. The bit pattern for a leaf node encodes the number of bursts and their start times for the video stream corresponding to that node. For example, in a tree of depth 3, there will be eight bursts, and the bit pattern 000 will correspond to the first burst. The bit pattern 011 means that the video stream is assigned to the fourth burst in a frame of eight bursts. For leaf nodes at levels less than the depth of the tree, the bit pattern is padded on the left with "x"s. For example, if a leaf node has the bit pattern x01 in a tree of depth 3, this means that the node is at level 2 from the root. It also means that this node should be assigned the two bursts: 001 and 101. Notice that the first burst assigned to any video stream is equal to the numerical value of its bit pattern with any "x" bits set to zero. The algorithm computes this value and refers to it as the offset. The algorithm then computes successive bursts for the stream relative to this offset.

6.3.2.2 Illustrative Example

Consider four video streams distributed over three different classes: two streams in class I with $r_1 = r_2 = 256$ kbps, one in class II with $r_3 = 512$ kbps, and one in class III with $r_4 = 1024$ kbps. Let the wireless medium bandwidth be $R = 2048$ kbps and the receiver buffer size be $B = 1$ Mb. As explained later, p^* is given by $p^* = B/r_1 = 4$ s. The algorithm divides each frame into $R/r_1 = 8$ bursts, and assigns 1, 1, 2, 4 bursts to video streams 1, 2, 3, 4, respectively. The algorithm constructs a binary tree bottom-up as shown in Figure 6.8. Four leaf nodes are created, each representing a video stream with a key value equal to the number of bursts that should be allocated to that video stream. Notice that the leaf nodes are logically placed at different levels based on their key values. The algorithm repeatedly merges nodes with the same key values until it creates the root node with key value $8 = R/r_1$. Then, the algorithm constructs the schedule by traversing the tree from each leaf node to the root and assigning the bit patterns shown in Figure 6.8 to the leaf nodes. Using these bit patterns, the offset for each node

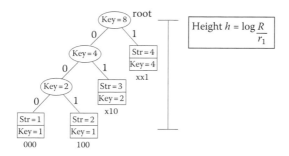

FIGURE 6.8
An illustrative example for the P2OPT algorithm. (From Hefeeda, M. and Hsu, C., On burst transmission scheduling in mobile TV broadcasting networks, *IEEE/ACM Tran. Netw.*, 18(2), 616. Copyright 2010 IEEE. With permission.)

is computed and the bursts are assigned. The resulting burst schedule is

$$\mathbb{T} = \{(0.0,1),(0.5,4),(1.0,3),(1.5,4),(2.0,2),(2.5,4),(3.0,3),(3.5,4)\},$$ where the

first element in the parentheses is the start time for sending the burst and the second element indicates the video stream.

We have presented a rather simple example for illustration, but the P2OPT algorithm is quite flexible concerning the bit rates of individual video streams. In fact, network operators can multicast each video stream s ($2 < s < S$) at an average bit rate $r_s = 2^i \times r_1$ for *any* $i \in \{0,1,2,3, ...\}$. For instance, by setting $i = 0$ for all video streams, network operators can compute the optimal streaming schedule for video streams with a uniform bit rate. Therefore, the P2OPT algorithm provides a systematic way to construct optimal burst schedules for mobile video streaming networks that transmit video streams at a uniform bit rate, which is currently a common practice.

6.3.2.3 Analysis of the Algorithm

We prove the correctness, efficiency, and optimality of our algorithm in the following two theorems.

Theorem 6.2 (Correctness and efficiency). The burst scheduling algorithm P2OPT in Figure 6.7 returns a conflict-free schedule with no buffer overflow/ underflow instances, if one exists, and P2OPT has a worst-case time complexity of $O(S \log S)$, where S is the number of video streams.

Proof: We prove the correctness part in two steps. First, observe that a burst schedule produced by P2OPT is conflict free because the algorithm assigns a unique bit pattern to each video stream that specifies the bursts that are allocated to that video stream. Moreover, the bit pattern is padded on the left with zero or more "x"s, which guarantees that video stream s is assigned r_s/r_1 equally spaced bursts. Hence, if P2OPT returns a schedule, then this schedule is conflict free with no buffer overflow/underflow instances.

Second, we prove that if P2OPT fails to return a schedule, then there exists no feasible schedule for the given video streams. P2OPT merges only nodes at the same binary tree level, and nodes at lower levels have strictly smaller key values than nodes at higher levels. Therefore, P2OPT merges all nodes in the tree bottom-up and creates at most one dummy node at each level. Moreover, line 12 of P2OPT ensures that the key value of each node is the number of time slots that are consumed by the node (for a leaf node) or by all leaf nodes in its subtree (for an internal node). P2OPT returns no feasible solution for a given problem if a full binary tree with height $h = \log(R/r_1)$ is built and $|P| > 1$ in line 16. Let z be the first merged node with key value R/r_1. Then z is the last merged node to be added to P before returning from line 16. Since $|P| > 1$, there must be at least one other node $w \neq z$ in P. Furthermore, w must have a key value no less than $(1/2)sR/r_1$, otherwise w would have been merged before the children of z. Now we count the number of time slots consumed by real (non-dummy) leaf nodes in the subtrees beneath w and z. The level of w in P is either at least as high as the level of z (case I) or it is at a level lower than the level of z (case II). In case I, we know that $w.\text{key} \geq z.\text{key} = R/r_1$ and that w is a real leaf node because z is the first merged node at its level. Since P2OPT guarantees that at most one dummy leaf node exists at each level, the total number of time slots occupied by dummy leaf nodes in z's subtree cannot exceed

$$\sum_{i=0}^{\log R/r_1 - 1} 2^i = \frac{R}{r_1} - 1.$$

This shows that the subtrees of w and z together consume at least

$$2\frac{R}{r_1} - \left(\frac{R}{r_1} - 1\right) = \frac{R}{r_1} + 1$$

time slots. Since $R/r_1 + 1$ exceeds the total number of available time slots R/r_1, there exists no feasible schedule. The proof for case II is similar.

For time complexity, P2OPT can be efficiently implemented using a binary heap, which can be initialized in time $O(S)$. There is at most one dummy leaf node at each level so there are at most $\log(R/r_1)$ dummy leaf nodes. The while loop in lines 4–15 iterates at most $O(S + \log(R/r_1)) = O(S)$ times, because $\log(R/r_1)$ can be considered to be a constant for practical encoding bit rates. Since each iteration takes $O(\log S)$ steps, the while loop takes $O(S \log S)$ steps. Constructing the burst schedule in lines 17–24 takes $O(S)$ steps since the tree has at most $2S$ nodes. Thus, the time complexity of P2OPT is $O(S \log S)$.

The following theorem shows that the P2OPT algorithm produces optimal burst schedules in terms of maximizing energy saving.

Theorem 6.3 (Optimality). The frame duration $p* = B/r_1$ computed by P2OPT, where B is the receiver buffer size, maximizes the average energy saving γ over all video streams.

Proof: We use the fact that we established in the proof of Theorem 6.1: the problem of maximizing energy saving is equivalent to the problem of minimizing the total number of bursts in each frame for a given frame length p. To maximize the energy saving γ, we have to minimize the ratio of the wireless interface on time to the frame length. We again divide the wireless interface on time into burst and overhead durations. Notice that the ratio of burst duration to frame length is constant for any feasible schedule because each video stream is transmitted at a fixed bit rate. Since each burst incurs overhead T_o s, following the definition of γ, our burst scheduling problem is reduced to the problem of minimizing the total number of bursts normalized by the frame length p.

We first prove by contradiction that no frame length $p < p*$ can result in higher energy saving. Assume that a feasible schedule $\mathbb{T}_1$ results in higher energy saving with frame length $p_1 < p*$ than the schedule $\mathbb{T}*$ with frame length $p*$ that is produced by P2OPT. To outperform $\mathbb{T}*$, $\mathbb{T}_1$ must have fewer bursts than $\mathbb{T}*$, i.e., $|\mathbb{T}_1| < |\mathbb{T}*|$. However, $\mathbb{T}_1$ must lead to some buffer overflow, because fewer bursts are equivalent to longer bursts, and P2OPT has fully utilized (filled up) the receiver buffer B for all bursts. This contradicts the assumption that $\mathbb{T}_1$ is a feasible schedule.

We next consider $p > p*$. Assume that $\mathbb{T}_2$ is a feasible schedule that results in better energy saving with frame length $p_2 > p*$. Let $k = \lceil p_2 / p* \rceil$. Define T′ by repeating $\mathbb{T}*$, which is produced by P2OPT, k times. Since burst schedule T′ also fills the receiver buffer in all bursts, and $\mathbb{T}_2$ has fewer (and therefore longer) bursts than T′, the same argument as earlier shows that $\mathbb{T}_2$ must lead to some buffer overflow. This contradicts the assumption that is a feasible schedule.

6.3.2.4 Trade-Off between Energy Saving and Switching Delay

We first compute the optimal average energy saving γ. Since video stream s has an inter-burst duration of B/r_s and the burst length is B/R, the average energy saving is as follows:

$$\gamma = \frac{\sum_{s=1}^{S} \left(1 - r_s(1/R + T_o/B)\right)}{S} = 1 - \sum_{s=1}^{S} r_s \frac{1/R + T_o/B}{S}. \tag{6.1}$$

Next, we analyze another important metric in mobile video streaming networks: the channel switching delay d is the time after requesting a change of video stream that a user waits before starting to view the selected video stream.

Channel switching delay is composed of several parts, and the frame refresh delay and time slicing delay are the two dominant contributors [45,46]. The frame refresh delay is controlled in the application layer and is orthogonal to our burst scheduling problem. The time slicing delay refers to the time period between locking on to the wireless medium and the beginning of the first burst of that video stream. Since the time slicing delay is a *by-product* of the time slicing-based energy saving scheme, we consider time slicing delay in this chapter. We assume that all other parts of channel switching delays are constant as they are outside of the scope of this chapter. With this assumption, the average channel switching delay resulting from the optimal schedule is as follows:

$$d^* = \frac{\left(\sum_{s=1}^{S} B/2r_s \right)}{S} = \left(\sum_{s=1}^{S} \frac{1}{r_s} \right)\left(\frac{B}{2S} \right). \tag{6.2}$$

Notice that there is a trade-off between γ and d and the main control parameter is the buffer size B. To examine this trade-off, we present an illustrative example. We consider a video streaming service with $R = 10$ Mbps and 25 video streams equally distributed into five classes of heterogeneous video stream bit rates (i.e., 1024, 512, 256, 128, and 64 kbps). We plot the energy saving and channel switching delay under different overhead durations in Figure 6.9. This figure shows that larger buffer sizes and smaller overhead durations lead to higher energy savings *and* also to higher channel switching delays.

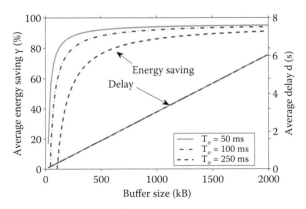

FIGURE 6.9
(See color insert.) The trade-off between energy saving and channel switching delay. (From Hefeeda, M. and Hsu, C., On burst transmission scheduling in mobile TV broadcasting networks, *IEEE/ACM Trans. Netw.*, 18(2), 615. Copyright 2010 IEEE. With permission.)

6.3.3 Evaluation and Analysis

We have evaluated the P2OPT algorithm using an actual implementation in a mobile video streaming test bed as well as by simulation. We present the results from the test bed in the following. Interested readers are directed to the work by Hefeeda and Hsu [31] for additional simulation and experimental results.

6.3.3.1 Setup of the Mobile Video Streaming Test Bed

The test bed that we used to evaluate our algorithm, shown in Figure 6.10, consists of a base station, mobile receivers, and data analyzers. The base station includes an RF signal modulator that produces DVB-H standard-compliant signals. The signals are amplified to around 0 dBm before transmission through a low-cost antenna to provide coverage to approximately 20 m for cellular phones. The base station has a multi-threaded design that enables the transmission of multiple video streams in real time on a low-end server.

The mobile receivers in our test bed are Nokia N96 cell phones that have DVB-H signal receivers and video players. Two DVB-H analyzers are included in the test bed system. The first one, a DiviCatch RF T/H tester [15], has a graphical interface for monitoring detailed information about the received signals, including burst schedules and burst jitters. The second analyzer, a dvbSAM [17], is used to access and analyze received data at the byte level to monitor the correctness of the received content.

We configured the modulator to use a 5 MHz radio channel with quadrature phase-shift keying (QPSK) modulation. This leads to 5.445 Mbps air medium bandwidth according to the DVB-H standard [21]. We concurrently

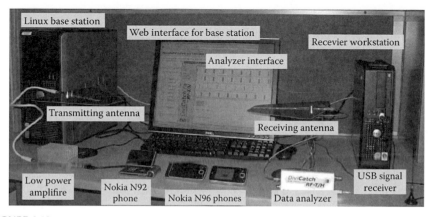

FIGURE 6.10
(See color insert.) The mobile video streaming test bed used in the evaluation. (From Hefeeda, M. and Hsu, C., Design and evaluation of a testbed for mobile TV networks, *ACM Transactions on Multimedia Computing, Communications, and Applications*, 8(1), 3:1. Copyright 2012 ACM. With permission.)

multicasted nine video streams using our P2OPT algorithm for 10 min. These video streams were classified into four classes: two streams at 64 kbps, three streams at 256 kbps, two streams at 512 kbps, and two streams at 1024 kbps. The receiver buffer size was 1 Mb. For each of these video streams, we set up a streaming server on the base station to send 1 kB IP packets at the specified bit rate. To conduct a statistically meaningful performance analysis, we collected detailed event logs from the base station. The logs contain the start and end times (in millisecond) of the transmission of every burst of data and its size.

We developed software utilities to analyze the logs for three performance metrics: bit rate, time spacing between successive bursts, and energy saving. We compute the bit rate for each video stream by considering the start times of two consecutive bursts and the burst size. We use the bit rate to verify that our burst scheduling algorithm leads to no buffer overflow/underflow instances. We compute the time spacing between bursts by first sorting bursts of all video streams based on their start times. Then, we sequentially compute the time spacing between the start time of a burst and the end time of the immediately previous burst. We use the time spacing to verify that there are no burst conflicts, as a positive time spacing indicates that bursts do not intersect with each other. We compute the energy saving for each video stream as the ratio between the wireless interface on time and off time. We assume that the overhead duration is $T_o = 100$ ms.

6.3.3.2 Experimental Validation of P2OPT Correctness

We first validate the correctness of our P2OPT algorithm using the actual test bed implementation. Figure 6.11 demonstrates the correctness of the P2OPT algorithm. In Figure 6.11a and b, we plot the cumulative data received by the receivers of two sample video streams as time progresses (other results are similar). We also show the consumed data over time. To account for the worst case, we assume that the receiver starts consuming (playing back) the video data immediately after receiving a burst. The two figures clearly show that there are (a) no buffer underflow instances as the consumption line never crosses the staircase curve representing received data, and (b) no buffer overflow instances as the distance between the data arrival and consumption curves never exceeds the buffer size (1 Mb). Notice that Figure 6.11a and b shows only short time periods for clarity, but since these short periods cover multiple frame periods and the burst scheduling is identical in successive frames, the results are the same for the whole streaming period.

In order to show that there are no burst conflicts, we plot the cumulative distribution function (CDF) of the time spacing between successive bursts in Figure 6.11c. This CDF curve is computed from all bursts of all video streams during the experiment period (10 min). Negative time spacing would indicate that two bursts are intersecting in time, i.e., a burst conflict. This figure clearly shows that our P2OPT algorithm results in no burst conflicts.

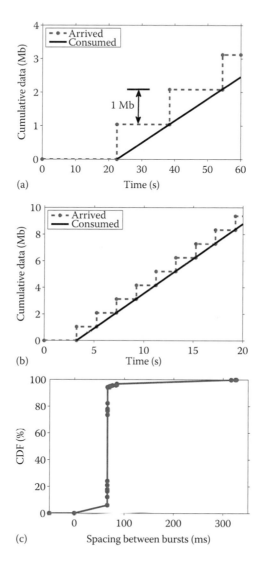

FIGURE 6.11
Experimental validation of the P2OPT algorithm: (a) and (b) show no overflow/underflow instances for Stream 1 (64 kbps) and Stream 6 (512 kbps), respectively, and (c) shows no burst conflicts among the bursts of all video streams. (From Hefeeda, M. and Hsu, C., Energy optimization in mobile TV broadcast networks, in *Proc. of IEEE International Conference on Innovations in Information Technology (Innovations'08)*, pp. 430–434, 2008. Copyright 2008 IEEE; Hefeeda, M. and Hsu, C., On burst transmission scheduling in mobile TV broadcasting networks, *IEEE/ACM Trans. Netw.*, 18(2), 619. Copyright 2010 IEEE. With permission.)

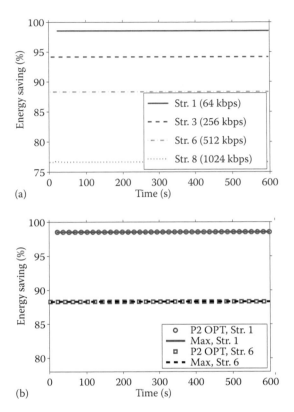

FIGURE 6.12
(a) Energy saving achieved by the P2OPT algorithm for individual video streams and (b) comparing the energy saving achieved by P2OPT to the absolute maximum saving. (From Hefeeda, M. and Hsu, C., On burst transmission scheduling in mobile TV broadcasting networks, *IEEE/ACM Trans. Netw.*, 18(2), 619–620. Copyright 2010 IEEE. With permission.)

6.3.3.3 Energy Saving Achieved by P2OPT

Next, we report the energy saving achieved by our algorithm. Figure 6.12a presents the energy saving of each video stream. We observe that the energy saving for low-bit-rate video streams can be as high as 99%, while it is only 76% for high bit rate video streams. This dramatic difference emphasizes the importance of transmitting video streams at heterogeneous bit rates: to maximize energy saving on mobile devices, a video stream should be sent at the lowest bit rate that fulfills its minimum quality requirements.

6.3.3.4 Optimality of P2OPT

Lastly, we verify that the energy saving achieved by our P2OPT algorithm is indeed optimal. To do this, we computed the absolute maximum energy saving that can be achieved by any algorithm for a given video stream. We computed this maximum by making the base station transmit only the given video stream. In this case, the base station easily maximizes the energy saving by allocating the largest burst that can fill the receiver's buffer. The wireless interface of the receiver was then turned off until the data of this burst are consumed. We repeated this experiment nine times: once for each video stream, and we computed the maximum possible energy saving. We then ran our algorithm to compute the burst schedule for the nine video streams, and we made the base station transmit all of them *concurrently*. We computed the energy saving for each video stream. Sample results for streams 1 and 6 are presented in Figure 6.12b; all other results are similar. The figure confirms the optimality of the P2OPT algorithm in terms of energy saving.

6.4 Bandwidth Optimization in Mobile Video Streaming

The focus of the previous section was energy optimization. Another important consideration in mobile video streaming networks is bandwidth optimization. Since the wireless spectrum in wide-area wireless networks is license based and expensive, maximizing the utilization of this spectrum is important. In this section, we study the problem of optimizing the use of bandwidth when transmitting several video streams from a wireless base station to a large number of mobile receivers. To achieve high bandwidth utilization, we employ the VBR model for encoding video streams in this section. Unlike the CBR model, the VBR model allows statistical multiplexing of video streams [32] and yields better perceived video quality [36]. However, the VBR model makes video transmission much more challenging than the CBR model in mobile video streaming networks [44].

VBR encoding of video streams results in higher-quality video by assigning more bits to complex frames and fewer bits to less complex frames. This results in more complicated requirements for applications of VBR streams [28]. Transmitting VBR video streams over a wireless channel while avoiding buffer overflow and underflow at mobile devices is a difficult problem [44]. Rate smoothing is one approach to reducing the complexity of this problem. The main idea in this approach is to transmit video streams in segments of CBR and let the receiver store some data ahead of time to avoid large changes in the transmission bit rate of video streams. This reduces the complexities resulting from variability of video stream bit rates. The performance of smoothing algorithms can be measured by different metrics like the peak bit rate, the number of changes in the bit rate, traffic variability, and the receiver buffer requirements. The minimum requirements of rate smoothing algorithms in terms of playback delay, lookahead time, and buffer size are discussed by Thiran et al. [56]. The smoothing algorithms proposed in [23,24,60] minimize the number of rate changes.

In Refs. [43,48], a smoothing algorithm is proposed to minimize traffic variability subject to a given receiver buffer size and startup delay. The algorithms by Lin et al. [37] reduce the complexity by controlling the transmission rate of a single video stream to produce a CBR stream. Ribas-Corbera et al. and Sen et al. [47,50] suggest smoothing algorithms to reduce the peak bandwidth requirements for video streams. In Ref. [35], a monotonically decreasing rate algorithm for smoothing the bit rates of video streams is discussed. This algorithm produces segments with monotonically decreasing bit rates for video streams to remove the resource allocation complexities resulting from upward adjustment. The worst case buffer requirement in this algorithm is unbounded, which makes it unsuitable for mobile receivers with limited buffer capacity. None of the smoothing algorithms provided earlier considers energy saving as a performance metric as these algorithms are not designed for mobile multicast/broadcast networks with limited-energy receivers. In Ref. [10], an online smoothing algorithm for bursts (SAB) is introduced that tries to minimize the percentage of lost frames due to bandwidth and buffer limitations. The algorithm calculates the minimum and maximum possible bit rates for video streams without experiencing buffer underflow and overflow instances. SAB transmits video data in bursts to achieve energy saving for mobile receivers; however, the algorithm considers the transmission of only one video stream.

The VBR transmission algorithms deployed in practice are simple heuristics. For example, in the Nokia mobile broadcast solution (MBS) [1,41], the operator determines a bit rate for each video stream and a time interval based on which bursts are transmitted. The time interval is calculated on the basis of the size of the receiver buffers and the largest bit rate among all video streams and is used for all video streams to avoid buffer overflow instances at the receivers. In each time interval, a burst is scheduled for each video stream based on its bit rate. It is difficult to assign bit rate values to VBR video streams for good performance while avoiding buffer underflow and overflow instances at the receivers.

A recent algorithm for transmitting VBR video streams to mobile devices without requiring joint video coders is presented by Hsu and Hefeeda [32]. This algorithm, called statistical multiplexing scheduling (SMS), performs statistical multiplexing of video streams. However, the SMS algorithm does not dynamically control the buffers of the receivers. The algorithm that we describe in this section uses dynamic control of buffers to improve the performance in terms of dropped video frames and energy consumption of mobile devices.

In this section, we describe an adaptive algorithm for the burst scheduling of multiple VBR streams and use experimental results to compare it to the SMS algorithm [32] and to the algorithm used in the Nokia MBS.

6.4.1 Problem Statement

To achieve the burst transmission of video streams, we need to create a transmission schedule that specifies for each stream the number of bursts, the size of each burst, and the start time of each burst. Note that only one burst can be transmitted on the broadcast channel at any time. The problem we address in this section is to design an algorithm to create a transmission schedule for bursts that yields better performance than current algorithms in the literature.

We study the problem of broadcasting S VBR video streams from a base station to mobile receivers in bursts over a wireless channel of bandwidth R kbps. The base station runs the transmission scheduling algorithm every Γ s; we say that Γ is the scheduling window. The base station receives the video data belonging to video streams from streaming servers and/or reads it from local video databases. The base station aggregates video data for Γ s. Then, it computes the required number of bursts for each stream s. We denote the size of burst k of video stream s by b_k^s kb, and the transmission start time for it by f_k^s s. The end time of the transmission for burst k of stream s is $f_k^s + b_k^s/R$ s.

After computing the schedule, the base station will start transmitting bursts in the next scheduling window. Each burst may contain multiple video frames. We denote the size of frame i of video stream s by l_i^s kb. Each video frame i has a decoding deadline, which is i/F, where F is the frame rate in fps. The goals of our scheduling algorithm are (a) maximize the number of frames delivered on time (before their decoding deadlines) for all video streams and (b) maximize the average energy saving for all mobile receivers. As in Section 6.3, we define the average energy saving to be $\gamma = \sum_{s=1}^{S} \gamma_s/S$, where γ_s is the fraction of time that the wireless interfaces of the receivers of stream s are turned off.

6.4.2 Transmission Algorithm

The algorithm that we describe in this section, called the adaptive data transmission (ADT) algorithm, solves the burst transmission problem for VBR video streams described earlier. The key idea of the algorithm is to *adaptively* control the buffer levels of mobile devices receiving different video streams. The buffer level of each mobile device is adjusted as a function of the bit rate of the video stream being received by that device. Since we consider VBR video streams, the bit rate of each stream is changing with time according to the visual characteristics of the video. This means that the buffer level of each mobile device is also changing with time. The receiver buffer levels are controlled through the sizes and timings of the bursts transmitted by the base station in each scheduling window. The sizes and timings of the transmitted bursts are computed by the ADT algorithm. This adaptive control of the receiver buffers provides flexibility to the base station that can be used to transmit more video data on time to mobile receivers.

6.4.2.1 Algorithm Description

The ADT algorithm defines control points at which it makes decisions about which stream should have access to the wireless medium and for how long it should have access. Control points for each video stream are determined separately based on a parameter α, where $0 < \alpha < 1$. This parameter is the fraction of a receiver's buffer capacity B that is played out between two control points. The parameter α can change dynamically between scheduling windows but is the same for all video streams. At a given control point, the base station selects a stream, computes the buffer level of the receivers for the selected stream, and can transmit data as long as there is no buffer overflow at the receivers.

For small values of α, control points are closer to each other (in time), which results in smaller bursts. This gives the base station more flexibility when deciding which video stream should be transmitted to meet its deadline. That is, the base station has more opportunities to adapt to the changing bit rates of the different VBR video streams being transmitted. For example, the base station can quickly transmit more bursts for a video stream experiencing high bit rate in the current scheduling window and fewer bursts for another stream with low bit rate in the current scheduling window. This dynamic adaptation increases the number of video frames that meet their deadlines from the high bit rate stream while not harming the low bit rate stream. However, smaller bursts may result in less energy saving for the mobile receivers because they may turn their wireless interfaces on and off more often. In each transition from off to on, the wireless interface incurs an overhead because it has to wake up shortly before the arrival of the burst to initialize its circuitry and lock onto the radio frequency of the wireless channel. We denote this overhead by T_o, which is on the order of milliseconds, depending on the wireless technology.

The ADT algorithm is to be run by the wireless base station to schedule the transmission of S video streams to mobile receivers. The algorithm can be called periodically in every scheduling window of length Γ s and whenever a change in the number of video streams occurs.

We define several variables that are used in the algorithm. Each video stream s is encoded at F fps. We assume that mobile receivers of video streams have a buffer capacity of B kb. We denote the size of frame i of video stream s by l_i^s kb. We denote the time by which the data in the buffer of a receiver of stream s is completely played out (so the buffer is completely drained) by d_s s. Thus, d_s is the deadline for receiving another burst of stream s. We use the parameter M kb to indicate the maximum size of a burst that could be scheduled for a video stream in our algorithm when there are no other limitations like buffer size. In some wireless network standards, there are limitations on the value of M. In this chapter, we assume that the buffer capacity is small, so the upper bound M on the burst size does not affect our algorithm in practice. A control point is a time when the scheduling algorithm decides which stream should be assigned a burst. A control point for video stream s is set every time the receivers of stream s have played out αB kb of video data.

The algorithm schedules burst in scheduling windows of Γ s. In each scheduling window, the scheduler schedules Γ s of data that will be played within that scheduling window. We assume that, during each scheduling window, the algorithm also has access to a small amount of data (as large as the size of the buffer) from the next scheduling window. Therefore, if all Γ s of video data for the video streams are scheduled within the current scheduling window and there is still some extra time remaining, the scheduler can schedule some bursts from the next scheduling window within the current scheduling window.

Let us assume that the algorithm is currently computing the schedule for the time window t_{start} to $t_{start} + \Gamma$ s. The algorithm defines the variable $t_{schedule}$ and sets it equal to t_{start}. Then, the algorithm computes bursts one by one and keeps incrementing $t_{schedule}$ until it reaches the end of the current scheduling window, i.e., $t_{start} < t_{schedule} < t_{start} + \Gamma$. For instance, if the algorithm schedules a burst of size 125 kb on a channel with 1 Mbps bandwidth, then the length of this burst will be 0.125 s and the algorithm increments $t_{schedule}$ by 0.125 s.

The number of video frames for video stream s that belong to the current scheduling window and are scheduled until $t_{schedule}$ is denoted by m_s. This gives the receiver of stream s a playout time of m_s/F s. Based on this, we can define the playout deadline for stream s at time $t_{schedule}$ as

$$d_s = t_{start} + \frac{m_s}{F}. \tag{6.3}$$

Let p_s denote the number of frames played out at the receivers of stream s by time $t_{schedule}$. Therefore, we can define p_s as $p_s = t_{schedule} \times F$. The next control point h_s of stream s after time $t_{schedule}$ will be when the receivers of stream s have played out αB kb of video data. We compute the number of frames g_s corresponding to this amount as follows:

$$\sum_{i=p_s+1}^{p_s+g_s} l_i^s \leq \alpha B < \sum_{i=p_s+1}^{p_s+g_s+1} l_i^s. \tag{6.4}$$

The control point h_s is then given by

$$h_s = t_{schedule} + \frac{g_s}{F}. \tag{6.5}$$

The algorithm uses α_{min} and α_{max} to define the range within which the value of α can vary. The operator presets these values based on the desired control over bandwidth and flexibility in energy saving.

The high-level pseudo-code of the ADT algorithm is given in Figure 6.13. The algorithm works as follows. The scheduler chooses a new video stream to receive a burst at each control point and also at any point when the buffers of the receivers of the currently scheduled burst are full (after scheduling

---

Adaptive Data Transmission (ADT) Algorithm

---

1.   **compute** control points and deadlines for each stream s
2.   **while** there is a video stream to transmit {
3.   **create** a new scheduling window from t_{start} to $t_{start} + \Gamma$
4.   **while** the current scheduling window is not complete and α could be improved {
5.   **pick** video stream s having the earliest deadline
6.   **schedule** a burst until the next control point or until the buffer is full
7.   **if** receivers of s are full, do not schedule any burst for until the next control point belonging to s
8.   **update** $t_{schedule}$ based on the length of scheduled burst
                  // gradually increase α if it was reduced in the previous scheduling window
9.   **if** $\alpha < \alpha_{max}$ and α was not reduced during the current scheduling window
10.   **update** α (increase linearly)
11.   **update** d_s and h_s (control point) for video stream s
12.   **if** there is a stream s' which is late and $\alpha > \alpha_{min}$ or if α could be improved
                  // go back within the scheduling window and reschedule bursts
13.   **reset** $t_{schedule}$ to t_{start}
14.   **update** α through binary search
15.   }
16.   }

FIGURE 6.13
The ADT burst scheduling algorithm.

the current burst). When deciding to select a stream, the algorithm finds the stream s', which has the closest deadline $d_{s'}$ and the stream s'' with the closest control point $h_{s''}$. Then the algorithm schedules a burst for stream s' until the control point $h_{s''}$ if this does not exceed the available buffer space at the receivers of s'. Otherwise, the size of the new burst for s' is set to the available buffer space. If the scheduled burst is as large as the available buffer space, which makes the buffer full, then the algorithm does not schedule any bursts for stream s' before its next control point. This lets the algorithm avoid small bursts for video streams when the buffers are nearly full and results in more energy saving for the receivers. The algorithm repeats the earlier steps until there are no more bursts to be transmitted from the video streams in the current scheduling window, or until $t_{schedule}$ exceeds $t_{start} + \Gamma$ and there remain data to be transmitted. The latter case means that some frames will not be transmitted. In this case, the algorithm tries to find a better schedule by increasing the number of possible control points to introduce more flexibility. This is done by decreasing the value of α.

In order to find the largest value of α that gives us the possibility to transmit all video streams, we do a binary search between α_{min} and α_{max}. The search considers only values between α_{min} and α_{max} with a constant resolution (in our case 0.05), so there are a constant number of choices of values for α. At each step of the binary search, we select a value for α and run the algorithm. If it was successful, we continue the binary search to try to find a larger α. Otherwise, we continue binary search to choose a smaller α. This process finds the largest value of α for which the algorithm can successfully schedule all video data within the current scheduling window. If no such α is found, then the binary search will result in the selection of α_{min}. It is important to note that the search space is of constant size, so the number of steps in the binary search is also constant, and the scheduling algorithm will be run a constant number of times to find the best α. If α is reduced in a scheduling window, then it will be gradually increased in the following scheduling windows based on a linear function. This means that after scheduling every burst, the value of α will be increased by a constant value. In our work, we have used 0.01 as this constant value.

6.4.2.2 Analysis of the ADT Algorithm

As shown in Section 6.3.1, the burst scheduling problem is NP-complete, so solving it optimally and in real time is probably not feasible. However, we can prove that the ADT algorithm produces near-optimal schedules in terms of energy saving for mobile receivers and that it is efficient enough to run in real time. Furthermore, we prove that the algorithm produces correct burst transmission schedules with no buffer overflow instances at the mobile receivers and no two bursts overlapping in time.

The following theorem shows that our algorithm produces near-optimal energy saving for mobile devices. The full proof of this theorem can be found in Ref. [53].

Theorem 6.4 (Energy saving). The difference between the optimal energy saving and the energy saving of the ADT algorithm is less than or equal to $(\bar{r}\,T_o/B)(2/\alpha_{min} - 1)$, where $\bar{r}$ is the average bit rate of all video streams, T_o is the wake-up overhead for the mobile network interfaces, B is the buffer size, and α_{min} is the minimum value of the control parameter used in the ADT algorithm.

Proof outline: The amount of energy saving is related to the number of the bursts scheduled for the video streams, and a new burst is scheduled only at a control point or when a burst completely fills its receivers' buffers. The control points that ADT uses for a video stream are at least αB bytes of data apart. We can show that at most one burst that fills its receivers' buffers occurs between two consecutive control points and that the total number of bursts for any stream s is at most $2r_sT/(\alpha_{min}B)$. This gives a lower bound on the total overhead that can be combined with the total time that the receivers are receiving actual data and the average bit rate of the video streams to obtain a lower bound of $\gamma = 1 - \bar{r}\left((2T_o/\alpha_{min}B) + (1/R)\right)$ on the energy saving of the ADT algorithm.

The size of a burst cannot exceed B bytes without causing a buffer over-flow. This gives a lower bound on the minimum number of bursts and the minimum total overhead time. Combining this with the total time that the receivers are receiving data and the average bit rate of the video streams gives an upper bound of $\gamma_{opt} \leq 1 - \bar{r}\left((T_o/B) + (1/R)\right)$ on the optimal energy saving.

The difference between the lower and upper bounds is $\left(\bar{r}\,T_o/B\right)(2/\alpha_{min} - 1)$.

According to Theorem 6.4, if energy saving is critical, and both T_o and the average bit rate of the video streams are large, then the operator can use α as a control parameter to adjust the system for more energy saving or for finer control over the bandwidth that results in a lower dropped frame rate.

The next theorem shows that our proposed algorithm is computationally efficient. The proof of this theorem can be found in Ref. [53].

Theorem 6.5 (Efficiency). The ADT scheduling algorithm runs in time O(NS), where N is the total number of control points and S is the number of video streams.

We note that our algorithm is linear in terms of S (number of streams) and N (number of control points). Thus, our algorithm can easily run in real time. Indeed, our experiments with the algorithm in the mobile video streaming test bed confirm that the algorithm runs in real time on a regular personal computer.

Next, we address the correctness of the proposed algorithm. We assume that $R \geq \sum_{s=1}^{S} r_s$ since a correct schedule without dropped frames is not possible otherwise.

Theorem 6.6 (Correctness). The ADT algorithm produces feasible burst schedules with no two bursts overlapping in time and no buffer overflow or underflow instances.

Proof: First, we show that there always exists a value of α that results in no buffer underflow instances. When the ADT algorithm is used, there will be time intervals during which bursts are scheduled, and there will be slack times when no bursts are scheduled as shown in Figure 6.14. The slack time could happen for two reasons. The first is that all receiver buffers have been completely filled, so no bursts will be scheduled for them before their next control points. Between two consecutive control points, αB kb of data will be played out, so the buffer levels of the receivers will be at least $(1 - \alpha)B$ during the slack time. The second situation when slack time could occur is at the end of a scheduling window when all video data for the window has been scheduled and there are no more data available. However, we have assumed that there is a small lookahead window that provides access to as much as a buffer of data from the next scheduling window, so this case will never happen and we will not run out of data at the end of a scheduling

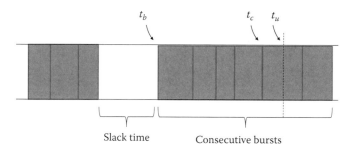

FIGURE 6.14
Scheduled bursts are free of buffer underflow instances.

window. Therefore, when ADT is used, a slack time period can occur for the first reason. Based on this, after a slack time and at the beginning of a period of consecutive bursts, the buffer levels of all receivers will be at least $(1 - \alpha)B$ kb. Now, by way of contradiction, assume that there is a case for which there is no value of α that can avoid buffer underflows. Assume that stream s is the first video stream that will have a buffer underflow in this case. Since all buffer levels are at least $(1 - \alpha)B$ kb during the slack time, buffer underflow for stream s happens during an interval containing consecutive scheduled bursts. Let t_b denote the beginning time of this interval and let t_u denote the time when buffer underflow occurs. Let t_c be the time of the last control point before t_u. Since buffer underflow occurs for stream s, this stream was not selected at time t_c for burst scheduling. Let s' be the stream that was selected at time t_c. Hence, the deadline for stream s' is before the deadline for stream s, which is t_u. Since there is no control point between time t_c and time t_u, and the control points for each stream are at least αB bytes apart, the distance between t_c and t_u is less than αB bytes. It follows that the buffer levels of both streams s and s' are less than αB kb at time t_c. Now, if we denote the sum of the buffer levels of all video streams at time t by e_t, we have $e_{t_c} < (S-2)B + 2\alpha B$ and $e_{t_b} \geq S(1-\alpha)B$. On the other hand, we have $e_{t_c} = e_{t_b} + R(t_c - t_b) - \sum_{s=1}^{S} r_s(t_c - t_b)$. Since $R \geq \sum_{s=1}^{S} r_s$, we will have $e_{t_c} \geq e_{t_b}$. Therefore, $(S - 2)B + 2\alpha B > S(1 - \alpha)B$. It follows that $\alpha > 2/(S+2)$, so any value of $\alpha < 2/(S+2)$ will avoid buffer underflow. This contradicts the assumption that there is no value of α that avoids buffer underflow.

At each scheduling time, we know the buffer level for each stream s by subtracting the size of the p_s frames that have been played out at the receiver from the total size of the m_s frames that have been scheduled for stream s. Based on this, we know the remaining capacity of the buffer, and therefore line 6 of the algorithm (Figure 6.13) guarantees that no buffer overflows occur at the receivers.

Finally, we show that no two scheduled bursts overlap in time. Each new burst is created starting at the current scheduling time $t_{schedule}$ in line 6 of the algorithm (Figure 6.13). After scheduling a burst in line 6, $t_{schedule}$ is updated in line 8 and moved forward as much as the length of the burst scheduled in line 6. Other than line 8, there is only one place where $t_{schedule}$ is updated in the algorithm and that is line 13. In line 13, $t_{schedule}$ is reset to the start time and the scheduling is started from scratch. Therefore, the ADT algorithm produces a feasible burst schedule for the base station.

Theorem 6.6 shows that the ADT algorithm will find a correct schedule if one exists and there are no constraints on the value chosen for α. If $R < \sum_{s=1}^{S} r_s$, then the available bandwidth of the wireless channel is insufficient to transmit all frames of all video streams and buffer underflows are inevitable. Buffer underflows are also possible if α_{min} is set too high or the distance d_α between the values of α considered by the binary search is too large. In particular, the algorithm must be able to find a value $a < 2/(S+2)$ to avoid underflows. If $2/(S+2) < \alpha_{min} + d_\alpha$, then ADT will not find α and underflows could occur. However, even if underflows occur, the schedules produced by ADT will have no buffer overflows and no pair of bursts overlapping. Furthermore, the values α_{min} and d_α are under the control of the operator.

6.4.3 Experimental Evaluation and Analysis

The ADT algorithm, the SMS algorithm [32], and the Nokia MBS [1,41] were implemented in the test bed described in Section 6.3.3 and integrated with the IP encapsulator of the transmitter. Our experiments show that ADT produces better bandwidth utilization resulting in fewer dropped frames than SMS and MBS and also achieves higher energy saving.

6.4.3.1 Experimental Setup

The overhead T_o was set to 100 ms and the modulator was set to a QPSK scheme and a 5 MHz radio channel. We fixed the maximum receiver buffer size B at 4 Mb (500 kB). We prepared a test set of 17 diverse VBR video streams to evaluate the algorithms. The different content of the streams (TV commercials, sports, action movies, documentaries) provided a wide range of video characteristics with average bit rates ranging from 25 to 600 kbps. The average bit rate of the video streams was 260 kbps. Each video stream was played at 30 fps and had length 566 s. We transmitted the 17 VBR video streams concurrently to the receivers and collected detailed statistics from the analyzers. Each experiment was repeated for each of the three algorithms (ADT, SMS, and MBS). In addition, for our ADT algorithm, we repeated the experiments with several values of the buffer adaptation parameter α. Preliminary versions of the results provided in the following text were first reported in Ref. [52]. More extensive experimental results appear in Ref. [53].

6.4.3.2 Results for Dropped Video Frames

Dropped frames are frames that are received at the mobile receivers after their decoding deadlines or not received at all. The number of dropped frames is an important quality of service metric as it impacts the visual quality and smoothness of the received videos. Figure 6.15a shows the cumulative total over all video streams of the numbers of dropped frames for ADT with fixed values of α ranging from 0.10 to 0.50, and for SMS and MBS. The figure clearly shows that ADT consistently drops significantly fewer

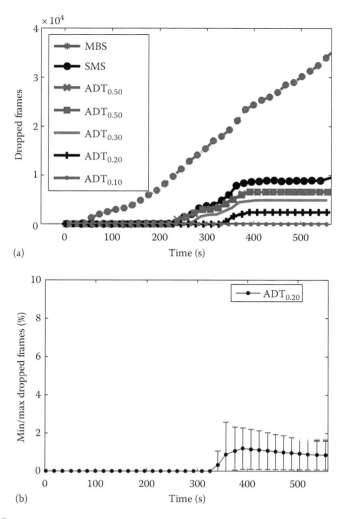

(a)

(b)

FIGURE 6.15
Video frames dropped by ADT. (a) Total number of dropped video frames and (b) minimum and maximum number of dropped video frames.

frames than the SMS and MBS algorithms. The figure also shows the effect of decreasing the value of α for the ADT algorithm. The total number of dropped frames decreases from 6630 with $\alpha=0.50$ to 2074 with $\alpha=0.20$. No frames are dropped when α is reduced to 0.10. On the other hand, the SMS algorithm is significantly worse with 9605 dropped frames. The results for MBS in Figure 6.15a were obtained by running the algorithm for each video stream with different assigned bit rates ranging from 0.25 times the average bit rate to 4 times the average bit rate of the video stream and then choosing the best result for each video stream. Even in this total of best cases, the number of dropped frames is more than 34,000. In practice, an operator heuristically chooses the assigned bit rates for video streams, so the results in practice likely will be worse.

We counted the number of dropped frames for each video stream to check whether the ADT algorithm improves the quality of some video streams at the expense of others. A sample of our results is shown in Figure 6.15b; results for other values of α are similar. The curves in the figure show the average over all streams of the percentage of dropped frames; each point on the curves is the average percentage of frames transmitted to that point in time that were dropped. The bars show the ranges over all video streams. As shown in the figure, the difference between the maximum and minimum dropped frame percentages at the end of the transmission period is small. We conclude that the ADT algorithm does not sacrifice the quality of service for some streams to achieve good aggregate results.

We further analyze the patterns of dropped video frames for each algorithm in more detail. We plot the total number of dropped frames during each 1 s interval across all video streams. Samples of our results are shown in Figure 6.16. For the ADT algorithm with $\alpha=0.50$, frames were dropped mostly during a 150 s period (between 220 and 370 s) because several video streams have high bit rate during this period. Reducing α to 0.20 permits finer control over the distribution of bandwidth among the video streams, and significantly fewer frames were dropped as shown in Figure 6.16b. Further reducing α to 0.10 eliminated all dropped frames as we have already seen in Figure 6.15a. The SMS algorithm on the other hand dropped up to 42% of the frames during the period in which the aggregate bit rate of all streams is high. Using the MBS algorithm results in dropping frames in a very wide time range. The results from these experiments confirm the performance benefits that our ADT algorithm achieves by dynamically adapting to the changing bit rate nature of VBR video streams by controlling the level of receivers' buffers through the parameter α.

6.4.3.3 Results for Energy Saving

We compute the average energy saving γ achieved across all video streams, which represents the average amount of time that the wireless interfaces are off. The results are shown in Figure 6.17a. The figure shows that the average

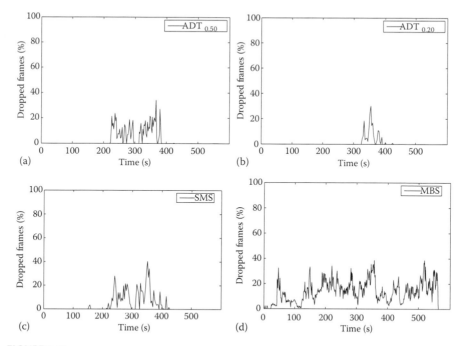

FIGURE 6.16
Dropped video frames over 1 s periods. (a) ADT with $\alpha=0.50$, (b) ADT with $\alpha=0.20$, (c) SMS, and (d) MBS.

saving resulting from the ADT algorithm when $\alpha=0.5$ is very close to the upper bound on optimal energy saving from the proof of Theorem 6.4 and is about 6.74% more than the average energy saving achieved by the SMS algorithm. Also our experiments show that the energy saving achieved by ADT is considerably higher than the MBS algorithm, which achieves an average energy saving of up to 41.14%.

The impact of changing α on the energy saving achieved by the ADT algorithm is shown in Figure 6.17b. The average over all video streams of the energy saving is 89.52% for the ADT algorithm when $\alpha=0.10$, which is better than the SMS algorithm. Increasing α to 0.50 increases the energy saving to 93.47%. The small improvement of 3.95% in energy saving is nontrivial, but it might not be large enough in many practical situations to offset the advantage of minimizing the number of dropped frames by setting $\alpha=0.10$.

Finally, we measured the energy saving for the receivers of each individual stream to check whether the ADT algorithm unfairly saves more energy for some receivers at the expense of others. A sample of our results is shown in Figure 6.17c. The figure confirms that ADT does not sacrifice energy saving in some streams to achieve good average energy saving.

In summary, the results in this section show that the proposed algorithm achieves higher energy saving than previous algorithms, that the saving is

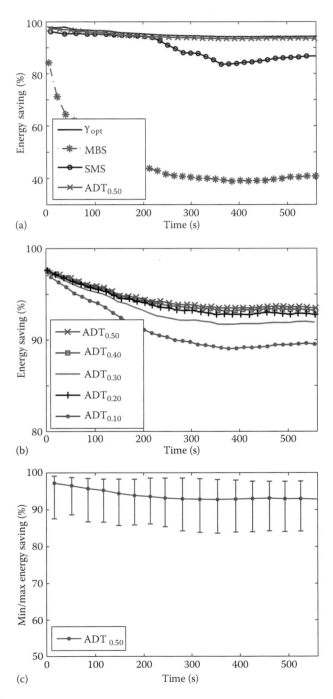

FIGURE 6.17
Energy saving with ADT. (a) Average energy saving, (b) Average energy saving with different values of α, and (c) min and max energy saving.

uniform across all mobile receivers, and that the saving is very close to the optimal energy saving.

6.5 Conclusions

In this chapter, we studied the scheduling problem in a mobile video streaming network in which a base station concurrently transmits multiple video streams over a multicast-enabled wireless network to many mobile devices. Multicast-enabled wireless networks allow the base station to send a common copy of video streams to all mobile devices within the range of the base station, thereby achieving high bandwidth efficiency. We presented systematic and provably optimal (and near-optimal) algorithms to maximize: (a) streaming quality, (b) energy saving, and (c) bandwidth efficiency. We achieved these objectives in two steps. We first considered video streaming networks of CBR streams and developed an algorithm to optimize streaming quality and energy saving. We then considered video streaming networks of VBR streams and designed an algorithm to optimize all three objectives.

In Section 6.3, we considered the problem of maximizing video quality and energy saving of mobile devices in a mobile video multicast network of CBR streams. To prevent playout glitches and achieve the optimum streaming quality, the base station must carefully construct burst schedules that have (a) no buffer violations, (b) no burst conflicts, and (c) high energy saving. We considered a problem in which every video stream can be coded at a different bit rate, and we formally proved that it is NP-complete. We then proposed a practical simplification, which allows video streams to be classified into multiple classes with power-of-two bit rate increments. We presented a burst scheduling algorithm, called P2OPT, to solve this practical simplification. We analytically proved that the P2OPT algorithm is optimal in terms of energy saving and that it runs in polynomial time. We conducted extensive experiments to validate the correctness and optimality of the P2OPT algorithm. The experimental results from a real mobile video streaming test bed indicate that the P2OPT algorithm achieves high energy saving: savings between 77% and 99% were observed for streams at different bit rates.

In Section 6.4, we studied the problem of multicasting several VBR video streams for high video quality, energy saving, and bandwidth efficiency. We presented a novel algorithm, called ADT, that dynamically adjusts the level of mobile device buffers according to the bit rates of the video streams being received by each device. By controlling the allocation of wireless bandwidth among the video streams, the ADT algorithm reduces the number of late frames while achieving the best possible statistical multiplexing gain. We also proved that the ADT algorithm computes feasible schedules whenever they exist. We analytically computed the gap between the energy saving resulting from the ADT algorithm and the optimal energy saving, and we showed that this gap is very small. We presented a proof-of-concept implementation of the ADT algorithm in a mobile video streaming test bed. We conducted extensive experiments using a number of VBR video streams with diverse visual characteristics and bit rates. Our results confirm that the ADT algorithm yields high energy saving for mobile receivers and reduces the number of video frames that miss their deadlines. The results also demonstrated that the ADT algorithm outperforms two state-of-the-art algorithms: one has been proposed in the literature and the other one is used in real networks.

Both the P2OPT and ADT algorithms are practical and can run in real time, as verified by our implementations in an actual mobile video streaming test bed that conforms to one of the international standards (DVB-H). P2OPT is suitable for network operators who do not have joint video coders and have to use CBR streams, while ADT is preferred by other network operators.

References

1. Private communication with Nokia's engineers managing mobile TV base stations, December 2008.
2. Multimedia Broadcast/Multicast Service (MBMS); Stage 1: User Services. Third Generation Partnership Project (3GPP) Standard TS 22.246 Ver. 9.0.0, 2009.
3. Improved video support for Packet Switched Streaming (PSS) and Multimedia Broadcast/Multicast Service (MBMS) Services (Release 9). Third Generation Partnership Project (3GPP) Standard TR 26.903 Ver. 9.0.0, 2010.
4. ATSC-Mobile DTV Standard, Part 2 RF/Transmission System Characteristics. ATSC Document A/153 Part 2:2009, 2009.
5. ATSC mobile DTV standard, 2009. *Open Mobile Video Coalition*, http://www.openmobilevideo.com/about-mobile-dtv/standards/ (Accessed August 22, 2012).
6. E. Balaguer, F. Fitzek, O. Olsen, and M. Gade. Performance evaluation of power saving strategies for DVB-H services using adaptive MPE-FEC decoding. In *Proceedings of IEEE International Symposium on Personal, Indoor and Mobile Radio Communications (PIMRC'05)*, pp. 2221–2226, Berlin, Germany, September 2005.
7. Y. Bartal, S. Leonardi, G. Shallom, and R. Sitters. On the value of preemption in scheduling. In *Workshop on Approximation Algorithms for Combinatorial Optimization Problems (APPROX'06)*, pp. 39–48, Barcelona, Spain, August 2006.
8. O. Braun and G. Schmidt. Parallel processor scheduling with limited number of preemptions. *SIAM Journal on Computing*, 32(3):671–680, 2003.
9. P. Brucker. *Scheduling Algorithms*, 4th edn. Springer, London, U.K., 2004.
10. P. Camarda, G. Tommaso, and D. Striccoli. A smoothing algorithm for time slicing DVB-H video transmission with bandwidth constraints. In *Proceedings of ACM International Mobile Multimedia Communications Conference (MobiMedia'06)*, pp. 6:1–6:5, Alghero, Italy, September 2006.

11. M. Chari, F. Ling, A. Mantravadi, R. Krishnamoorthi, R. Vijayan, G. Walker, and R. Chandhok. FLO physical layer: An overview. *IEEE Transactions on Broadcasting*, 53(1):145–160, March 2007.

12. S. Cho, G. Lee, B. Bae, K. Yang, C. Ahn, S. Lee, and C. Ahn. System and services of Terrestrial Digital Multimedia Broadcasting (T-DMB). *IEEE Transactions on Broadcasting*, 53(1):171–178, March 2007.

13. P. Chou. Streaming media on demand and live broadcast. In M. van der Schaar and P. Chou, eds., *Multimedia Over IP and Wireless Networks*, Chapter 14, pp. 453–502. Academic Press, St. Louis, MO, March 2007.

14. C. Cicconetti, L. Lenzini, E. Mingozzi, and C. Eklund. Quality of service support in IEEE 802.16 networks. *IEEE Network Magazine*, 20(2):50–55, March 2006.

15. Divi Catch RF-T/H transport stream analyzer, 2008. http://www.enensys.com/

16. DVB-H global mobile TV: FAQ, 2008. *DVB Mobile TV: FAQ, DVB Project Office.* http://dvb-h.org/faq.htm (Accessed August 22, 2012)

17. dvbSAM DVB-H solution for analysis, monitoring, and measurement, 2008. *ATSC, ATSC M/H, DVB-H, DVB-T, DVB-S, DVB-S2, DVB-C Analysis, Monitoring, Measurement, Test, Decontis GmbH.* http://www.decontis.com/ (Accessed August 22, 2012).

18. Radio broadcasting systems: Digital Audio Broadcasting (DAB) to mobile, portable and fixed receivers. European Telecommunications Standards Institute (ETSI) Standard EN 300 401 Ver. 1.3.3, May 2001.

19. Digital Video Broadcasting (DVB); framing structure, channel coding and modulation for digital terrestrial television. European Telecommunications Standards Institute (ETSI) Standard EN 300 744 Ver. 1.5.1, June 2004.

20. Digital Video Broadcasting (DVB); transmission system for handheld terminals (DVB-H). European Telecommunications Standards Institute (ETSI) Standard EN 302 304 Ver. 1.1.1, November 2004.

21. Digital Video Broadcasting (DVB); DVB-H implementation guidelines. European Telecommunications Standards Institute (ETSI) Standard EN 102 377 Ver. 1.3.1, May 2007.

22. G. Faria, J. Henriksson, E. Stare, and P. Talmola. DVB-H: Digital broadcast services to handheld devices. *Proceedings of the IEEE*, 94(1):194–209, January 2006.

23. W. Feng, F. Jahanian, and S. Sechrest. An optimal bandwidth allocation strategy for the delivery of compressed prerecorded video. *Multimedia Systems*, 5(5): 297–309, September 1997.

24. W. Feng and S. Sechrest. Critical bandwidth allocation for the delivery of compressed video. *Computer Communications*, 18(10):709–717, 1995.

25. FLO forum home page, 2008. *TR-47, Terrestrial and Non Terrestrial Mobile Multimedia Multicast, Telecommunications Industry Association*, http://www.floforum.org/ (Accessed August 22, 2012).

26. M. Garey and D. Johnson. *Computers and Intractability: A Guide to the Theory of NP-Completeness.* W. H. Freeman, Gordonsville, VA, 1979.

27. A. Ghosh, D. Wolter, J. Andrews, and R. Chen. Broadband wireless access with WiMAX/802.16: Current performance benchmarks and future potential. *IEEE Communications Magazine*, 43(2):129–136, February 2005.

28. M. Grossglauser, S. Keshav, and D. Tse. RCBR: A simple and efficient service for multiple time-scale traffic. *IEEE/ACM Transactions on Networking*, 5:741–755, December 1997.

29. F. Hartung, U. Horn, J. Huschke, M. Kampmann, T. Lohmar, and M. Lundevall. Delivery of broadcast services in 3G networks. *IEEE Transactions on Broadcasting*, 53(1):188–199, March 2007.

30. Z. He and D. Wu. Linear rate control and optimum statistical multiplexing for H.264 video broadcast. *IEEE Transactions on Multimedia*, 10(7):1237–1249, November 2008.

31. M. Hefeeda and C. Hsu. On burst transmission scheduling in mobile TV broadcast networks. *IEEE/ACM Transactions on Networking*, 18(2):610–623, April 2010.

32. C. Hsu and M. Hefeeda. On statistical multiplexing of variable-bit-rate video streams in mobile systems. In *Proceedings of ACM Multimedia'09*, pp. 411–420, Beijing, China, October 2009.

33. M. Jacobs, J. Barbarien, S. Tondeur, R. Van de Walle, T. Paridaens, and P. Schelkens. Statistical multiplexing using SVC. In *Proceedings of IEEE International Symposium on Broadband Multimedia Systems and Broadcasting (BMSB'08)*, pp. 1–6, Las Vegas, NV, March 2008.

34. M. Kornfeld and G. May. DVB-H and IP Datacast—Broadcast to handheld devices. *IEEE Transactions on Broadcasting*, 53(1):161–170, March 2007.

35. H. Lai, J. Lee, and L. Chen. A monotonic-decreasing rate scheduler for variable-bit-rate video streaming. *IEEE Transactions on Circuits and Systems for Video Technology*, 15(2):221–231, February 2005.

36. T. Lakshman, A. Ortega, and A. Reibman. VBR video: Tradeoffs and potentials. *Proceedings of the IEEE*, 86(5):952–973, May 1998.

37. J. Lin, R. Chang, J. Ho, and F. Lai. FOS: A funnel-based approach for optimal online traffic smoothing of live video. *IEEE Transactions on Multimedia*, 8(5):996–1004, October 2006.

38. G. May. The IP Datacast system—Overview and mobility aspects. In *Proceedings of IEEE International Symposium on Consumer Electronics (ISCE'04)*, pp. 509–514, Reading, U.K., September 2004.

39. Statistical multiplexing over IP, 2008. Statistical Multiplexing over IP StatmuxIP Solution, Motorola Inc. http://www.motorola.com/web/Business/Products/TV%20Video%20Distribution/Video%20Distribution/Multiplexers/SX-1000/_Documents/_Static%20files/SP_StatmuxIP.pdf (Accessed August 22, 2012).

40. R. Motwani, S. Phillips, and E. Torng. Nonclairvoyant scheduling. *Theoretical Computer Science*, 130(1):17–47, August 1994.

41. Nokia mobile broadcast solution, February 2009. http://web.archive.org/web/20091012075639/ (Accessed October 12, 2009), http://www.mobiletvforum.com/solutions/mbs/ (Accessed August 22, 2012).

42. S. Parkvall, E. Englund, M. Lundevall, and J. Torsner. Evolving 3G mobile systems: Broadband and broadcast services in WCDMA. *IEEE Communications Magazine*, 44(2):30–36, February 2006.

43. A. Reibman and A. Berger. Traffic descriptors for VBR video teleconferencing over ATM networks. *IEEE/ACM Transactions on Networking*, 3:329–339, June 1995.

44. M. Rezaei. Video streaming over DVB-H. In F. Luo, ed., *Mobile Multimedia Broadcasting Standards*, Chapter 4, pp. 109–131. Springer, New York, November 2009.

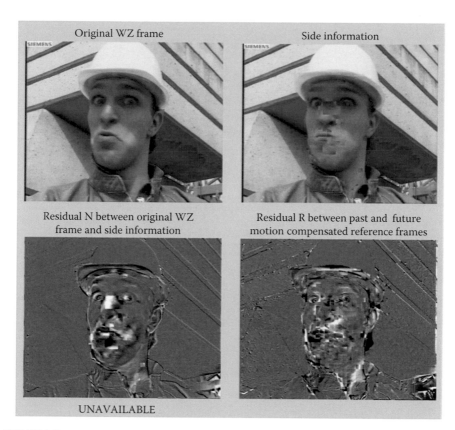

FIGURE 2.8

The motion-compensated residual (bottom-right) resembles the true residual between the original and the side information (bottom-left). (Since the true residual is unavailable during (de)coding, the motion-compensated residual can be used to estimate α.)

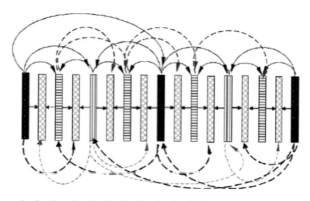

Image type:	I_1	B_3	B_2	B_3	B_1	B_3	B_2	B_3	P_1	B_3	B_2	B_3	B_1	B_3	B_2	B_3	I_2
Display order:	1	2	3	4	5	6	7	8	9	10	11	12	13	14	15	16	17
(a) Coding order:	1	6	5	7	4	9	8	10	3	13	12	14	11	16	15	17	2

FIGURE 3.10

(a) Dyadic hierarchical B-picture prediction structure using two frames and three frames including the third frame (dotted arrows).

(continued)

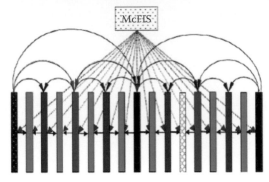

Image type: P_1 B_3 B_2 B_3 B_1 B_3 B_2 B_3 P_2 B_3 B_2 B_3 B_1 B_3 B_2 B_3 P_3

Display order: 1 2 3 4 5 6 7 8 9 10 11 12 13 14 15 16 17

(b) Coding order: 2 7 6 8 5 10 9 11 4 14 13 15 12 17 16 18 3

FIGURE 3.10 (continued)
(b) proposed triple frame referencing with the McFIS as the third frame.

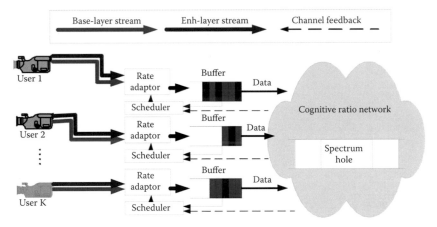

FIGURE 5.1
A WLAN system where each user is equipped with a size B buffer, a decentralized scheduler, and a rate adaptor for transmission control. The users transmit a scalable video payload in which enhancement layers provide quality refinements over the base layer bitstream.

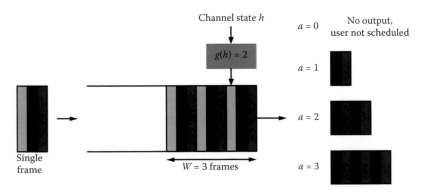

FIGURE 5.3
Example of the buffer control mechanism assumed in this chapter where $f_{in,k}=1$ ($k=1, 2, ..., K$). At every time slot, a new coded video frame enters the buffer. The buffer output depends on the scheduling algorithm involved, the buffer occupancy, and the channel quality. If a user is scheduled for transmission, then the action taken will extract a specific number l of video frame layers from up to N frames stored in the buffer.

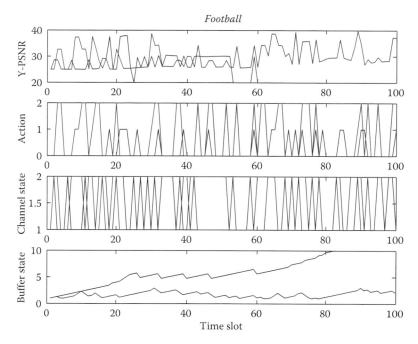

FIGURE 5.6

Result of the transmission of the *Football* sequence comparing the performance in terms of video PSNR and buffer utilization between the proposed switching control game policy (blue lines) and the myopic policy (red lines) with 80 ms delay constraint. The result shows that the proposed switching control game policy performs better than the myopic policy.

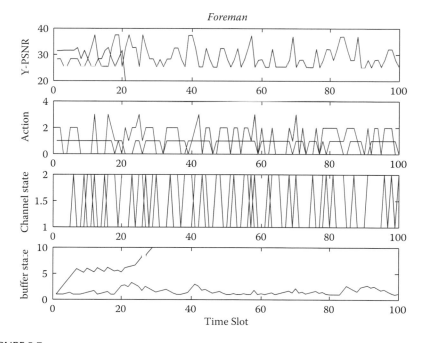

FIGURE 5.7

Result of the transmission of the *Foreman* sequence comparing the performance in terms of video PSNR and buffer utilization between the proposed switching control game policy (blue lines) and the myopic policy (red lines) with 80 ms delay constraint. The result shows that the proposed switching control game policy performs better than the myopic policy.

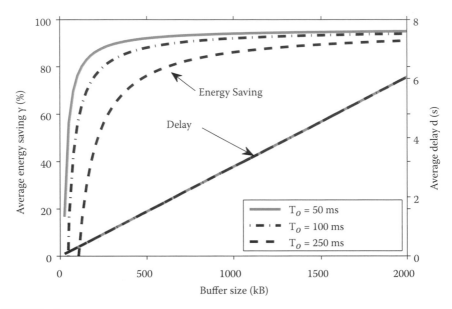

FIGURE 6.9
The trade-off between energy saving and channel switching delay. (From Hefeeda, M. and Hsu, C., On burst transmission scheduling in mobile TV broadcasting networks, *IEEE/ACM Trans. Netw.*, 18(2), 615. Copyright 2010 IEEE. With permission.)

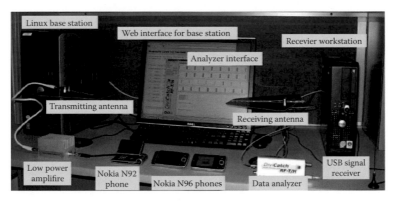

FIGURE 6.10
The mobile video streaming test bed used in the evaluation. (From Hefeeda, M. and Hsu, C., Design and evaluation of a testbed for mobile TV networks, *ACM Transactions on Multimedia Computing, Communications, and Applications*, 8(1), 3:1. Copyright 2012 ACM. With permission.)

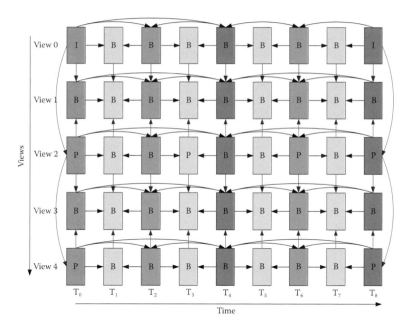

FIGURE 12.3
Multiview coding structure with temporal/interview prediction.

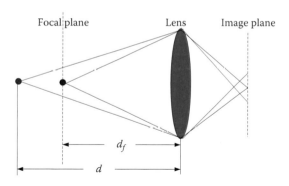

FIGURE 13.3
Focus and defocus for the thin lens.

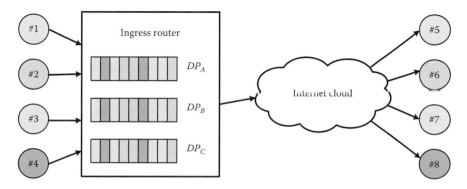

FIGURE 14.5
DP queues in the ingress router. (From Jahaniaval, A., Video quality enhancement through end-to-end distortion optimization and enriched video traces, M.Sc. thesis, University of Guelph, Guelph, Ontario, Canada, pp. 511–516, 2010.)

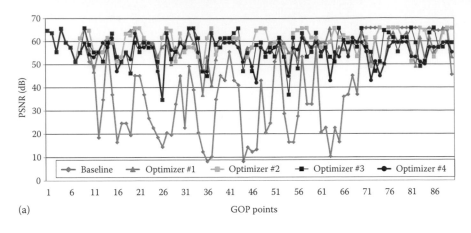

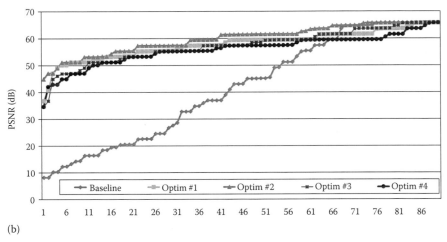

FIGURE 14.8
Received PSNR for Stream 1. (a) Received PSNR for Stream 1 (Akiyo). (b) Reordered with ascending PSNR value. (From Jahaniaval, A., Video quality enhancement through end-to-end distortion optimization and enriched video traces, M.Sc. thesis, University of Guelph, Guelph, Ontario, Canada, pp. 511–516, 2010.)

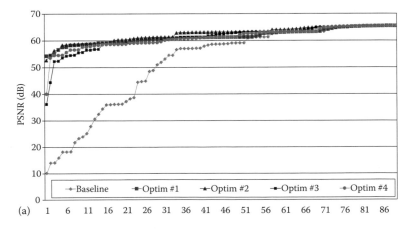

FIGURE 14.9
Received PSNR values in ascending order for (a) Stream 2 (Carphone).

(*continued*)

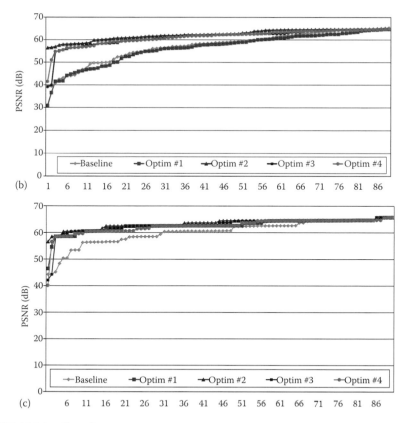

FIGURE 14.9 (continued)
Received PSNR values in ascending order for (b) Stream 3 (Foreman), (c) Stream 4 (Silent). (From Jahaniaval, A., Video quality enhancement through end-to-end distortion optimization and enriched video traces, M.Sc. thesis, University of Guelph, Guelph, Ontario, Canada, 2010.)

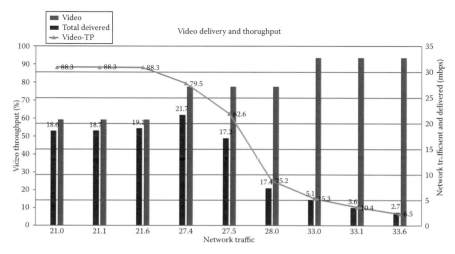

FIGURE 15.22
Video QoS—traffic delivered and throughput.

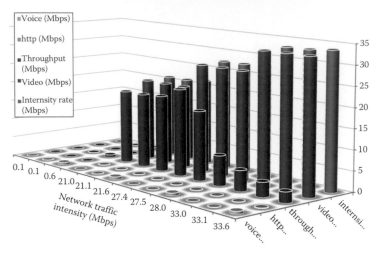

FIGURE 15.23
Network traffic components—intensities and throughput.

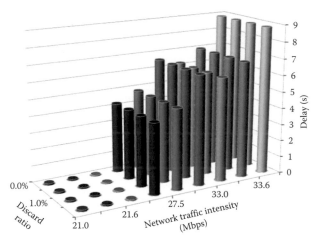

FIGURE 15.24
Average network delay against MAN packet discard ratio and network traffic.

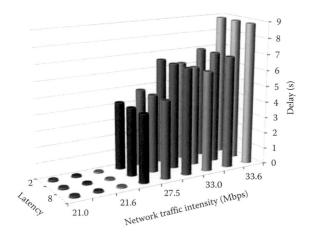

FIGURE 15.25
The average network delay against MAN latency and network traffic intensity.

45. M. Rezaei, I. Bouazizi, and M. Gabbouj. Joint video coding and statistical multiplexing for broadcasting over DVB-H channels. *IEEE Transactions on Multimedia*, 10(7):1455–1464, December 2008.

46. M. Rezaei, M. Hannuksela, and M. Gabbouj. Tune-in time reduction in video streaming over DVB-H. *IEEE Transactions on Broadcasting*, 53(1):320–328, March 2007.

47. J. Ribas-Corbera, P. Chou, and S. Regunathan. A generalized hypothetical reference decoder for H.264/AVC. *IEEE Transactions on Circuits and Systems for Video Technology*, 13(7):674–687, July 2003.

48. J. Salehi, S. Zhang, J. Kurose, and D. Towsley. Supporting stored video: Reducing rate variability and end-to-end resource requirements through optimal smoothing. *IEEE/ACM Transactions on Networking*, 6:397–410, August 1998.

49. U. Schwiegeishohn. Preemptive weighted completion time scheduling of parallel jobs. In *Proceedings of European Symposium on Algorithms (ESA'96)*, pp. 39–51, Barcelona, Spain, September 1996.

50. S. Sen, J. Rexford, J. Dey, J. Kurose, and D. Towsley. Online smoothing of variable-bit-rate streaming video. *IEEE Transactions on Multimedia*, 2(1):37–48, March 2000.

51. E. Stanley and J. Sutton. *Public Response to Alerts and Warnings on Mobile Devices*. The National Academics, Washington, DC, April 2011.

52. F. Tabrizi, J. Peters, and M. Hefeeda. Adaptive transmission of variable-bit-rate video streams to mobile devices. In *Proceedings of IFIP Networking 2011*, Lecture Notes in Computer Science, pp. 213–224, Springer-Verlag, Valencia, Spain, 2011.

53. F. M. Tabrizi. Adaptive transmission of variable-bit-rate video streams to mobile devices. Master's thesis, School of Computing Science, Simon Fraser University, Surrey, British Columbia, Canada, April 2011.

54. M. Tagliasacchi, G. Valenzise, and S. Tubaro. Minimum variance optimal rate allocation for multiplexed H.264/AVC bitstreams. *IEEE Transactions on Image Processing*, 17(7):1057–1143, July 2008.

55. M. Takada and M. Saito. Transmission system for ISDB-T. *Proceedings of the IEEE*, 94(1):251–256, January 2006.

56. P. Thiran, J. Le Boudec, and F. Worm. Network calculus applied to optimal multimedia smoothing. In *Proceedings of IEEE INFOCOM'01*, pp. 1474–1483, Anchorage, AK, April 2001.

57. G. van der Auwera, P. David, and M. Reisslein. Traffic characteristics of H.264/AVC variable bit rate video. *IEEE Communications Magazine*, 46(11):164–174, November 2008.

58. F. Wang, A. Ghosh, C. Sankaran, P. Fleming, F. Hsieh, and S. Benes. Mobile WiMAX systems: Performance and evolution. *IEEE Communications Magazine*, 46(10):41–49, October 2008.

59. X. Yang, Y. Song, T. Owens, J. Cosmas, and T. Itagaki. Performance analysis of time slicing in DVB-H. In *Proceedings of Joint IST Workshop on Mobile Future and Symposium on Trends in Communications (SympoTIC'04)*, pp. 183–186, Bratislava, Slovakia, October 2004.

60. J. Zhang and J. Hui. Applying traffic smoothing techniques for quality of service control in VBR video transmissions. *Computer Communications*, 21(4):375–389, 1998.

61. Q. Zhang, F. Fitzek, and M. Katz. Cooperative power saving strategies for IP-services supported over DVB-H networks. In *Proceedings of IEEE Wireless Communications and Networking Conference (WCNC'07)*, pp. 4107–4111, Hong Kong, China, March 2007.

62. C. Hsu and M. Hefeeda. Using simulcast and scalable video coding to efficiently control channel switching delay in mobile TV broadcast networks. *ACM Transactions on Multimedia Computing, Communications, and Applications*, 7(2):8:1–8:29, February 2011.

63. C. Hsu and M. Hefeeda. Statistical multiplexing of variable-bit-rate videos streamed to mobile devices. *ACM Transactions on Multimedia Computing, Communications, and Applications*, 7(2):12:1–12:23, February 2011.

64. C. Hsu and M. Hefeeda. Broadcasting video streams encoded with arbitrary bit rates in energy-constrained mobile TV networks. *IEEE/ACM Transactions on Networking*, 18(3):681–694, June 2010.

65. M. Hefeeda and C. Hsu. Energy optimization in mobile TV broadcast networks. In *Proc. of IEEE International Conference on Innovations in Information Technology (Innovations'08)*, pp. 430–434, Al Ain, United Arab Emirates, December 2008.

66. C. Hsu and M. Hefeeda. Time slicing in mobile TV broadcast networks with arbitrary channel bit rates. In *Proc. of IEEE INFOCOM'09*, pp. 2231–2239, Rio de Janeiro, Brazil, April 2009.

67. M. Hefeeda and C. Hsu. Design and evaluation of a testbed for mobile TV networks. *ACM Transactions on Multimedia Computing, Communications, and Applications*, 8(1):3:1–3:25, January 2012.

7

Resource Allocation for Scalable Videos over Cognitive Radio Networks

Donglin Hu and Shiwen Mao

CONTENTS

7.1 Introduction

Video content delivery over wireless networks is expected to grow drastically in the coming years. The compelling need for ubiquitous video content access will significantly stress the capacity of existing and future wireless networks. To meet this critical demand, the cognitive radio (CR) technology provides an effective solution that can exploit co-deployed networks and aggregate underutilized spectrum for future video-aware wireless networks.

CR was motivated by the FCC's spectrum measurement study, where a significant amount of the licensed spectrum is found to remain underutilized [1,2]. A CR is an advanced radio device that enables dynamic spectrum access (DSA). It represents a paradigm change in spectrum regulation and access, from exclusive use by licensed (or primary users) to sharing spectrum and allowing dynamic access for unlicensed (or secondary users), with the objective of enhancing spectrum utilization and achieving high-throughput capacity.

The high potential of CRs has attracted substantial interest. The mainstream CR research has focused on developing effective spectrum sensing and access techniques (e.g., see Refs. [1,2]). Although considerable advances have been achieved, the important problem of guaranteeing application performance has not been well studied. We find that video streaming can make excellent use of the enhanced spectrum efficiency in CR networks. Unlike data, where each bit should be delivered, video is loss tolerant and rate adaptive. They are highly suited for CR networks, where the available bandwidth depends on primary user transmission behavior. Graceful degradation of received video quality can be achieved with scalable video coding as spectrum opportunities evolve over time.

CR is an evolving concept with various network models and levels of cognitive functionality [1,2]. IEEE 802.22 wireless regional area network (WRAN) is the first CR standard for reforming broadcast TV bands, where a base station (BS) controls medium access for customer-premises equipments [3]. Therefore, we first consider multicasting scalable videos in such an infrastructure-based CR network. The spectrum consists of multiple channels, each allocated to a primary network. The CR network is colocated with the primary networks, where a CR BS seeks spectrum opportunities for multicasting multiple video streams, each to a group of secondary subscribers. The problem is to exploit spectrum opportunities for minimizing video distortion, while keeping the collision rate with primary users below a prescribed threshold. We consider scalable video coding, such as fine-grained scalability (FGS) and medium grain scalable (MGS) videos [4,5]. We model the problem of CR video multicast over the licensed channels as a mixed-integer nonlinear programming (MINLP) problem and then develop a sequential fixing (SF) algorithm and a greedy algorithm to solve the MINLP, while the latter has a low computational complexity and a proved optimality gap [6].

We then tackle the problem of video over multi-hop CR networks, e.g., a wireless mesh network with CR-enabled nodes. This problem is more challenging than the problem presented earlier due to the lack of infrastructure support. We assume that each secondary user is equipped with two transceivers. To model and guarantee end-to-end video performance, we adopt the amplify-and-forward approach for video data transmission, which is well studied in the context of cooperative communications [7]. This is equivalent to setting up a "virtual tunnel" through a multi-hop multichannel path. The challenging problem, however, is how to set up the virtual tunnels, while the available channels at each relay evolve over time due to primary user transmissions. The formulated MINLP problem is first solved using a centralized SF algorithm, which provides upper and lower bounds for the achievable video quality. We then apply dual decomposition to develop a distributed algorithm and prove its optimality as well as the convergence condition [8].

The rest of the chapter is organized as follows. We review related work in Section 7.2 and present preliminaries in Section 7.3. We examine video over infrastructure-based CR networks in Section 7.4 and over multi-hop CR networks in Section 7.5. Simulation results are presented in Section 7.6. We conclude the chapter in Section 7.7 with a discussion of open problems.

7.2 Related Work

The high potential of CRs has attracted considerable interest from industry, government, and academia [1,9]. The mainstream CR research has been focused on spectrum sensing and DSA issues. For example, the impact of spectrum sensing errors on the design of spectrum access schemes has been addressed in several papers [10–13]. The approach of iteratively sensing a selected subset of available channels has been developed in the design of CR MAC protocols [10,14–16]. The optimal trade-off between the two kinds of sensing errors is investigated comprehensively and addressed in depth in Ref. [11].

A few recent works [17–19] have studied multi-hop CR networks. The authors formulate cross-layer optimization problems considering factors from the physical layer (PHY) up to the transport layer. The dual decomposition technique [20,21] is adopted to develop distributed algorithms. We choose similar methodology in our work and apply it to the more challenging problem of real-time video streaming.

The important problem of QoS provisioning in CR networks has been studied in a few papers (e.g., see Refs. [12,14,16,22]), where the objective is still focused on the so-called network-centric metrics such as maximum throughput and delay [12,14,16]. In Refs. [14,16], MAC protocols are developed for CR networks with throughput guarantees and bounded interference to primary users. In Ref. [12], an interesting delay-throughput trade-off for a multi-cell CR network is derived, while the goal of primary user protection is achieved by stabilizing a virtual "collision queue." In Ref. [22], a game-theoretic framework is described for resource allocation for multimedia transmissions in spectrum agile wireless networks. In this interesting work, each wireless station participates in a resource management game, which is coordinated by a network moderator. A mechanism-based resource management scheme determines the amount of transmission opportunities to be allocated to various users on different frequency bands such that certain global system metrics are optimized.

The problem of video over CR networks has been addressed only in a few recent papers. In Ref. [23], a priority virtual queue model is adopted for wireless CR users to select channel and maximize video qualities. In Ref. [24], the impact of system parameters residing in different network layers is jointly considered to achieve the best possible video quality for CR users. The problem is formulated as a minimax problem and solved with a dynamic programming approach. Ali and Yu [25] jointly optimize video parameter with spectrum sensing and access strategy. A rate-distortion model is adopted to optimize the intra-mode selection and source-channel rate with a partially observable Markov decision process formulation. In Ref. [26], video encoding rate, power control, relay selection, and channel allocation are jointly considered for video over cooperative CR networks. The problem is formulated as a mixed-integer nonlinear problem and solved by a solution algorithm based on a combination of the branch and bound framework and convex relaxation techniques. In our prior work, we studied the problem of scalable video multicast in an infrastructure-based CR network in Refs. [6,27], multiuser scalable video streaming over a multi-hop CR network in Ref. [8], and multiuser downlink video streaming over a CR femtocell network in Refs. [28,29], where effective algorithms that achieve optimality or with bounded performance are developed. We also investigate the problem of combing cooperative relay with CR for multiuser downlink video streaming in Ref. [30], where interference alignment is incorporated to facilitate concurrent transmissions of multiple video packets.

Video multicast, as one of the most important multimedia services, has attracted considerable interest from the research community. Layered video multicast has been studied in the context of mobile ad hoc networks [31,32] and infrastructure-based wireless networks [4,33]. A greedy algorithm is presented in Ref. [33] for layered video multicast in WiMAX networks with a proven optimality gap. Heuristic algorithms are proposed in Refs. [31,32] to establish disjoint multicast trees.

7.3 System Model and Preliminaries

7.3.1 Primary Network

We consider a spectrum band consisting of M orthogonal channels with identical bandwidth [34]. We assume that the M channels are allocated to K primary networks, which cover different service areas. A primary network can use any of the M channels without interfering with other primary networks. We further assume that the primary systems use a synchronous slot structure as in prior work [1,14]. Due to primary user transmissions, the occupancy of each channel evolves following a discrete-time Markov process, as validated by recent measurement studies [1,14,35].

In primary network k, the status of channel m in time slot t is denoted by $S_m^k(t)$ with idle (i.e., $S_m^k(t) = 0$) and busy (i.e., $S_m^k(t) = 1$) states. Let λ_m^k and μ_m^k be the transition probability of remaining in state 0 and that from state 1 to 0, respectively, for channel m in primary network k. The utilization of channel m in primary network k, denoted by $\eta_m^k = \Pr\left(S_m^k = 1\right)$, is

$$\eta_m^k = \lim_{T \to \infty} \frac{1}{T} \sum_{t=1}^{T} S_m^k(t) = \frac{1 - \lambda_m^k}{1 - \lambda_m^k + \mu_m^k}. \tag{7.1}$$

Note that in infrastructure-based CR networks, we assume that there is only one primary network (i.e., $K=1$) since video streams are multicast to single-hop CR users. In infrastructure-based CR networks introduced in Section 7.4, we adopt N as the number of licensed channels since M is denoted as the number of modulation-coding (MC) schemes.

7.3.2 Infrastructure-Based CR Networks

We consider a CR BS that multicasts G real-time videos to G multicast groups, each of which have N_g users, $g = 1,2,\ldots,G$. The BS seeks spectrum opportunities in the N channels to serve CR users. In each time slot t, the BS selects a set of channels $\mathcal{A}_1(t)$ to sense and a set of channels $\mathcal{A}_2(t)$ to access. Without loss of generality, the BS has $|\mathcal{A}_1(t)|$ transceivers such that it can sense $|\mathcal{A}_1(t)|$ channels simultaneously. The combination of a time slot and channel, termed a *tile*, is the basic unit for resource allocation.

We adopt the same time slot structure as in Refs. [1,15]. At the beginning of each time slot, the BS senses channels in $A_1(t)$ and then chooses a set of available channels for opportunistic transmissions based on sensing results. After a successful transmission, the BS will receive an ACK from the user with the highest SNR in the target multicast group. Without loss of generality, we assume that each CR network user can access all the available channels with channel bonding/aggregation techniques [36,37].

7.3.3 Multi-Hop CR Networks

We also consider a multi-hop CR network that is colocated with the primary networks, within which S real-time videos are streamed among N CR nodes. Let U^k denote the set of CR nodes that are located within the coverage of primary network k. A video session l may be relayed by multiple CR nodes if source z_l is not a one-hop neighbor of destination d_l. We assume a *common control channel* for the CR network [14]. We also assume that the timescale of the primary channel process (or, the time slot durations) is much larger than the broadcast delays on the control channel, such that feedbacks of channel information can be received at the source nodes in a timely manner.

The time slot structure is the same as that in the case of infrastructure-based CR networks. In the sensing phase, one transceiver of a CR node is used to sense one of the M channels, while the other is tuned to the control channel to exchange channel information with other CR users. Each video source computes the optimal path selection and channel assignment based on sensing results. In the transmission phase, the channels assigned to a video session l at each link along the path form a virtual "tunnel" connecting source z_l and destination d_l. Similarly, each node can use one or more channels to communicate with other nodes using the channel bonding/aggregation techniques [36,37]. When multiple channels are available on all the links along a path, multiple tunnels can be established and used simultaneously for a video session. In the acknowledgment phase, the destination sends ACK to the source for successfully received video packets through the same tunnel.

We adopt amplify-and-forward for video transmission [7]. During the transmission phase, one transceiver of the relay node receives video data from the upstream node on one channel, while the other transceiver of the relay node amplifies and forwards the data to the downstream node on a different, orthogonal channel. There is no need to store video packets at the relay nodes. Error detection/correction will be performed at the destination node. As a result, we can transmit through the tunnel a block of video data with minimum delay and jitter in one time slot.

7.3.4 Spectrum Sensing

During the sensing process, two types of sensing errors may occur. A *false alarm* may lead to wasting a spectrum opportunity and a *miss detection* may cause collision with primary users. In a multi-hop CR network, the sensing results from various users may be different. Denote H_0 as the hypothesis that channel m in primary network k is idle, and H_1 the hypothesis that channel m in primary network k is busy in time slot t. The conditional probability that channel m is available in primary network k, denoted by $a_m^k(t)$, can be derived as [8]

$$a_m^k(t) = \Pr(H_0 \mid W_i^m = \theta_i^m, i \in \mathcal{U}_m^k, \pi_m^k)$$

$$= \left[1 + \left(\varphi_m^k \right)^{u_m^k} \left(\phi_m^k \right)^{|\mathcal{U}_m^k| - u_m^k} \frac{\Pr(H_1 \mid \pi_m^k)}{\Pr(H_0 \mid \pi_m^k)} \right]^{-1}. \tag{7.2}$$

where
  θ_i^m represents a specific sensing result (0 or 1)
  $\mathcal{U}_m^k$ is the subset of users in $\mathcal{U}^k$ (i.e., the set of CR nodes that are located within the coverage of primary network k) that sense channel m
  μ_m^k is the number of users in $\mathcal{U}_m^k$ observing channel m is idle
  π_m^k represents the history of channel m in primary network k
  φ_m^k and ϕ_m^k are defined as

$$\begin{cases} \varphi_m^k = \dfrac{P(W_i^m = 0 \mid H_1)}{P(W_i^m = 0 \mid H_0)} = \dfrac{\delta_m}{1 - \epsilon_m}, & \text{when } \theta_i^m = 0 \\[3mm] \phi_m^k = \dfrac{P(W_i^m = 1 \mid H_1)}{P(W_i^m = 1 \mid H_0)} = \dfrac{1 - \delta_m}{\epsilon_m}, & \text{when } \theta_i^m = 1. \end{cases} \tag{7.3}$$

Based on the Markov chain channel model, we have (7.4), which can be recursively expanded:

$$\begin{cases} \Pr(H_0 \mid \pi_m^k) = \lambda_m^k a_m^k(t-1) + \mu_m^k \left[1 - a_m^k(t-1) \right] \\ \Pr(H_1 \mid \pi_m^k) = 1 - \Pr(H_0 \mid \pi_m^k). \end{cases} \tag{7.4}$$

7.3.5 Video Performance Measure

Both FGS and MGS videos are highly suited for dynamic CR networks. With FGS or MGS coding, each video l is encoded into one base layer with rate R_l^b and one enhancement layer with rate R_l^e. The total bit rate for video l is $R_l = R_l^b + R_l^e$.

We consider peak-signal-noise-ratio (PSNR) (in dB) of reconstructed videos. As in prior work [4,6], the average PSNR of video l, denoted as Q_l, can be estimated as follows:

$$Q_l(R_l) = Q_l^b + \beta_l(R_l - R_l^b) = Q_l^0 + \beta_l R_l, \qquad (7.5)$$

where

Q_l^b is the resulting PSNR when the base layer is decoded alone

β_l is a constant depending on the video sequence and codec setting, and $Q_l^0 = Q_l^b - \beta_l R_l^b$

We verified the model (7.5) with several test video sequences using the MPEG-4 FGS codec and the H.264/SVC MGS codec and found it highly accurate.

Due to the real-time nature, we assume that each *group of pictures* (GoPs) must be delivered during the next GoP window, which consists of N_G time slots. Beyond that, overdue data from the current GoP will be useless and will be discarded. In infrastructure-based network, G video stream are multicast to G groups of CR users, so we use g to denote the group index instead of video session index l.

7.4 Scalable Video over Infrastructure-Based CR Networks

In this section, we examine the problem of video over infrastructure-based CR networks. We consider cross-layer design factors such as scalable video coding, spectrum sensing, opportunistic spectrum access, primary user protection, scheduling, error control, and modulation. We propose efficient optimization and scheduling algorithms for highly competitive solutions and prove the complexity and optimality bound of the proposed greedy algorithm.

7.4.1 Spectrum Access

At the beginning of each time slot t, the CR BS senses the M channels and compute $a_n(t)$ for each channel n. Based on spectrum sensing results, the BS determines which channels to access for video streaming. We adopt an opportunistic spectrum access approach, aiming to exploit unused spectrum while probabilistically bounding the interference to primary users.

Let $\gamma_n \in (0,1)$ be the *maximum allowed collision probability* with primary users on channel n, and $p_n^{tr}(t)$ the *transmission probability* on channel n for the BS in time slot t. The probability of collision caused by the BS should be kept below γ_n, i.e., $p_n^{tr}(t)\left[1-a_n(t)\right]\leq\gamma_n$. In addition to primary user protection, another important objective is to exploit unused spectrum as much as possible. The transmission probability can be determined by jointly considering both objectives, as

$$p_n^{tr}(t) = \begin{cases} \min\left\{1, \dfrac{\gamma_n}{1-a_n(t)}\right\}, & \text{if } 0 \leq a_n(t) < 1 \\ 1, & \text{if } a_n(t) = 1. \end{cases} \qquad (7.6)$$

If $p_n^{tr}(t)=1$, channel n will be accessed deterministically. If $p_n^{tr}(t)=\gamma_n/\left[1-a_n(t)\right]<1$, channel n will be accessed opportunistically with probability $p_n^{tr}(t)$.

7.4.2 Modulation-Coding Schemes

At the PHY layer, we consider various modulation and channel coding combination schemes. Without loss of generality, we assume several choices of modulation schemes, such as QPSK, 16-QAM, and 64-QAM, combined with several choices of forward error correction (FEC) schemes, e.g., with rates 1/2, 2/3, and 3/4. We consider M unique combinations of modulation and FEC schemes, termed *modulation-coding* schemes, in this chapter.

Under the same channel condition, different MC schemes will achieve different data rates and symbol error rates. Adaptive modulation and channel coding allow us to exploit user channel variations to maximize video data rate under a given residual bit error rate constraint. When a user has a good channel, it should adopt an MC scheme that can support a higher data rate. Conversely, it should adopt a low-rate MC scheme when the channel condition is poor. Let $\{MC_m\}_{m=1,...,M}$ be the list of available MC schemes indexed according to their data rates in the increasing order. We assume slow fading channels with coherence time larger than a time slot. Each CR user measures its own channel and feedbacks measurements to the BS when its channel quality changes. At the beginning of a time slot, the BS is able to collect the number $n_{g,m}$ of users in each multicast group g who can successfully decode MC_m signals for $m=1,2,...,M$

Since the base layer carries the most important data, the most reliable MC scheme $MC_{b(g)}$ should be used, where $b(g)=\min_i\{i:\ n_{g,i}=N_g\}$, for all g. Without loss of generality, we assume that the base layer is always transmitted using MC_1. If a user's channel is so poor that it cannot decode the MC_1 signal, we consider it disconnected from the CR network. We further divide the enhancement layer into M sub-layers, where sub-layer m has rate $R_{g,m}^e$ and uses MC_m. Assuming that MC_m can carry $b_{g,m}$ bits of video g in one tile, we denote the number of tiles for sub-layer m of video g as $l_{g,m} \geq 0$. We have

$$R_g^e = \sum_{m=1}^{M} R_{g,m}^e = \sum_{m=1}^{M} b_{g,m} l_{g,m}.$$

7.4.3 Proportional Fair Resource Allocation

Since we consider video quality in this chapter, we define the utility for user i in group g as $U_{g,i} = \log Q_{g,i} = \log\left(Q_g^b + \beta_g R_g^e(i)\right)$, where $R_g^e(i)$ is the received enhancement layer rate of user i in group g.

The total utility for group g is $U_g = \sum_{i=1}^{N_g} U_{g,i}$. Intuitively, a lower layer should use a lower (i.e., more reliable) MC scheme. This is because if a lower layer is lost, a higher layer cannot be used at the decoder even if it is correctly received. Considering the user classification based on their MC schemes, we can rewrite U_g as follows [33]:

$$U_g = \sum_{k=1}^{M} (n_{g,k} - n_{g,k+1}) \log\left(Q_g^b + \beta_g \sum_{m=1}^{k} R_{g,m}^e \right), \tag{7.7}$$

where $n_{g,M+1} = 0$. The utility function of the entire CR video multicast system is $U = \sum_{g=1}^{G} U_g$. Maximizing U will achieve *proportional fairness* among the video sessions [38].

7.4.4 Enhancement Layer Partitioning and Tile Allocation

As a first step, we need to determine the effective rate for each enhancement layer $R_g^e \leq \bar{R}_g^e$. We also need to determine the optimal partition of each enhancement layer. Recall that the base layers are transmitted using MC_1 first in each GoP window. The *remaining* available tiles can then be allocated to the enhancement layers. We assume that the number of tiles used for the enhancement layers in a GoP window, T_e, is known at the beginning of the GoP window. For example, we can estimate T_e by computing the total average "idle" intervals of all the N channels based on the channel model, decreased by the number of tiles used for the base layers (i.e., $R_g^b / b_{g,1}$). We then split the enhancement layer of each video g into M sub-layers, each occupying $l_{g,m}$ tiles when coded with MC_m, $m = 1, 2, \ldots, M$.

Letting $\vec{l} = [l_{1,1}, l_{1,2}, \ldots, l_{1,M}, l_{2,1}, \ldots, l_{G,M}]$ denote the *tile allocation vector*, we formulate an optimization problem OPT-Part as follows:

$$\text{Maximize:} U(\vec{l}) = \sum_{g=1}^{G} \sum_{k=1}^{M} (n_{g,k} - n_{g,k+1}) \times \log\left[Q_g^b + \beta_g \sum_{m=1}^{k} b_{g,m} l_{g,m} \right] \tag{7.8}$$

$$\text{Subject to:} \sum_{g=1}^{G} \sum_{m=1}^{M} l_{g,m} \leq T_e \tag{7.9}$$

$$\sum_{m=1}^{M} b_{g,m} l_{g,m} \leq \bar{R}_g^e, \quad g \in [1,\ldots,M] \tag{7.10}$$

$$l_{g,m} \geq 0, m \in [1,\ldots,M], \quad g \in [1,\ldots,G]. \tag{7.11}$$

OPT-Part is solved at the beginning of each GoP window to determine the optimal partition of the enhancement layer for channel coding and modulation. The objective is to maximize the overall system utility by choosing optimal values for the $l_{g,m}$'s for given amount of spectrum opportunities. We can derive the effective video rates as $R_g^e = \sum_{m=1}^{M} b_{g,m} l_{g,m}$. The formulated problem is an MINLP problem, which is generally NP-hard [33]. In the following, we present two algorithms for computing near-optimal solutions to problem OPT-Part: (a) a *sequential fixing* (SF) algorithm based on a linear relaxation of (7.8) and (b) a *greedy algorithm* GRD1 with proven optimality gap.

7.4.4.1 A Sequential Fixing Algorithm

With this algorithm, the original MINLP is first linearized to obtain a linear programming (LP) relaxation. Then we iteratively solve the LP, while fixing one integer variable in every iteration [18,39]. We use the *reformulation–linearization technique* (RLT) to obtain the LP relaxation [40]. RLT is a technique that can be used to produce LP relaxations for a nonlinear, nonconvex polynomial programming problem. This relaxation will provide a tight upper bound for a maximization problem. Specifically, we linearize the logarithm function in (7.8) over some suitable, tightly bounded interval using a polyhedral outer approximation comprising a convex envelope in concert with several tangential supports. We further relax the integer constraints, i.e., allowing the $l_{g,m}$'s to take fractional values. Then we obtain an upper-bounding LP relaxation that can be solved in polynomial time. Due to lack of space, we refer interested readers to Ref. [40] for a detailed description of the technique.

We next solve the LP relaxation iteratively. During each iteration, we find the $l_{\hat{g},\hat{m}}$, which has the minimum value for $\left(\lceil l_{\hat{g},\hat{m}} \rceil - l_{\hat{g},\hat{m}}\right)$ or $\left(l_{\hat{g},\hat{m}} - \lfloor l_{\hat{g},\hat{m}} \rfloor\right)$ among all fractional $l_{g,m}$'s, and round it up or down to the nearest integer. We next reformulate and solve a new LP with $l_{\hat{g},\hat{m}}$ fixed. This procedure repeats until all the $l_{g,m}$'s are fixed. The complete SF algorithm is given in Table 7.1. The complexity of SF depends on the specific LP algorithm (e.g., the *simplex method* can be used with a polynomial-time average-case complexity) [41].

TABLE 7.1

The Sequential Fixing Algorithm

1:	Use RLT to linearize the original problem;
2:	Solved the LP relaxation;
3:	Suppose $l_{\hat{g},\hat{m}}$ is the integer variable with the minimum $\left(\left\lceil l_{\hat{g},\hat{m}} \right\rceil - l_{\hat{g},\hat{m}}\right)$ or $\left(l_{\hat{g},\hat{m}} - \left\lfloor l_{\hat{g},\hat{m}} \right\rfloor\right)$ value among all $l_{g,m}$ variables that remain to be fixed, round it up or down to the nearest integer;
4:	**if** all $l_{g,m}$'s are fixed, got to Step 6;
5:	**else** reformulate a new relaxed LP with the newly fixed $l_{g,m}$ variables, and go to Step 2;
6:	Output all fixed $l_{g,m}$ variables and $R_g^e = \sum_{m=1}^{M} b_{g,m} l_{g,m}$;

Source: Hu, D., Mao, S., Hou, Y., and Reed, J., Fine grained scalability video multicast in cognitive radio networks, *IEEE J. Sel. Areas Commun.*, 28(3), 334–344, Copyright 2010 IEEE. With permission.

7.4.4.2 A Greedy Algorithm

Although SF can compute a near-optimal solution in polynomial time, it does not provide any guarantee on the optimality of the solution. In the following, we describe a greedy algorithm, termed GRD1, which exploits the inherent priority structure of layered video and MC schemes and has a proven optimality bound.

The complete greedy algorithm is given in Table 7.2, where $R = \sum_{g=1}^{G} \bar{R}_g^e$ is the total rate of all the enhancement layers and $\vec{e}_i$ is a *unit vector* with "1" at the ith location and "0" at all other locations. In GRD1, all the $l_{g,m}$'s are initially set to 0. During each iteration, one tile is allocated to the $\hat{m}$th sub-layer of video $\hat{g}$. In step 4, $l_{\hat{m},\hat{g}}$ is chosen to be the one that achieves the largest increase in terms of the "normalized" utility (i.e., $[U(\vec{l} + \vec{e}_{g,m}) - U(\vec{l})]/[b_{g,m} + R/T_e]$) if it is assigned with an additional tile. Lines 6, 7, and 8 check if the assigned rate exceeds the maximum rate $\bar{R}_g^e$. GRD1 terminates when either all the available tiles are used or when all the video data are allocated with tiles. In the latter case, all the videos are transmitted at full rates. We have the following theorem for GRD1. The proof can be found in Ref. [6].

Theorem 7.1 The greedy algorithm GRD1 shown in Table 7.2 has complexity $O(MGT_e)$. It guarantees a solution that is within a factor of $(1 - e^{-1/2})$ of the global optimal solution.

7.4.4.3 A Refined Greedy Algorithm

GRD1 computes $l_{g,m}$'s based on an estimate of network status $\vec{s}(t)$ in the next T_{GoP} time slots. Due to channel dynamics, the computed $l_{g,m}$'s may not be

TABLE 7.2

Greedy Algorithm (GRD1)

1:	Initialize $l_{g,m}=0$ for all g and m;
2:	Initialize $A=\{1,2,\ldots,G\}$;
3:	**while** $\left(\sum_{g=1}^{G}\sum_{m=1}^{M}l_{g,m}\leq T_e \text{ and } A \text{ is not empty}\right)$
4:	Find $l_{\hat{g},\hat{m}}$ that can be increased by one:
	$\vec{e}_{\hat{g},\hat{m}}=\arg\max_{g\in A,m\in[1,\ldots,M]}\left\{\dfrac{U(\vec{l}+\vec{e}_{g,m})-U(\vec{l})}{b_{g,m}+R/T_e}\right\}$;
5:	$\vec{l}=\vec{l}+\vec{e}_{\hat{g},\hat{m}}$;
6:	**if** $\left(\sum_{m}b_{\hat{g},m}l_{\hat{g},m}>\bar{R}_{\hat{g}}^{e}\right)$
7:	$\vec{l}=\vec{l}-\vec{e}_{\hat{g},\hat{m}}$;
8:	Delete $\hat{g}$ from A;
9:	**end if**
10:	**end while**

Source: Hu, D., Mao, S., Hou, Y., and Reed, J., Fine grained scalability video multicast in cognitive radio networks, *IEEE J. Sel. Areas Commun.*, 28(3), 334–344, Copyright 2010 IEEE. With permission.

exactly accurate, especially when T_{GoP} is large. We next present a refined greedy algorithm, termed GRD2, which adjusts the $l_{g,m}$'s based on more accurate estimation of the channel status.

GRD2 is executed at the beginning of every time slot. It estimates the number of available tiles $T_e(t)$ in the next T_{est} successive time slots, where $1\leq T_{est}\leq T_{GoP}$ is a design parameter depending on the coherence time of the channels. Such estimates are more accurate than that in GRD1 since they are based on recently received ACKs and recent sensing results. Specifically, we estimate $T_e(t)$ using the belief vector $\vec{a}(t)$ in time slot t. Recall that $a_n(t)$'s are computed based on the channel model, feedback, sensing results, and sensing errors. For the next time slot, $a_n(t+1)$ can be estimated as $\hat{a}_n(t+1)=\lambda_n a_n(t)+\mu_n[1-a_n(t)]=(\lambda_n-\mu_n)a_n(t)+\mu_n$. Recursively, we can derive $\hat{a}_n(t+\tau)$ for the next τ time slots.

$$\hat{a}_n(t+\tau)=(\lambda_n-\mu_n)^{\tau}a_n(t)+\mu_n\frac{1-(\lambda_n-\mu_n)^{\tau}}{1-(\lambda_n-\mu_n)}. \tag{7.12}$$

At the beginning section of a GoP window, all the base layers will be firstly transmitted. We start the estimation after all the base layers have been successfully received (possibly with retransmissions). The number of available tiles in the following T_{est} time slots can be estimated as $T_e(t) = \sum_{n=1}^{N} \sum_{\tau=0}^{\tau_{min}} \hat{a}_n(t+\tau)$, where $\hat{a}_n(t+0) = a_n(t)$ and $t_{min} = \min\{T_{est} - 1, T_{GoP} - (t \bmod T_{GoP})\}$. $T_e(t)$ may not be an integer, but it does not affect the outcome of GRD2.

We then adjust the $l_{g,m}$'s based on $T_e(t)$ and $N_{ack}(t)$, the number of ACKs received in time slot t. If $T_e(t) + N_{ack}(t-1) > T_e(t-1) + N_{ack}(t-2)$, there are more tiles that can be allocated, and we can increase some of the $l_{g,m}$'s. On the other hand, if $T_e(t) + N_{ack}(t-1) < T_e(t-1) + N_{ack}(t-2)$, we have to reduce some of the $l_{g,m}$'s. Due to layered videos, when we increase the number of allocated tiles, we need to consider only $l_{g,m}$ for $m = m', m'+1,\ldots,M$, where $MC_{m'}$ is the highest MC scheme used in the previous time slot. Similarly, when we reduce the number of allocated tiles, we need to consider only $l_{g,m}$ for $m = m', m'+1,\ldots,M$.

The refined greedy algorithm is given in Table 7.3. For time slot t, the complexity of GRD2 is $O(MGK)$, where $K = |N_{ack}(t-1) - N_{ack}(t-2) + T_e(t) - T_e(t-1)|$. Since $K \ll T_e$, the complexity of GRD2 is much lower than GRD1, suitable for execution in each time slot.

7.4.5 Tile Scheduling in a Time Slot

Once the enhancement layer is partitioned, coded with channel codes, and modulated, in each time slot t, we need to schedule the remaining tiles for transmission on the N channels. We define Inc (g,m,i) to be the increase in the group utility function $U(g)$ after the ith tile in the sub-layer using MC_m is successfully decoded. It can be shown that

$$\text{Inc}(g,m,i) = \sum_{k=m}^{M} (n_{g,k} - n_{g,k+1}) \times \log\left[1 + \frac{\beta_g b_{g,m}}{Q_g^b + \beta_g \sum_{u=1}^{m-1} b_{g,u} I_{g,u} + (i-1)\beta_g b_{g,m}}\right].$$

Inc (g,m,i) can be interpreted as the *reward* if the tile is successfully received.

Letting $c_n(t)$ be the probability that the tile is successfully received, we have $c_n(t) = p_n^{tr}(t) a_n(t)$. Our objective of tile scheduling is to maximize the expected reward, i.e.,

$$\text{Maximize: } E[\text{Reward}(\vec{\xi})] = \sum_{n=1}^{N} c_n(t) \cdot \text{Inc}(\xi_n), \tag{7.13}$$

where
$\vec{\xi} = \{\xi_n\}_{n=1,\ldots,N}$
ξ_n is the tile allocation for channel n, i.e., representing the three-tuple $\{g, m, i\}$

TABLE 7.3

Refined Greedy Algorithm (GRD2) for Each Time Slot

1:    Initialize $l_{g,m} = 0$ for all g and m;

2:    Initialize $A = \{1,2, \ldots, G\}$;

3:    Initialize $N_{ack}(0) = 0$;

4:    Estimate $T_e(1)$ based on the Markov Chain channel model;

5:    Use GRD1 to find all $l_{g,m}$'s based on $T_e(1)$;

6:    **while** $t = 2$ to T_{GoP}

7:    Estimate $T_e(t)$;

8:    **if** $[T_e(t) + N_{ack}(t-1) < T_e(t-1) + N_{ack}(t-2)]$

9:    **while** $\left[\displaystyle\sum_{g=1}^{G} \sum_{m=1}^{M} l_{g,m} > T_e(t) + N_{ack}(t-2) \right]$

10:    Find $l_{\hat{g},\hat{m}}$ that can be reduced by 1: $\vec{e}_{\hat{g},\hat{m}} = \arg\min_{\forall g, m \in \{m', \ldots, M\}} \left\{ \dfrac{U(\vec{l}) - U(\vec{l} - \vec{e}_{g,m})}{b_{g,m} + R/T_e} \right\}$;

11:    $\vec{l} = \vec{l} - \vec{e}_{\hat{g},\hat{m}}$;

12:    **if** $(\hat{g} \notin A)$

13:    Add $\hat{g}$ to A;

14:    **end if**

15:    **end while**

16:    **end if**

17:    **if** $[T_e(t) + N_{ack}(t-1) > T_e(t-1) + N_{ack}(t-2)]$

18:    **while** $\left[\displaystyle\sum_{g=1}^{G} \sum_{m=1}^{M} l_{g,m} \le T_e(t) + N_{ack}(t-1) \text{ and } A \text{ is not empty} \right]$

19:    Find $l_{\hat{g},\hat{m}}$ that can be increased by 1

$$\vec{e}_{\hat{g},\hat{m}} = \arg\max_{g \in A, m \in \{m', \ldots, M\}} \left\{ \dfrac{U(\vec{l} + \vec{e}_{g,m}) - U(\vec{l})}{b_{g,m} + R/T_e} \right\};$$

20:    $\vec{l} = \vec{l} + \vec{e}_{\hat{g},\hat{m}}$;

21:    **if** $\left(\displaystyle\sum_{m} b_{\hat{g},m} l_{\hat{g},m} > \bar{R}_{\hat{g}}^e \right)$

22:    $\vec{l} = \vec{l} - \vec{e}_{\hat{g},\hat{m}}$;

23:    Delete $\hat{g}$ from A;

24:    **end if**

25:    **end while**

26:    **end if**

27:    Update $N_{ack}(t-1)$;

28:    **end while**

TABLE 7.4

Algorithm for Tile Scheduling in a Time Slot

1:	Initialize m_g to the lowest MC that has not been ACKed for all g;
2:	Initialize i_g to the first packet that has not been ACKed for all g;
3:	Sort $\{c_n(t)\}$ in decreasing order. Let the sorted channel list be indexed by j;
4:	**while** ($j = 1$ to N)
5:	Find the group having the maximum increase in $U(g)$: $\hat{g} = \arg \max_{v_g} \mathrm{Inc}(g, m_g, i_g)$;
6:	Allocate the tile on channel j to group $\hat{g}$;
7:	Update m_g and i_g;
8:	**end while**

Source: Hu, D., Mao, S., Hou, Y., and Reed, J., Fine grained scalability video multicast in cognitive radio networks, *IEEE J. Sel. Areas Commun.*, 28(3), 334–344, Copyright 2010 IEEE. With permission.

The TSA algorithm is shown in Table 7.4, which solves the aforementioned optimization problem. The complexity of TSA is $O(N \log N)$. We have the following theorem for TSA.

Theorem 7.2 $E[Reward]$ defined in (7.13) is maximized if Inc (ξ_i) > Inc (ξ_j) when $c_i(t) > c_j(t)$ for all i and j.

7.5 Scalable Video over Multi-Hop CR Networks

In this section, we examine the problem of video over multi-hop CR networks. We model streaming of concurrent videos as an MINLP problem, aiming to maximize the overall received video quality and fairness among the video sessions, while bound the collision rate with primary users under spectrum sensing errors. We solve the MINLP problem using a centralized SF algorithm and derive upper and lower bounds for the objective value. We then apply dual decomposition to develop a distributed algorithm and prove its optimality and convergence conditions.

7.5.1 Spectrum Access

During the transmission phase of a time slot, a CR user determines which channel(s) to access for transmission of video data based on spectrum sensing results. Let κ_m^k be a threshold for spectrum access: channel m is considered idle if the estimate a_m^k is greater than the threshold, and busy otherwise. The availability of channel m in primary network k, denoted as A_m^k, is

$$A_m^k = \begin{cases} 0, & a_m^k \geq \kappa_m^k \\ 1, & \text{otherwise.} \end{cases} \tag{7.14}$$

For each channel m, we can calculate the probability of collision with primary users as follows:

$$\Pr(A_m^k = 0 \mid H_1) = \sum_{i \in \psi_m^k} \binom{|\mathcal{U}_m^k|}{i} (1 - \delta_m)^{|\mathcal{U}_m^k| - i} (\delta_m)^i, \tag{7.15}$$

where set ψ_m^k is defined as

$$\psi_m^k = \left\{ i \left| \left[1 + \varphi_m^i \phi_m^{|\mathcal{U}_m^k| - i} \frac{\Pr(H_1 \mid \pi_m^k)}{\Pr(H_0 \mid \pi_m^k)} \right]^{-1} \geq \kappa_m^k \right. \right\}. \tag{7.16}$$

For nonintrusive spectrum access, the collision probability should be bounded with a prescribed threshold γ_m^k. A higher spectrum access threshold κ_m^k will reduce the potential interference with primary users but increase the chance of wasting transmission opportunities. For a given collision tolerance γ_m^k, we can solve $\Pr(A_m^k = 0 \mid H_1) = \gamma_m^k$ for κ_m^k. The objective is to maximize CR users' spectrum access without exceeding the maximum collision probability with primary users.

Let $\Omega_{i,j}$ be the set of available channels at link $\{i, j\}$. Assuming $i \in \mathcal{U}^k$ and $j \in \mathcal{U}^{k'}$, we have

$$\Omega_{i,j} = \left\{ m \middle| A_m^k = 0 \text{ and } A_m^{k'} = 0 \right\}. \tag{7.17}$$

7.5.2 Link and Path Statistics

Due to the amplify-and-forward approach for video data transmission, there is no queuing delay at intermediate nodes. Assume that each link has a fixed delay $\omega_{i,j}$ (i.e., processing and propagation delays). Let $\mathcal{P}_l^A$ be the set of all possible paths from z_l to d_l. For a given delay requirement T_{th}, the set of feasible paths $\mathcal{P}_l$ for video session l can be determined as follows:

$$\mathcal{P}_l = \left\{ \mathcal{P} \middle| \sum_{\{i,j\} \in \mathcal{P}} \omega_{i,j} \leq T_{th}, \mathcal{P} \in \mathcal{P}_l^A \right\}. \tag{7.18}$$

Let p_{ij}^m be the packet loss rate on channel m at link $\{i, j\}$. A packet is successfully delivered over link $\{i, j\}$ if there is no loss on all the channels that were used for transmitting the packet. The link loss probability $p_{i,j}$ can be derived as thus:

$$p_{i,j} = 1 - \prod_{m \in \mathcal{M}} (1 - p_{i,j}^m)^{I_m}, \tag{7.19}$$

where

$\mathcal{M}$ is a set of licensed channels

I_m is an indicator: $I_m = 1$ if channel m is used for the transmission, and $I_m = 0$ otherwise

Assuming independent link losses, the end-to-end loss probability for path $\mathcal{P}_l^h \in \mathcal{P}_l$ can be estimated as follows:

$$p_l^h = 1 - \prod_{\{i,j\} \in \mathcal{P}_l^h} (1 - p_{i,j}). \tag{7.20}$$

7.5.3 Problem Statement

We also aim to achieve fairness among the concurrent video sessions. It has been shown that *proportional fairness* can be achieved by maximizing the sum of logarithms of video PSNRs (i.e., utilities). Therefore, our objective is to maximize the overall system utility, i.e.,

$$\text{Maximize: } \sum_l U_l(R_l) = \sum_l \log(Q_l(R_l)). \tag{7.21}$$

The problem of video over multi-hop CR networks consists of path selection for each video session and channel scheduling for each CR node along the chosen paths. We define two sets of index variables. For channel scheduling, we have

$$x_{i,j,m}^{l,h,r} = \begin{cases} 1, & \text{at link } \{i, j\}, \text{ if channel } m \text{ is assigned to tunnel } r \text{ in path } \mathcal{P}_l^h \\ 0, & \text{otherwise.} \end{cases} \tag{7.22}$$

For path selection, we have

$$y_l^h = \begin{cases} 1, & \text{if video session } l \text{ selects path } \mathcal{P}_l^h \in \mathcal{P}_l \\ 0, & \text{otherwise.} \end{cases} \tag{7.23}$$

Note that the indicators, $x_{i,j,m}^{l,h,r}$ and y_l^h, are not independent. If $y_l^h = 0$ for path $\mathcal{P}_l^h$, all the $x_{i,j,m}^{l,h,r}$'s on that path are 0. If link $\{i, j\}$ is not on path $\mathcal{P}_l^h$, all its $x_{i,j,m}^{l,h,r}$'s are also 0. For link $\{i, j\}$ on path $\mathcal{P}_l^h$, we can choose only those available channels in set $\Omega_{i,j}$ to schedule video transmission. That is, we have $x_{i,j,m}^{l,h,r} \in \{0,1\}$ if $m \infty \Omega_{i,j}$, and $x_{i,j,m}^{l,h,r} = 0$ otherwise. In the rest of the chapter, we use $\mathbf{x}$ and $\mathbf{y}$ to represent the vector forms of $x_{i,j,m}^{l,h,r}$ and y_l^h, respectively.

As discussed, the objective is to maximize the expected utility sum at the end of N_G time slots, as given in (7.21). Since $\log(Q_l(E\ [R_l(0)]))$ is a constant, (7.21) is equivalent to the sum of utility increments of all the time slots, as

$$
\begin{aligned}
& \sum_l \log(Q_l(E[R_l(N_G)])) - \log(Q_l(E[R_l(0)])) \\
& = \sum_t \sum_l \{\log(Q_l(E[R_l(t)])) - \log(Q_l(E[R_l(t-1)]))\}
\end{aligned}
\tag{7.24}
$$

Therefore, (7.21) will be maximized if we maximize the expected utility increment during each time slot, which can be written as follows:

$$
\begin{aligned}
& \sum_l \log(Q_l(E[R_l(t)])) - \log(Q_l(E[R_l(t-1)])) \\
& = \sum_l \sum_{h \in \mathcal{H}} y_l^h \log\left(1 + \rho_l^t \sum_r \sum_m x_{z_l, z_l', m}^{l,h,r}(1 - p_{l,h}^r)\right),
\end{aligned}
$$

where
z_l' is the next hop from z_l on path $\mathcal{P}_l^h$
$p_{l,h}^r$ is the packet loss rate on tunnel r of path $\mathcal{P}_l^h$
$Q_l^{t-1} = Q_l(E[R_l(t-1)])$
$\rho_l^t = \beta_1 L_p / (N_G T_s Q_l^{t-1})$

From (7.19) and (7.20), the end-to-end packet loss rate for tunnel r on path $\mathcal{P}_l^h$ is as follows:

$$
p_{l,h}^r = 1 - \prod_{\{i,j\} \in \mathcal{P}_l^h} \prod_{m \in \mathcal{M}} (1 - p_{i,j}^m)^{x_{i,j,m}^{l,h,r}}.
\tag{7.25}
$$

We assume that each tunnel can include only one channel on each link. When there are multiple channels available at each link along the path, a CR source node can set up multiple tunnels to exploit the additional bandwidth. We then have the following constraint:

$$\sum_m x_{i,j,m}^{l,h,r} \le 1, \quad \forall \{i,j\} \in \mathcal{P}_l^h. \tag{7.26}$$

Considering availability of the channels, we further have

$$\sum_r \sum_m x_{i,j,m}^{l,h,r} \le |\Omega_{i,j}|, \quad \forall \{i,j\} \in \mathcal{P}_l^h, \tag{7.27}$$

where $|\Omega_{i,j}|$ is the number of available channels on link $\{i, j\}$ defined in (7.17).

As discussed, each node is equipped with two transceivers: one for receiving and the other for transmitting video data during the transmission phase. Hence, a channel cannot be used to receive and transmit data simultaneously at a relay node. We have the following for each channel m:

$$\sum_r x_{i,j,m}^{l,h,r} + \sum_r x_{j,k,m}^{l,h,r} \le 1, \quad \forall\, m, l, \forall\, h \in \mathcal{P}_l, \forall\, \{i,j\}, \{j,k\} \in \mathcal{P}_l^h. \tag{7.28}$$

Let n_l^h be the number of tunnels on path $\mathcal{P}_l^h$. For each source z_l and each destination d_l, the number of scheduled channels is equal to n_l^h. We have for each source node

$$\sum_r \sum_m x_{z_l, z_l^r, m}^{l,h,r} = n_l^h y_l^h, \quad \forall\, h \in \mathcal{P}_l, \forall\, l. \tag{7.29}$$

Let d_l^r be the last hop to destination d_l on path $\mathcal{P}_l^h$; we have for each destination node

$$\sum_r \sum_m x_{d_l^r, d_l, m}^{l,h,r} = n_l^h y_l^h, \quad \forall\, h \in \mathcal{P}_l, \forall\, l. \tag{7.30}$$

At a relay node, the number of channels used to receive data is equal to that of channels used to transmit data, due to flow conservation and amplify-and-forward. At relay node j for session l, assume $\{i, j\} \in \mathcal{P}_l^h$ and $\{j, k\} \in \mathcal{P}_l^h$. We have,

$$\sum_r \sum_m x_{i,j,m}^{l,h,r} = \sum_r \sum_m x_{i,j,m}^{l,h,r}, \quad \forall\, h \in \mathcal{P}_l, \forall\, l, \forall\, \{i,j\}, \{j,k\} \in \mathcal{P}_l^h. \tag{7.31}$$

We also consider hardware-related constraints on path selection. We summarize such constraints in the following general form for ease of presentation:

$$\sum_l \sum_{h \in P_l} w_{l,h}^g y_l^h \leq 1, \quad \forall g. \tag{7.32}$$

To simplify exposition, we choose at most one path in P_l for video session l. Such a single path routing constraint can be expressed as $\sum_h y_l^h \leq 1$, which is a special case of (7.32), where $w_{l,h}^1 = 1$ for all h, and $w_{l',h}^g = 0$ for all $g \neq 1, l' \neq l$, and h. We can also have $\sum_h y_l^h \leq \xi$ to allow up to ξ paths for each video session. In order to achieve optimality in the general case of multipath routing, an optimal scheduling algorithm should be designed to dispatch packets to paths with different conditions (e.g., different number of tunnels and delays).

There are also disjointedness constraints for the chosen paths. This is because each CR node is equipped with two transceivers, and both will be used for a video session if it is included in a chosen path. Such disjointedness constraint is also a special case of (7.32) with the following definition for $w_{l',h}^g$ for each CR node g:

$$w_{l,h}^g = \begin{cases} 1, & \text{if node } g \in \text{path } P_l^h \\ 0, & \text{otherwise,} \end{cases} \tag{7.33}$$

Finally, we formulate the problem of multi-hop CR network video streaming (OPT-CRV) as follows:

$$\text{Maximize: } \sum_l \sum_{h \in P_l} y_l^h \log\left(1 + \rho_l^t \sum_r \sum_m x_{zl,zl,m}^{l,h,r}(1 - p_{l,h}^r)\right) \tag{7.34}$$

Subject to: (1.22)–(1.32).

7.5.4 Centralized Algorithm and Performance Bounds

Problem OPT-CRV is in the form of MINLP (without continuous variables), which is NP-hard in general. We first describe a centralized algorithm to derive performance bounds in this section and then present a distributed algorithm based on dual decomposition in the next section.

We first obtain a relaxed *nonlinear programming* (NLP) version of OPT-CRV. The binary variables $x_{i,j,m}^{l,h,r}$ and y_l^h are relaxed to take values in [0,1]. The integer variables n_l^h are treated as nonnegative real numbers. It can be shown that the relaxed problem has a concave object function and the constraints are convex. This relaxed problem can be solved using a constrained nonlinear optimization problem solver. If all the variables are integer in the solution,

TABLE 7.5

Sequential Fixing Algorithm for Problem OPT-CRV

1:   Relax integer variables $x_{i,j,m}^{l,h,r}$, y_l^h, and n_l^h;

2:   Solve the relaxed problem using a constrained NLP solver;

3:   **if** (there is y_l^h not fixed)

4:   Find the largest $y_{l'}^{h'}$, where $[l', h'] = \arg\max\{y_l^h\}$, and fix it to 1;

5:   Fix other y_l^h's according to constraint (7.32);

6:   Go to Step 2;

7:   **end if**

8:   **if** (there is $x_{i,j,m}^{l,h,r}$ not fixed)

9:   Find the largest $x_{i',j',m'}^{l',h',r'}$, where $[i', j', m', l', h', r'] = \arg\max\{x_{i,j,m}^{l,h,r}\}$, and set it to 1;

10: Fix other $x_{i,j,m}^{l,h,r}$'s according to the constraints;

11: **if** (there is other variable that is not fixed)

12: Go to Step 2;

13: **else**

14: Fix n_l^h's based on **x** and **y**;

15: Exit with feasible solution {**x**, **y**, **n**};

16: **end if**

17: **end if**

Source: Hu, D. and Mao, S., Streaming scalable videos over multi-hop cognitive radio networks, *IEEE Trans. Wireless Commun.*, 9(11), 3501–3511, Copyright 2010 IEEE. With permission.

then we have the exact optimal solution. Otherwise, we obtain an infeasible solution, which produces an upper bound for the problem. This is given in lines 1–2 in Table 7.5.

We then develop an *SF algorithm* for solving OPT-CRV. The pseudo-code is given in Table 7.5. SF iteratively solves the relaxed problem, fixing one or more integer variables after each iteration [6,18]. In Table 7.5, lines 3–7 fix the path selection variables y_l^h, and lines 8–16 fix the channel scheduling variables $x_{i,j,m}^{l,h,r}$ and tunnel variables n_l^h. The tunnel variables n_l^h can be computed using (7.29) after $x_{i,j,m}^{l,h,r}$ and y_l^h are solved. When the algorithm terminates, it produces a feasible solution that yields a lower bound for the objective value.

7.5.5 Dual Decomposition and Distributed Algorithm

SF is a centralized algorithm requiring global information. It may not be suitable for multi-hop wireless networks, although the upper and lower bounds provide useful insights on the performance limits. In this section, we develop a distributed algorithm for problem OPT-CRV and analyze its optimality and convergence performance.

7.5.5.1 Decompose Problem OPT-CRV

Since the domains of $x_{i,j,m}^{l,h,r}$ defined in (7.26)–(7.31) for different paths do not intersect with each other, we can decompose problem OPT-CRV into two subproblems. The first subproblem deals with channel scheduling for maximizing the expected utility on a chosen path $\mathcal{P}_l^h$. We have the *channel scheduling* problem (OPT-CS) as follows:

$$H_l^h = \max_x \sum_r \sum_m x_{zl,zl,m}^{l,h,r}(1 - p_{l,h}^r)$$

(7.35)

Subject to: $(1.26)–(1.31), x_{zl,zl,m}^{l,h,r} \in \{0,1\}, \quad$ for all l, h, r, m.

In the second part, optimal paths are selected to maximize the overall objective function. Letting $F_l^h = \log(1 + \rho_l^T H_l^h)$, we have the following *path selection* problem (OPT-PS):

$$\text{Maximize:} \quad f(\mathbf{y}) = \sum_l \sum_h F_l^h y_l^h$$

$$\text{Subject to:} \quad \sum_l \sum_{h \in \mathcal{P}_l} w_{l,h}^g y_l^h \leq 1, \text{ for all } g$$

(7.36)

$$y_l^h \in \{0,1\}, \text{ for all } l, h.$$

7.5.5.2 Channel Scheduling

We have the following result for assigning available channels at a relay node.

Theorem 7.3 Consider three consecutive nodes along a path, denoted as nodes i, j, and k. Idle channels 1 and 2 are available at link $\{i, j\}$ and idle channels 3 and 4 are available at link $\{j, k\}$. Assume the packet loss rates of the four channels satisfy $p_{i,j}^1 > p_{i,j}^2$ and $p_{j,k}^3 > p_{j,k}^4$. To set up two tunnels, assigning channels $\{1, 3\}$ to one tunnel and channels $\{2, 4\}$ to the other tunnel achieves the maximum expectation of successful transmission on path section $\{i, j, k\}$.

According to Theorem 7.3, a greedy approach, which always chooses the channel with the lowest loss rate at each link when setting up tunnels along a path, produces the optimal overall success probability. More specifically, when there is only one tunnel to be set up along a path, the tunnel should consist of the most reliable channels available at each link along the path. When there are multiple tunnels to set up along a path, tunnel 1 should consist of the most reliable channels that are available at each link; tunnel 2 should consist of the second most reliable links available at each link; and so forth.

Define the set of loss rates of the available channels on link $\{i, j\}$ as $\Lambda_{i,j} = \{p^m_{i,j} \mid m \in \Omega_{i,j}\}$. The greedy algorithm is given in Table 7.6, with which each video source node solves problem OPT-CS for each feasible path. Lines 2–3 in Table 7.6 check if there are more channels to assign and the algorithm terminates if no channel is left. In lines 4–10, links with only one available channel are assigned to tunnel r and the neighboring links with the same available channels are removed due to constraint (7.28). In lines 11–17, links with more than two channels are grouped to be assigned later. In lines 18–20, the available channel with the lowest packet loss rate is assigned to tunnel r at each unallocated link, according to Theorem 7.3. To avoid co-channel interference, the same channel on neighboring links is removed as in lines 21–33.

7.5.5.3 Path Selection

To solve problem OPT-PS, we first relax binary variables y^h_l to allow them take real values in $[0,1]$ and obtain the following *relaxed path selection* problem (OPT-rPS):

$$\text{Maximize:} \quad f(\mathbf{y}) = \sum_l \sum_h F^h_l y^h_l$$

$$\text{Subject to:} \quad \sum_l \sum_{h \in P_l} w^g_{l,h} y^h_l \leq 1, \text{ for all } g \tag{7.37}$$

$$0 \leq y^h_l \leq 1, \text{ for all } h,l.$$

We then introduce positive Lagrange multipliers e_g for the path selection constraints in problem OPT-rPS and obtain the corresponding *Lagrangian function*:

$$\mathcal{L}(\mathbf{y}, \mathbf{e}) = \sum_l \sum_h F^h_l y^h_l + \sum_g e_g \left(1 - \sum_l \sum_h w^g_{l,h} y^h_l \right)$$

$$= \sum_l \sum_h \left(F^h_l y^h_l - \sum_g w^g_{l,h} y^h_l e_g \right) + \sum_g e_g \tag{7.38}$$

$$= \sum_l \sum_h \mathcal{L}^h_l(y^h_l, \mathbf{e}) + \sum_g e_g.$$

Problem (7.38) can be decoupled since the domains of y^h_l's do not overlap. Relaxing the coupling constraints, it can be decomposed into two levels. At the lower level, we have the following subproblems, one for each path P^h_l:

TABLE 7.6

Greedy Algorithm for Channel Scheduling

1:	Initialization: tunnel $r = 1$, link $\{i, j\}$'s from z_l to d_l;
2:	**if** $(|\Lambda_{i,j}| == 0)$
3:	Exit;
4:	**else if** $(|\Lambda_{i,j}| == 1)$
5:	Assign the single channel in $\Lambda_{i,j}$, m', to tunnel r;
6:	Check neighboring link $\{k, i\}$;
7:	**if** $(p_{k,i}^{m'} \in \Lambda_{k,i})$
8:	Remove $p_{k,i}^{m'}$ from $\Lambda_{k,i}$, $i \leftarrow k$, $j \leftarrow i$ and go to Step 2;
9:	**else**
10:	Go to Step 13;
11:	**end if**
12:	**else**
13:	Put $\Lambda_{i,j}$ in set Λ_l^h;
14:	**if** (node j is not destination d_l)
15:	$i \leftarrow j$, $j \leftarrow v$;
16:	Go to Step 2;
17:	**end if**
18:	**end if**
19:	**while** (Λ_l^h is not empty)
20:	Find the maximum value $p_{i',j'}^{m'}$ in set $\Lambda_l^h \{i', j', m'\} = \arg\min\{p_{i,j}^m\}$;
21:	Assign channel m' to tunnel r;
22:	Remove set $\Lambda_{i',j'}$ from set Λ_l^h;
23:	Check neighboring link $\{k, i\}$ and $\{j, v\}$;
24:	**if** $(p_{k,i}^{m'} \in \Lambda_{k,i}$ and $\Lambda_{k,i} \in \Lambda_l^h)$
25:	Remove $p_{k,i}^{m'}$ from $\Lambda_{k,i}$;
26:	**if** ($\Lambda_{k,i}$ is empty)
27:	Exit;
28:	**end if**
29:	**end if**
30:	**if** $(p_{j,v}^{m'} \in \Lambda_{j,v}$ and $\Lambda_{j,v} \in \Lambda_l^h)$
31:	Remove $p_{j,v}^{m'}$ from $\Lambda_{j,v}$;
32:	**if** ($\Lambda_{j,v}$ is empty)
33:	Exit;
34:	**end if**
35:	**end if**
36:	**end while**
37:	Compute the next tunnel: $r \leftarrow r + 1$ and go to Step 2;

Source: Hu, D. and Mao, S., Streaming scalable videos over multi-hop cognitive radio networks, *IEEE Trans. Wireless Commun.*, 9(11), 3501–3511, Copyright 2010 IEEE. With permission.

$$\max_{0 \le y_l^h \le 1} \mathcal{L}_l^h(y_l^h, \mathbf{e}) = F_l^h y_l^h - \sum_g w_{l,h}^g y_l^h e_g. \tag{7.39}$$

At the higher level, by updating the dual variables e_g, we can solve the *relaxed dual problem*:

$$\min_{\mathbf{e} \ge 0} \ q(\mathbf{e}) = \sum_l \sum_h \mathcal{L}_l^h \left(\left(y_l^h\right)^*, \mathbf{e} \right) + \sum_g e_g, \tag{7.40}$$

where $\left(y_l^h\right)^*$ is the optimal solution to (7.39). Since the solution to (7.39) is unique, the relaxed dual problem (7.40) can be solved using the following *subgradient method* that iteratively updates the Lagrange multipliers [21]:

$$e_g(\tau+1) = \left[e_g(\tau) - \alpha(\tau) \left(1 - \sum_l \sum_h w_{l,h}^g y_l^h \right) \right]^+, \tag{7.41}$$

where
τ is the iteration index
$\alpha(\tau)$ is a sufficiently small positive step size
$[x]^+$ denotes $max\{x, 0\}$

The pseudo-code for the distributed algorithm is given in Table 7.7.

TABLE 7.7

Distribution Algorithm for Path Selection

1:	Initialization: set $\tau = 0$, $e_g(0) > 0$ and step size $s \in [0,1]$;
2:	Each source locally solves the lower level problem in (7.39);
	if $(F_l^h - \sum_g d_{l,h}^g e_g(\tau)) > 0)$ $y_l^h = y_l^h + s$, $y_l^h = \min\{y_l^h, 1\}$;
	else $y_l^h = y_l^h - s$, $y_l^h = \max\{y_l^h, 0\}$;
3:	Broadcast solution $y_l^h(\mathbf{e}(\tau))$;
4:	Each source updates $\mathbf{e}$ according to (7.41) and broadcasts $\mathbf{e}(\tau+1)$ through the common control channel;
5:	$\tau \leftarrow \tau+1$ and go to Step 2 until termination criterion is satisfied;

Source: Hu, D. and Mao, S., Streaming scalable videos over multi-hop cognitive radio networks, *IEEE Trans. Wireless Commun.*, 9(11), 3501–3511, Copyright 2010 IEEE. With permission.

7.5.5.4 Optimality and Convergence Analysis

The distributed algorithm in Table 7.7 iteratively updates the dual variables until they converge to stable values. In this section, we first show that the solution obtained by the distributed algorithm is also optimal for the original path selection problem OPT-PS. We then derive the convergence condition for the distributed algorithm. The proofs can be found in Ref. [6].

Lemma 7.1 The optimal solution for the relaxed primal problem OPT-rPS in (7.37) is also feasible and optimal for the original problem OPT-PS in (7.36).

Lemma 7.2 If the relaxed primal problem OPT-rPS in (7.37) has an optimal solution, then the relaxed dual problem (7.40) also has an optimal solution, and the corresponding optimal values of the two problems are identical.

We have Theorem 7.4 on the optimality of the path selection solution, which follows naturally from Lemmas 7.1 and 7.2.

Theorem 7.4 The optimal solution to the relaxed dual problems (7.39) and (7.40) is also feasible and optimal to the original path selection problem OPT-PS given in (7.36).

As discussed, the relaxed dual problem (7.40) can be solved using the *subgradient method* that iteratively updates the Lagrange multipliers. We have the following theorem on the convergence of the distributed algorithm given in Table 7.7.

Theorem 7.5 Let $\mathbf{e}^*$ be the optimal solution. The distributed algorithm in Table 7.7 converges if the step sizes $\alpha(\tau)$ in (7.41) satisfy the following condition:

$$0 < \alpha(\tau) < \frac{2\left[q(\mathbf{e}(\tau)) - q(\mathbf{e}^*)\right]}{||G(\tau)||^2}, \quad \text{for all } \tau, \tag{7.42}$$

where $G(\tau)$ is the gradient of $q(\mathbf{e}(\tau))$.

Since the optimal solution $\mathbf{c}^*$ is not known a priori, we use the following approximation in the algorithm: $\alpha(\tau) = [q(\mathbf{e}(\tau)) - \bar{q}(\tau)]/||G(\tau)||^2$, where $\bar{q}(\tau)$ is the current estimate for $q(\mathbf{e}^*)$. We choose the mean of the objective values of the relaxed primal and dual problems for $\bar{q}(\tau)$.

7.6 Simulation Results

We implement the proposed algorithms with a combination of C and MATLAB®, and evaluate their performance with simulations. In the following, we first present the simulation results on scalable video over infrastructure-based CR networks in Section 7.6.1 and then present the simulation study on scalable video over multi-hop CR networks in Section 7.6.2.

7.6.1 Scalable Video over Infrastructure-Based CR Networks

With our customized simulator, the LPs are solved using the MATLAB Optimization Toolbox and the remaining parts are written in C. For the results reported later, we have $N = 12$ channels (unless otherwise specified). The channel parameters λ_n and μ_n are set between (0, 1). The maximum allowed collision probability γ_n is set to 0.2 for all the channels unless otherwise specified.

The CR BS multicasts three common intermediate format (CIF, 352×288) video sequences to three multicast groups, i.e., *Bus* to group 1, *Foreman* to group 2, and *Mother and Daughter* to group 3. The $n_{1,m}$'s are {42, 40, 36, 30, 22, 12} (i.e., 42 users can decode MC_1 signal, 40 users can decode MC_2 signal, and so forth); the $n_{2,m}$'s are {51, 46, 40, 32, 23, 12} and the $n_{3,m}$'s are {49, 44, 40, 32, 24, 13}. The bits carried in one tile using the MC schemes are 1, 1.5, 2, 3, 5.3, and 6 kb/s, respectively. We choose $T_{GoP} = 150$ and $T_{est} = 10$, sensing interval $W = 3$, false alarm probability $\varepsilon_n = 0.3$, and miss-detection probability $\delta_n = 0.25$ for all n, unless otherwise specified.

In every simulation, we compare three schemes: (a) a simple heuristic scheme that equally allocates tiles to each group (equal allocation); (b) a scheme based on SF, and (c) a scheme based on the greedy algorithm GRD2 (greedy algorithm). These schemes have increasing complexity in the order of equal allocation, greedy algorithm, and SF. They differ on how to solve problem OPT-Part, while the same tile scheduling algorithm and opportunistic spectrum access scheme are used in all the schemes. The C/MATLAB code is executed in a Dell Precision Workstation 390 with an Intel Core 2 Duo E6300 CPU working at 1.86 GHz and a 1066 MB memory. For number of channels ranging from $N = 3$ to $N = 15$, the execution times of equal allocation and greedy algorithm are about a few milliseconds, while SF takes about 2 s.

In Figure 7.1, we plot the average PSNR among all users in each multicast group. We find that the greedy algorithm achieves the best performance, with up to 4.2 dB improvements over equal allocation and up to 0.6 dB improvements over SF. We find that SF achieves a lower PSNR than equal allocation for group 3, but higher PSNRs for groups 1 and 2. This is because equal allocation does not consider channel conditions and fairness. It achieves better performance for group 3 at the cost of much lower PSNRs for groups 1 and 2.

Then, we demonstrate the impact of user channel variations (i.e., due to mobility) in Figure 7.2. We choose a tagged user in group 1 and assume that its channel condition changes every 20 GoPs. The highest MC scheme that the tagged user can decode is changed according to the following sequence: MC3, MC5, MC4, MC6, MC5, and MC3. All other parameters remain the same as in the previous experiments. In Figure 7.2, we plot the average PSNRs

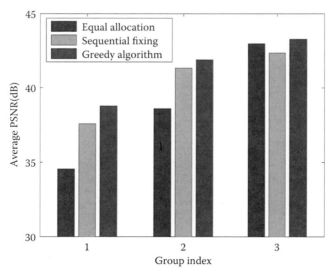

FIGURE 7.1
Average PSNR of the three multicast group users. (From Hu, D., Mao, S., Hou, Y., and Reed, J., Fine grained scalability video multicast in cognitive radio networks, *IEEE J. Sel. Areas Commun.*, 28(3), 334–344, Copyright 2010 IEEE. With permission.)

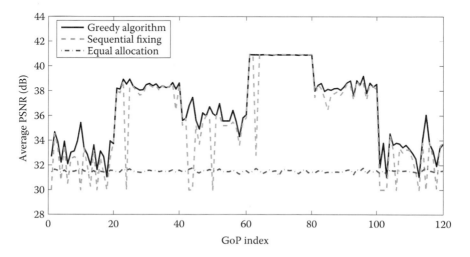

FIGURE 7.2
Per GoP average PSNRs of a tagged user in group 1, when its channel condition varies over time. (From Hu, D., Mao, S., Hou, Y., and Reed, J., Fine grained scalability video multicast in cognitive radio networks, *IEEE J. Sel. Areas Commun.*, 28(3), 334–344, Copyright 2010 IEEE. With permission.)

for each GoP at this user that are obtained using the three algorithms. We observe that both greedy algorithm and SF can quickly adapt to changing channel conditions. Both algorithms achieve received video qualities commensurate with the channel quality of the tagged user. We also find that the video quality achieved by greedy algorithm is more stable than that of SF, while the latter curve has some deep fades from time to time. This is due to the fact that greedy algorithm has a proven optimality bound, while SF does not provide any guarantee. The equal allocation curve is relatively constant for the entire period since it does not adapt to channel variations.

7.6.2 Scalable Video over Multi-Hop CR Networks

For the results reported in this section, we have $K=3$ primary networks and $M=10$ channels. There are 56, 55, and 62 CR users in the coverage areas of primary networks 1, 2, and 3, respectively. The $|\mathcal{U}_m^1|$'s are [5 4 6 4 8 7 5 6 7 4] (i.e., five users sense channel 1, four users sense channel 2, and so forth); the $|\mathcal{U}_m^2|$'s are [4 6 5 7 6 5 3 8 5 6], and the $|\mathcal{U}_m^3|$'s are [8 6 5 4 7 6 8 5 6 7]. We choose $L_p=100$, $T_s=0.02$, and $N_G=10$. The channel utilization is $\eta_m^k = 0.6$ for all the channels. The probability of false alarm is $\epsilon_m^k = 0.3$ and the probability of miss detection is $\delta_m^k = 0.2$ for all m and k, unless otherwise specified. Channel parameters λ_m^k and μ_m^k are set between (0,1). The maximum allowed collision probability γ_m^k is set to 0.2 for all the M channels in the three primary networks.

We consider three video sessions, each streaming a video in the CIF (352×288), i.e., *Bus* to destination 1, *Foreman* to destination 2, and *Mother and Daughter* to destination 3. The frame rate is 30 fps, and a GOP consists of 10 frames. We assume that the duration of a time slot is 0.02 s and each GOP should be delivered in 0.2 s (i.e., 10 time slots).

We compare four schemes in the simulations: (a) the upper-bounding solution by solving the relaxed version of problem OPT-CRV using an NLP solver, (b) the proposed distributed algorithm in Tables 7.6 and 7.7, (c) the SF algorithm given in Table 7.5, which computes a lower-bounding solution, and (d) a greedy heuristic where at each hop, the link with the most available channels is used.

To demonstrate the convergence of the distributed algorithm, we plot the traces of the four Lagrangian multipliers in Figure 7.3. We observe that all the Lagrangian multipliers converge to their optimal values after 76 iterations. We also plot the control overhead as measured by the number of distinct broadcast messages for $e_i(\tau)$ using the y-axis on the right-hand side. The overhead curve increases linearly with the number of iterations and gets flat (i.e., no more broadcast message) when all the Lagrangian multipliers converge to their optimal values.

We examine the impact of spectrum sensing errors in Figure 7.4. Six sensing error combinations are tested: $\{\epsilon_m, \delta_m\}$ as follows: {0.1, 0.5}, {0.2, 0.3},

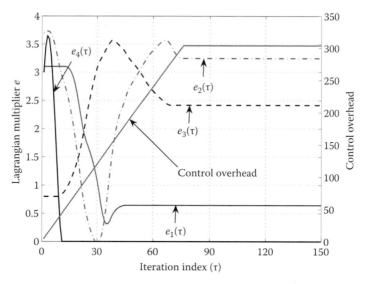

FIGURE 7.3
Convergence of the distributed algorithm. (From Hu, D. and Mao, S., Streaming scalable videos over multi-hop cognitive radio networks, *IEEE Trans. Wireless Commun.*, 9(11), 3501–3511, Copyright 2010 IEEE. With permission.)

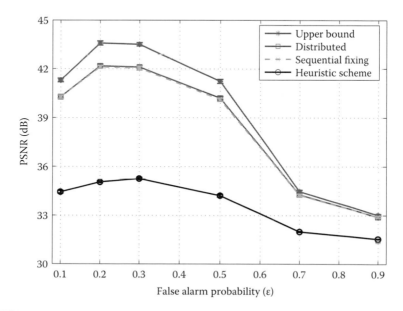

FIGURE 7.4
Average PSNRs versus spectrum sensing error. (From Hu, D. and Mao, S., Streaming scalable videos over multi-hop cognitive radio networks, *IEEE Trans. Wireless Commun.*, 9(11), 3501–3511, Copyright 2010 IEEE. With permission.)

{0.3, 0.2}, {0.5, 0.11}, {0.7, 0.06}, and {0.9, 0.02}. The average PSNR values of the Foreman session are plotted. Each point in the figure is the average of 10 simulation runs, with 95% confidence intervals plotted as error bars in the figures, which are all negligible. It is interesting to see that the best video quality is achieved when the false alarm probability ε_m is between 0.2 and 0.3. Since the two error probabilities are correlated, increasing one will generally decrease the other. With a larger ε_m, CR users are more likely to waste spectrum opportunities that are actually available, leading to lower bandwidth for videos and poorer video quality, as shown in Figure 7.4. On the other hand, a larger δ_m implies more aggressive spectrum access and more severe interference to primary users. Therefore, when ε_m is lower than 0.2 (and δ_m is higher than 0.3), the CR nodes themselves also suffer from the collisions and the video quality degrades.

7.7 Conclusion and Future Directions

In this chapter, we investigated two challenging problems of enabling video over CR networks: (a) an IEEE 802/22 WRAN-like, infrastructure-based CR network where the BS coordinates spectrum sensing and access, and (b) a multi-hop CR network, such as a wireless mesh network with CR-enabled nodes. The problem formulation considered spectrum sensing and sensing errors, spectrum access and primary user protection, video quality, and fairness. For the first problem, we proposed effective optimization and scheduling algorithms and proved the complexity and optimality bound for the proposed greedy algorithm. For the second problem, we applied dual decomposition to derive a distributed algorithm and analyzed its optimality and convergence performance.

Since video over CR networks is a relatively new problem area, there are many interesting open problems to be investigated. To name a few, interested readers may consider the following problems:

- *Spectrum process model*: We assumed that each licensed channel evolves over time following a two-state discrete-time Markov process, and the primary user activities on different channels are independent. However, these assumptions may not hold true in some CR networks. For example, the primary user transmissions may not be adequately modeled as discrete-time Markov processes, or even not by a continuous-time Markov process, while the primary transmissions on different channels may be statistically correlated. It would be interesting to investigate spectrum sensing and access under the more general spectrum process models [42]. The accuracy of the sensing process could be improved by exploiting the sensing results from adjacent channels and historic sensing results.

- *Admission control*: Admission control is an essential mechanism for QoS provisioning. We assumed that the lowest video quality requirement for CR users can be guaranteed in this chapter. However, the network capacity for CR users is dependent on both the primary user transmissions and the random channel fading. Such that for a given set of system parameters, there is a limit on the total number of video sessions that could be supported. In a reliable network (e.g., a wired network), admission control can be achieved by setting a threshold for the user with the lowest PSNR. However, in a CR network, the problem is more complicated due to primary user activity and channel condition. A simple yet efficient admission control mechanism that considers these two types of uncertainties would be useful for guaranteeing video quality in CR networks.
- *Testbed development and validation*: We presented a theoretical framework for video streaming in CR network and demonstrated the performance with simulations. It would be interesting to develop a proof-of-concept implementation and testbed, such that the system performance can be demonstrated under a realistic environment [34,43]. Such a testbed can not only validate the theoretical results, but also reveal new practical constraints that should be considered in the model, as well as identifying new research problems.

Acknowledgment

This work is supported in part by the U.S. National Science Foundation (NSF) under Grants CNS-0953513, ECCS-0802113, IIP-1127952, and DUE-1044021, and through the NSF Wireless Internet Center for Advanced Technology at Auburn University. Any opinions, findings, and conclusions or recommendations expressed in this material are those of the authors and do not necessarily reflect the views of the foundation.

References

1. Q. Zhao and B. Sadler, A survey of dynamic spectrum access, *IEEE Signal Process. Mag.*, 24(3), 79–89, May 2007.
2. Y. Zhao, S. Mao, J. Neel, and J. H. Reed, Performance evaluation of cognitive radios: Metrics, utility functions, and methodologies, *Proc. IEEE*, 97(4), 642–659, April 2009.

3. IEEE, Draft standard for wireless regional area networks part 22: Cognitive wireless RAN medium access control (MAC) and physical layer (PHY) specifications: Policies and procedures for operation in the TV bands, IEEE P802.22 Draft Standard (D0.3), May 2007.

4. M. van der Schaar, S. Krishnamachari, S. Choi, and X. Xu, Adaptive cross-layer protection strategies for robust scalable video transmission over 802.11 WLANs, *IEEE J. Sel. Areas Commun.*, 21(10), 1752–1763, December 2003.

5. M. Wien, H. Schwarz, and T. Oelbaum, Performance analysis of SVC, *IEEE Trans. Circuits Syst. Video Technol.*, 17(9), 1194–1203, September 2007.

6. D. Hu, S. Mao, Y. Hou, and J. Reed, Fine grained scalability video multicast in cognitive radio networks, *IEEE J. Sel. Areas Commun.*, 28(3), 334–344, April 2010.

7. N. Laneman, D. Tse, and G. Wornell, Cooperative diversity in wireless networks: Efficient protocols and outage behavior, *IEEE Trans. Inf. Theory*, 50(11), 3062–3080, November 2004.

8. D. Hu and S. Mao, Streaming scalable videos over multi-hop cognitive radio networks, *IEEE Trans. Wireless Commun.*, 9(11), 3501–3511, November 2010.

9. I. Akyildiz, W. Lee, M. Vuran, and S. Mohanty, NeXt generation/dynamic spectrum access/cognitive radio wireless networks: A survey, *Computer Netw. J.*, 50(9), 2127–2159, September 2006.

10. D. Hu and S. Mao, Design and analysis of a sensing error-aware MAC protocol for cognitive radio networks, in *Proc. IEEE GLOBECOM'09*, Honolulu, HI, November/December 2009, pp. 5514–5519

11. Y. Chen, Q. Zhao, and A. Swami, Joint design and separation principle for opportunistic spectrum access in the presence of sensing errors, *IEEE Trans. Inf. Theory*, 54(5), 2053–2071, May 2008.

12. R. Urgaonkar and M. Neely, Opportunistic scheduling with reliability guarantees in cognitive radio networks, *IEEE Trans. Mobile Comput.*, 8(6), 766–777, June 2009.

13. T. Shu and M. Krunz, Throughput-efficient sequential channel sensing and probing in cognitive radio networks under sensing errors, in *Proc. ACM MobiCom'09*, Beijing, China, September 2009, pp. 37–48.

14. H. Su and X. Zhang, Cross-layer based opportunistic MAC protocols for QoS provisionings over cognitive radio wireless networks, *IEEE J. Sel. Areas Commun.*, 26(1), 118–129, January 2008.

15. Q. Zhao, S. Geirhofer, L. Tong, and B. Sadler, Opportunistic spectrum access via periodic channel sensing, *IEEE Trans. Signal Process.*, 36(2), 785–796, February 2008.

16. D. Hu and S. Mao, A sensing error aware mac protocol for cognitive radio networks, *ICST Transact. Mobile Commun. Appl.*, 1(1), 2012, in press.

17. Y. Hou, Y. Shi, and H. Sherali, Optimal spectrum sharing for multi-hop software defined radio networks, in *Proc. IEEE INFOCOM'07*, Anchorage, AK, April 2007, pp. 1–9.

18. Y. Hou, Y. Shi, and H. Sherali, Spectrum sharing for multi-hop networking with cognitive radios, *IEEE J. Sel. Areas Commun.*, 26(1), 146–155, January 2008.

19. Z. Feng and Y. Yang, Joint transport, routing and spectrum sharing optimization for wireless networks with frequency-agile radios, in *Proc. IEEE INFOCOM'09*, Rio de Janeiro, Brazil, April 2009, pp. 1665–1673.

20. D. Palomar and M. Chiang, A tutorial on decomposition methods for network utility maximization, *IEEE J. Sel. Areas Commun.*, 24(8), 1439–1451, August 2006.

21. D. P. Bertsekas, *Nonlinear Programming*, Athena Scientific, Belmont, MA, 1995.
22. A. Fattahi, F. Fu, M. van der Schaar, and F. Paganni, Mechanism-based resource allocation for multimedia transmission over spectrum agile wireless networks, *IEEE J. Sel. Areas Commun.*, 25(3), 601–612, April 2007.
23. H.-P. Shiang and M. van der Schaar, Dynamic channel selection for multi-user video streaming over cognitive radio networks, in *Proc. IEEE ICIP'08*, San Diego, CA, October 2008, pp. 2316–2319.
24. H. Luo, S. Ci, and D. Wu, A cross-layer design for the performance improvement of real-time video transmission of secondary users over cognitive radio networks, *IEEE Trans. Circuits Syst. Video Technol.*, 21(8), 1040–1048, August 2011.
25. S. Ali and F. Yu, Cross-layer qos provisioning for multimedia transmissions in cognitive radio networks, in *Proc. IEEE WCNC'09*, Budapest, Hungary, April 2009, pp. 1–5.
26. Z. Guan, L. Ding, T. Melodia, and D. Yuan, On the effect of cooperative relaying on the performance of video streaming applications in cognitive radio networks, in *Proc. IEEE ICC'11*, Kyoto, Japan, June 2011, pp. 1–6.
27. D. Hu, S. Mao, and J. Reed, On video multicast in cognitive radio networks, in *Proc. IEEE INFOCOM'09*, Rio de Janeiro, Brazil, April 2009, pp. 2222–2230.
28. D. Hu and S. Mao, Resource allocation for medium grain scalable videos over femtocell cognitive radio networks, in *Proc. IEEE ICDCS'11*, Minneapolis, MN, June 2011, pp. 258–267.
29. D. Hu and S. Mao, On medium grain scalable video streaming over cognitive radio femtocell networks, *IEEE J. Sel. Areas Commun.*, 30(3), 641–651, April 2012.
30. D. Hu and S. Mao, Cooperative relay with interference alignment for video over cognitive radio networks, in *Proc. IEEE INFOCOM'12*, Orlando, FL, March 2012, pp. 2014–2022.
31. S. Mao, X. Cheng, Y. Hou, and H. Sherali, Multiple description video multicast in wireless ad hoc networks, *ACM/Kluwer Mobile Netw. Appl. J.*, 11(1), 63–73, January 2006.
32. W. Wei and A. Zakhor, Multiple tree video multicast over wireless ad hoc networks, *IEEE Trans. Circuits Syst. Video Technol.*, 17(1), 2–15, January 2007.
33. S. Deb, S. Jaiswal, and K. Nagaraj, Real-time video multicast in WiMAX networks, in *Proc. IEEE INFOCOM'08*, Phoenix, AZ, April 2008, pp. 1579–1587.
34. J. Jia, Q. Zhang, and X. Shen, HC-MAC: A hardware-constrained cognitive MAC for efficient spectrum management, *IEEE J. Sel. Areas Commun.*, 26(1), 106–117, January 2008.
35. A. Motamedi and A. Bahai, MAC protocol design for spectrum-agile wireless networks: Stochastic control approach, in *Proc. IEEE DySPAN'07*, Dublin, Ireland, April 2007, pp. 448–451.
36. C. Corderio, K. Challapali, D. Birru, and S. Shankar, IEEE 802.22: An introduction to the first wireless standard based on cognitive radios, *J. Commun.*, 1(1), 38–47, April 2006.
37. H. Mahmoud, T. Yücek, and H. Arslan, OFDM for cognitive radio: Merits and challenges, *IEEE Wireless Commun.*, 16(2), 6–14, April 2009.
38. A. M. F. Kelly and D. Tan, Rate control in communication networks: Shadow prices, proportional fairness and stability, *J. Oper. Res. Soc.*, 49(3), 237–252, March 1998.

39. Y. T. Hou, Y. Shi, and H. D. Sherali, Optimal base station selection for any cast routing in wireless sensor networks, *IEEE Trans. Veh. Technol.*, 55(3), 813–821, May 2006.
40. S. Kompella, S. Mao, Y. Hou, and H. Sherali, On path selection and rate allocation for video in wireless mesh networks, *IEEE/ACM Trans. Netw.*, 11(1), 63–73, January 2006.
41. M. S. Bazaraa, J. J. Jarvis, and H. D. Sherali, *Linear Programming and Network Flows*, 4th edn., John Wiley & Sons, Inc., New York, 2010.
42. X. Xiao, K. Liu, and Q. Zhao, Opportunistic spectrum access in self similar primary traffic, *EURASIP J. Adv. Signal Process.*, Vol. 2009, Article ID 762547, 8 pages, doi:10.1155/2009/762547, 2009.
43. Y. Huang, P. Walsh, Y. Li, and S. Mao, A GNU Radio testbed for distributed polling service-based medium access control, in *Proc. IEEE MILCOM'11*, Baltimore, MD, November 2011, pp. 1–6.

8

Cooperative Video Provisioning in Mobile Wireless Environments

Paolo Bellavista, Antonio Corradi, and Carlo Giannelli

CONTENTS

8.1 Introduction

In the last years, the wireless Internet has received evergrowing attention resulted from the unprecedented and mass-market spread of mobile devices: smartphones, tablets, and netbooks are widely and increasingly exploited to access the Internet from everywhere, in a ubiquitous manner. Initially, mobile devices were primarily used to surf the web and access contents managed by traditional servers connected to the fixed Internet via widebandwidth links. In other words, users exploited their mobile devices similar to what they were used to do with their desktops, with the only but notable differences that, in the case of mobile devices, (a) users were not tied to fixed locations and (b) devices usually had limited hardware/software capabilities, e.g., making harder to render HTML pages designed for high-resolution and large displays.

However, mobile computing has quickly moved toward more powerful service provisioning scenarios where devices not only play the role of clients that access remote resources while in mobility, but also generate and provide interesting content to other nodes. Mobile nodes generating audio/video real-time streams are a notable example, where communication endpoints can even simultaneously play the roles of both content generators and consumers. However, in relatively traditional solutions, inter-node communication is still supported by infrastructure-based networking, and multimedia stream delivery can exploit special-purpose infrastructure components, e.g., proxy nodes that transcode audio/video streams to fit client constraints dynamically.

More recently, mobile environments are shifting toward novel and more challenging paradigms for communication and service provisioning, which aim at connecting wireless devices one another directly, without any constraint to be connected to the traditional Internet or to exploit support components deployed on either the infrastructure or server-side. In particular, nowadays cooperative mobile networks are attracting significant academic/industrial attention and are providing promising preliminary results: the specific property of cooperative mobile networks is that they are originated from the willingness of social interaction of people via impromptu interconnection of the mobile personal devices they carry (e.g., smartphones, PDAs, and iPADs) (Feeney et al. 2001). Services take advantage of inter-device cooperation to dispatch packets toward their destination, thus potentially enabling the full exploitation of all the potential of all the available wireless connectivity opportunities, for instance, by allowing resource connectivity sharing and immediate connectivity offers in regions with difficult coverage (Ferreira et al. 2010, Salameh et al. 2009, Wu et al. 2007). In our vision, cooperative mobile networks will make soon another relevant evolution step: they will support multi-hop application-oriented scenarios where cooperative trusted users will collaborate to invoke/provide services to the community, such as sharing locally stored JPEG pictures, delivering live multimedia, and offering underutilized flat-rate Internet connectivity. In other words, we envision a collaborative provisioning environment where mobile communications are fully enabled also in the absence of a networking infrastructure: Cooperating nodes can generate multimedia flows, route them toward their destination, and eventually collaborate to adapt them at runtime, e.g., in response to the detection of bandwidth degradation.

Let us point out from the beginning that, on the one hand, the user-centric nature of cooperative mobile networks relaxes the constraint of having infrastructure-based communication support (e.g., expensive cellular coverage anywhere). On the other hand, it naturally yields to very heterogeneous, uncoordinated, and dynamic networking environments where, for instance, any node can create its self-administered layer 2 links. In addition, to fully exploit the potential of these networks, nodes should be able to take advantage of different wireless interfaces simultaneously (multi-homing) to join/create multiple IP networks (via both ad hoc and infrastructure-based connectivity). These user-centric networks should be autonomously created, configured, and destroyed by collaborating users in a completely decentralized way and should be managed with very lightweight coordination (or without any coordination at all). Moreover, to be scalable, cooperative mobile networks call for the challenging issue of the effective management of end-to-end multi-hop communications with local and partial topology awareness.

Also due to the number of technical challenges (heterogeneity, lack of coordination, high dynamicity, and so on) to address to enable effective use of cooperative mobile networks, as better detailed in the following, up to now social collaboration has been limited to very simple applications, such as message/status sharing. However, because mobile devices are becoming general-purpose media players and personal data carriers, we strongly claim that there will be the need, in the near future, of supporting more complex resource sharing and providing more challenging collaborative applications. In this perspective, a crucial role will be played by collaborative streaming of live multimedia.

8.1.1 Application Case and Problem Statement

To practically clarify the targeted application case with an example, consider a hybrid scenario including the characteristics of both traditional wireless Internet and cooperative mobile networking. For instance, a student (Alice) is watching a soccer match on her smartphone via an HSDPA device (Figure 8.1). The quality of the visualized stream depends on many factors, ranging from client-side static capabilities, e.g., screen size and CPU/ memory, to server-side dynamic state, e.g., number of clients requiring the stream, and client-to-server connectivity quality, e.g., wireless connectivity

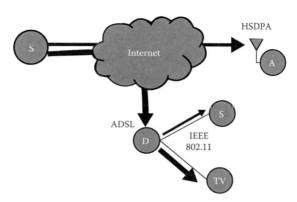

FIGURE 8.1
Multimedia stream delivery with differentiated quality.

quality is greatly influenced by interferences and competing traffic due to concurrent access by multiple mobile clients (Bellavista et al. 2009). In order to achieve a quality level suitable for the final user, the multimedia player on Alice's smartphone can negotiate the stream quality with the remote server at start-up, e.g., selecting the stream version with the best quality that fits client capabilities, and also during service provisioning, e.g., dynamically switching stream versions according to currently available bandwidth. In this way, Alice's smartphone can receive the stream at a quality compatible with its low capabilities, while the same multimedia content is sent at full quality to desktop nodes that connect to the Internet via traditional ADSL. Moreover, desktop nodes can forward the stream to other nodes in the subnet, eventually transcoding and tailoring it at runtime to fit codec and hardware capabilities of nearby devices, ranging from high-definition UPnP-enabled TVs to smartphones connected to the local gateway via IEEE 802.11.

In addition, Alice can decide to play an active collaborative role by working to provide online redistribution (namely recast) of the received stream to her friends' devices in the neighborhood, even if they are not equipped with HSDPA. To this purpose, Alice's smartphone starts streaming the match, possibly via multi-hop paths consisting of heterogeneous ad hoc links. Bob receives the match with automatically downscaled quality level (lower-frame frequency) because the path between Alice and Bob has limited bandwidth, e.g., due to the fact it consists of low-quality hops (Figure 8.2a). The stream is downscaled only if and when required, thus permitting to nodes connected to Alice via high-quality paths to receive the multimedia stream at full quality. To this purpose, nodes in the Alice-to-Bob path have to (a) collaboratively interact to dispatch packets along the path toward the destination and (b)

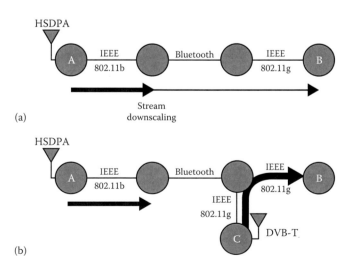

FIGURE 8.2
Collaborative content distribution/adaptation (a) and path reconfiguration (b).

actively monitor the path quality to tailor traversing streams depending on current connectivity capabilities. In the meanwhile, Carl joins the network and provides the same multimedia content at higher quality via its local DVB-T device (Figure 8.2b). Bob's smartphone perceives that Carl's netbook provides the same content at higher quality and via a better collaborative path; therefore, it switches streaming from Alice's laptop to Carl's netbook.

This chapter will focus on this articulated and state-of-the-art scenario of cooperative video provisioning in mobile wireless environments. The primary idea is to exploit the spontaneous and opportunistic collaboration of mobile wireless devices, which cooperate to gather, monitor, and tailor multimedia content depending on dynamically changing capabilities of their communication channels. In particular, the objective of this chapter is twofold. On the one hand, it provides an insight of the state-of-the-art literature about collaborative mobile networking, ranging from static server-to-client interaction for the selection of the proper multimedia flow at service start-up, to more challenging forms of peer-to-peer cooperation involving dynamic downscaling/upscaling of multimedia flows at service provisioning time. On the other hand, it reports about the RAMP case study, demonstrating that node cooperation can properly and effectively achieve multimedia streaming adaptation, with minimum impact on computing/memory resources available at intermediary participating nodes.

The structure of the chapter is as follows. Section 8.2 presents how the traditional Internet has evolved toward a more dynamic scenario with ubiquitous wireless connectivity, eventually involving even direct communication among devices. In particular, it focuses on multimedia stream adaptation techniques suitable for these environments. Section 8.3 outlines the primary related work in the field, classifying state-of-the-art contributions along three main guidelines: (a) traditional end-to-end stream provisioning/adaptation that involves only server and client; (b) more articulated and distributed solutions based on proxies, with the objective of improving the quality perceived by final users; and (c) challenging proposals based on the dynamic interaction of peers along the path between stream providers and final users. Section 8.4 details our RAMP middleware solution, aiming at supporting the provisioning of multimedia content in collaborative mobile networks. Finally, Section 8.5 reports conclusive remarks and points out some hot topics, which are promising research directions for the near future.

8.2 Multimedia Streaming in Mobile Wireless Environments

The traditional Internet has already evolved to a mobile one characterized by handheld devices capable of receiving streams via wireless links, while content providers are still usually fixed and reachable via the traditional

Internet. The widespread utilization of peer-to-peer multimedia services, e.g., audio/video communication over IP, calls for more complex provisioning environments where both streaming endpoints are connected via wireless links and mobile. Also the distinction between content providers and consumers is becoming blurred: the trend is that any node tends to be producer and consumer of multimedia content, e.g., by sharing user-generated multimedia content composed by text, pictures, videos, or a mix of them. As already sketched in the introduction, the upcoming communication trend goes beyond, toward more challenging scenarios where endpoints are connected via multi-hop wireless paths, not only exploiting special-purpose infrastructure equipment, e.g., IEEE 802.11s mesh routers, but also mobile nodes collaboratively cooperating to dispatch packets, e.g., as in our RAMP middleware for collaborative mobile networking (Bellavista et al. 2012).

In particular, we identify four main communication paradigms for mobile wireless streaming, spanning from the traditional wireless Internet to more challenging mobile environments based on peer-to-peer communications (see Figure 8.3):

1. *Only client endpoints are connected via wireless hops* (Figure 8.3a): This is the traditional wireless Internet environment, where mobile nodes (e.g., node A) access the Internet via one wireless hop toward an infrastructure of access points, e.g., by exploiting IEEE 802.11 or HSDPA interfaces. Typically, services are provided by fixed servers (e.g., node S) deployed on nodes with relatively high-computing capabilities and connected to the Internet via highly reliable and large-bandwidth wired links; clients are the only mobile entities and are connected to the Internet via wireless hops with low/medium connectivity quality. For instance, users can view videos on their mobile nodes connecting to YouTube and alike, as they are used to do from their desktop computers, even if with a reduced content resolution and screen size.

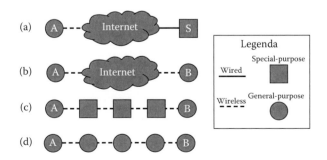

FIGURE 8.3
The proposed categorization of communication paradigms for mobile streaming: (a) only clients are wireless, (b) clients and servers are wireless, (c) also intermediary nodes are wireless, and (d) intermediary nodes are wireless, mobile, and collaborative.

2. *Both endpoints use wireless hops* (Figure 8.3b): The widespread diffusion of mobile devices has recently leveraged the adoption of applications where both endpoints of audio/video streams are connected via wireless hops. In addition, client and server roles may be blurred, since both nodes can offer and take advantage of services and resources in a peer-to-peer way. Several VoIP applications are a notable example in this category: both endpoints typically exploit wireless interfaces to connect to the Internet and to transmit/ receive audio and video streams; there is no streaming server playing the specific role of content generator and deployed on a high performance and reliable node, as it was used in more traditional web-based servers for streaming download.

3. *Also intermediary hops are wireless* (Figure 8.3c), e.g., *in wireless mesh networks* (Camp and Knightly 2008): In the last years, the deployment of large wireless networks covering small-to-medium urban areas, e.g., university campuses and city downtowns, has moved the attention to the proper management of wireless connectivity among nodes via multi-hop paths composed of wireless links provided by mesh routers, eventually without any need of connecting to the Internet. For instance, mesh networks can be adopted to set up medium-scale remote surveillance systems, via proper deployment of wireless routers along the covered area and to collect video streams toward a control point. Mesh-based solutions make easier and less expensive the dynamic extension of the wireless coverage area, e.g., by adding or relocating wireless mesh routers.

4. *Intermediary nodes are wireless and mobile* (Figure 8.3d), e.g., *collaborative mobile networks*: In collaborative mobile networks, devices discover and interact with one another opportunistically and without any prior mutual knowledge, by exploiting any wireless opportunity currently available in a best-effort way (e.g., Wi-Fi or Bluetooth ad hoc links and UMTS infrastructure-based ones). In particular, social/group-related behavior and the ever-increasing willingness to share rich user-generated contents, also pertaining to the personal sphere, call for a user-centric communication paradigm shift (as shown also by recent standardization efforts for point-to-point Wi-Fi connectivity [Wi-Fi Alliance 2009]), where the ad hoc interconnection of mobile devices in direct visibility plays a central role. For instance, it should be used for conference/class streaming with attendees/students spread on multiple rooms or for DVB-T program redistribution by (the generally few) nodes equipped with DVB-T receivers, as rapidly sketched in Section 8.1.

Such scenarios require novel multimedia management/adaptation techniques, aware of specific characteristics of mobile/wireless environments (e.g., heterogeneous performance between wired and wireless subpaths) and dynamically exploiting the resources offered by intermediary nodes (e.g., in collaborative mobile networks, intermediary nodes cooperate not only to dispatch packets but also to dynamically tailor the traversing multimedia streams). In particular, we classify the techniques for multimedia content provisioning, monitoring, and adaptation in three main categories, depending on where streaming quality assessment/tailoring are performed: *end-to-end*; *proxy-based* (one single- and special-purpose intermediary); *collaborative* (several general-purpose intermediary nodes).

End-to-end adaptation techniques represent the basic and currently most adopted mechanism to adapt the multimedia content. Servers and/or clients are aware that the same content can be provided at different quality levels and interact to select the suitable quality to fit the currently available capabilities. In the simplest case, the final user has the opportunity to select among multiple choices, e.g., low, medium, and high quality; once a quality level is selected, the server delivers the content to the client, despite that client device characteristics and available bandwidth actually fit to the case for the whole streaming duration. For instance, a user can select a movie in HD format from her smartphone connected to the Internet via UMTS, thus almost certainly receiving the stream jerkily, if receiving it at all. Moreover, the same multimedia content can be statically tailored at different quality levels and stored on the server-side, e.g., by generating offline multiple files with different codecs and bitrates. Or it can be dynamically adapted by need and on the fly during service provisioning with per-client granularity, thus finely tuning the content in relation to client capabilities; this produces the nonnegligible overhead of possibly transcoding/adapting the same multimedia content several times.

Slightly more sophisticated solutions involve *content distributors and clients in the quality selection process*. On the one hand, content distributors take into consideration server-side information to provide the set of available choices. For instance, the streaming server can provide the list of available quality levels and related codecs, eventually avoiding the delivery of HD streams in the case of heavy load due to many concurrent requests from multiple users. On the other hand, clients rendering the multimedia content to final users can exploit local resource information in order to discard unsuitable choices, e.g., by comparing the characteristics of potentially available streams with the maximum resolution of the local display. In this case, the final user achieves the notable advantage of not requiring any knowledge about multimedia content formats and device hardware/software capabilities, delegating to the underlying system the proper selection of the most suitable quality level in a completely transparent way. At the same time, the server can increase the amount of concurrently served clients by adapting the set of provided streams in relation to its capabilities. However, in this case, once service distribution is started, it is not possible to modify the content quality anymore; for instance, in case of abrupt bandwidth fluctuations, the final user may perceive frequent service disruptions, requiring to manually and explicitly restart the streaming with lower quality.

Wireless environments are usually characterized by highly varying connectivity quality. For instance, user movements or interferences may cause frequent packet loss, by negatively affecting the quality perceived by final users. For this reason, it is not possible to consider only static characteristics at service initialization time. To this purpose, recent proposals for mobile wireless streaming aim at *assessing connectivity quality and tailor multimedia content at service provisioning time*. The most notable advantage is the capability of adapting content quality dynamically, by fully exploiting the currently available capabilities. At service request time, content distributor and client select the proper quality level, then they properly renegotiate content quality at runtime in relation to, e.g., increased/decreased server-side load, better/worse connectivity bandwidth, or client-side stream delay/jitter.

The dynamic reconfiguration of multimedia content is particularly effective when coupled with *proxy-based solutions*. In this case, end-to-end paths are typically characterized by a reliable, powerful wired subpath, e.g., from the server to the wired/wireless edge, and an unreliable and bandwidth-limited wireless one, e.g., from a UMTS base station to the client on a smartphone. Proxy-based solutions decouple the content delivery process by carefully considering this difference: proxies are special-purpose components specifically designed to support multimedia content delivery at the proper quality level. They are typically deployed on the infrastructure-side, gather the content from the server, and provide it to wireless clients considering the unreliable nature of their adjacent wireless links. For instance, proxies can retrieve multimedia content at full quality and then deliver it to clients of the same UMTS base station in a differentiated manner, considering that different smartphones may have different hardware/software capabilities and access the network with different bandwidth. In addition, proxies can eventually pre-fetch large chunks of multimedia content and/or transcode it in order to fit client capabilities. In this way, it is possible to achieve the notable advantage of lowering the load on servers because multimedia content is requested only once and delivered to multiple users by exploiting different codecs and bitrates.

In the case of collaborative mobile networks, each intermediary node can perform quality assessment and content tailoring similar to the proxies provided earlier, by achieving *highly collaborative solutions*. Proxies are usually deployed on special-purpose equipment with the main goal of supporting content delivery and adaptation. On the contrary, intermediary nodes of collaborative mobile networks share part of their computational/memory/connectivity resources to support content delivery/tailoring in a best-effort way. In collaborative environments, users are willing to cooperate to allow content dispatching, but they also require limiting local resource consumption, e.g., avoiding excessive consumption of local computing resources for multimedia content transcoding. At the same time, since the same path is split into multiple segments, each one corresponding to a couple of adjacent intermediary nodes, it is possible to monitor local subpaths with finer granularity if compared with proxy-based solutions. For instance, it is possible to tailor the multimedia content only in correspondence of a Bluetooth link with limited capabilities, while maintaining higher quality in other parts of the served path.

8.3 Related Work

This section classifies the state-of-the-art literature contributions about multimedia content adaptation for mobile/wireless streaming by taking into consideration the definitions and concepts presented in Section 8.2. Existing approaches are organized in three categories: *end-to-end*, based on quality selection performed at content distributor, typically choosing among few coarse-grained possibilities, eventually by following client indications; *proxy-based*, performing dynamic adaptation typically at the wired/wireless edge (Gene Cheung et al. 2005); and *collaborative* tailoring, where clients are more involved in quality assessment and eventually adapt multimedia streams in a collaborative and peer-to-peer way.

The remainder of the section focuses, for each category, on the related work considered as the most relevant and exemplar, with the main goal of reporting few but notable solutions that are also useful to fully understand the categorization that we have originally introduced in the previous section. For the sake of presentation clarity, in the following, all the reported solutions are associated with one single category in our classification (in relation to the solutions characteristics we deem more relevant), even if minor relationships with other categories may exist.

8.3.1 End-to-End Solutions

Traditional stream tailoring solutions for fixed and mobile multimedia are mainly based on end-to-end mechanisms, involving only the server offering the multimedia content and the client rendering it to the user.

Adaptive bitrate streaming describes a basic and effective mechanism to provide content adaptation via differentiated multimedia streams to HTTP clients, by avoiding asking final users for the explicit selection of the requested quality level (Adobe 2010). In particular, the client actively probes local computing resources and available bandwidth, and accordingly selects the multimedia content available on the server-side at the most suitable bitrate. In this way, adaptive bitrate streaming succeeds in fitting the provided quality to current computing/networking capabilities. Moreover, the adoption of HTTP as the transport protocol easily supports the delivery of both live and on-demand streams to very differentiated clients. However, bitrate options have to be statically decided, and modifications of video bitrate are possible only at the beginning of a multimedia fragment (temporal subset of the provided stream). There is no possibility of gracefully and dynamically increasing/decreasing multimedia bitrate at runtime, in a fine-grained way and only when actually needed.

In order to provide a more powerful and sophisticated mechanism to dynamically tailor multimedia content at provisioning time, research activities in scalable video coding (SVC) for H.264 have mainly focused on multilayered codecs, where the same multimedia resource can be provided to clients in the form of different subsets of layers (a base layer plus one or more enhancement ones) (Schwarz et al. 2007). In particular, temporal scalability approaches provide clients with a different set of frames by varying the desired frame rate; in spatial scalability, enhancement layers enable higher spatial resolution, e.g., by augmenting horizontal or vertical frame resolutions; in quality scalability, enhancement layers permit to improve the rendering process. SVC has the notable advantage of easily allowing fine-tuned adaptation of multimedia content on the fly, by overcoming many issues associated with solutions that exploit a small set of predefined quality levels, such as adaptive bitrate streaming. In particular, SVC permits to store multimedia content on the server-side in only one format: The same multimedia content is tailored, only at provisioning time, in relation to client requirements. Moreover, content adaptation can be performed only when actually required, thus overcoming limitations of coarse-grained per-fragment quality selection of adaptive bitrate streaming. Note that the great flexibility of SVC makes its adoption suitable not only for end-to-end solutions, but also for proxy-based and collaborative ones.

By focusing on mobile wireless environments, some relevant work has explored quality management via dynamic flow transformation. The main goal is to assess end-to-end quality in order to select the most suitable quality level. The work by Frojdh et al. (2006) specifically considers 2.5G and 3G cellular networks, by evaluating end-to-end paths based on information sent by the client. The main goal is to send the proper amount of multimedia content to the client in order to avoid both client-side buffer overflow (e.g., sending the stream at a bitrate higher than rendering capabilities) and underflow (e.g., sending the stream at a bitrate higher than the currently available bandwidth). To this purpose, the client provides the server with RTCP feedback about the free space available on the client-side buffer, the amount of lost packets, RTT, and the sequence numbers of received packets. Then, the server is in charge of dynamically selecting the proper quality and downscaling/upscaling the delivered multimedia stream.

The work by Zhi-Jin et al. (2005) is another example of multimedia content adaptation driven by the conditions of the buffer on the receiver side. In this case, content adaptation is performed by providing differentiated priority to different stream layers: The basic layer has greatest priority and is scheduled to be transmitted first, while enhancement layers are sent only if there is enough bandwidth.

The work by Guenkova-Luy et al. (2004) adopts a more general approach, which aims at facilitating the QoS negotiation process not only at service start-up, but also when notable events may vary end-to-end characteristics, such as vertical handovers. The proposed solution adopts SIP to transfer control data and XML to describe system characteristics and quality parameters. The proposed management process considers a wide set of parameters: quality perceived by the user, application-level stream characteristics such as frame rate and size, transport-level parameters such as media codec and network-access technology, and specific characteristics of end systems, e.g., memory and CPU.

Finally, the work by Wanghong Yuan et al. (2006) represents an interesting solution aiming at offering the best user experience possible by adapting not only the multimedia stream but also the behavior of the client at the hardware and operating system layers. The goal is to coordinate the adaptation process with the application, operating system, and hardware layers in order to improve user satisfaction while minimizing the imposed overhead. For instance, the proposed solution takes into consideration that processors can run at a selectable speed in a discrete set, that codec operations are usually long-living and CPU intensive, and that different video frames may have different computational complexity and different importance in the rendering process.

8.3.2 Proxy-Based Solutions

In mobile environments, the necessity of properly and differently managing the heterogeneity of performance in the wired and wireless links that usually compose paths between servers and clients has generally emerged. Traditional solutions have managed content delivery to mobile and wired clients in the same manner, e.g., simply considering the available bandwidth in the end-to-end path equal to its bottleneck, i.e., the wireless hop. However, the unreliable and ephemeral nature of wireless connectivity pushes for novel solutions explicitly considering the peculiarities of mobile wireless networks. In particular, proxy-based solutions represent the most spread and widely accepted architecture proposal for delivering multimedia streams with adaptation capabilities, by taking into primary consideration the specific differences and characteristics between the wired and the wireless subpaths.

In fact, proxies reside at the edge between wired and wireless path segments and actively interact with content providers/consumers to dynamically assess stream quality and tailor multimedia contents, e.g., via selective frame dropping. On the one hand, they take full advantage of reliable and powerful connectivity toward the Internet to get multimedia content from servers, possibly at full quality. On the other hand, they consider the intermittent and highly variable characteristics of the wireless subpath to properly and dynamically adapt delivered content, e.g., upscaling/downscaling video quality in relation to wireless interferences and the actually achievable throughput. For instance, a proxy can get a multimedia stream from a web server only once and then redistribute the same stream to multiple mobile devices with differentiated quality levels. In this way, it achieves the twofold benefit of lowering the traffic load on the server-side and providing the multimedia stream to users in a per-client fine-grained fashion.

To this purpose, Liu et al. (2010) propose components at the wired/wireless borders that manage multimedia streams to maximize quality while reducing the needed throughput over limited wireless links. Proxies divide streams into different layers, by grouping frames with the same priority; then, they can separately handle different layers, by dropping part of the streams with per-layer per-client granularity. Jeong et al. (2007) focus on home network environments, with the specific goal of transcoding multimedia content on residential gateways between high-quality and bandwidth-eager digital video format and the more efficient low-quality MPEG4 one. Thus, while Liu et al. (2010) and Jeong et al. (2007) have demonstrated their effectiveness in access point-based network infrastructures, their architecture are only partially distributed and only border nodes manage streams, e.g., depending on explicit client feedback.

Ji Shen et al. (2004) delineate an architecture composed of several proxies interacting the one with the other in order to provide multimedia content to wireless devices. The main purpose is to overcome typical issues associated with wireless communication, such as hardware/software heterogeneity of mobile devices, reduced capabilities (preventing from on-the-fly content transcoding on the client side), and intermittent connectivity. Proxies deployed close to wireless devices interact to negotiate stream quality on behalf of clients, eventually by buffering part of the content in the case of abrupt connectivity disruption; the buffered content can be delivered immediately once the wireless link has been recovered.

The SProxy project aims at increasing the quality perceived by final users while lowering the load on web servers that provide multimedia content (Chen et al. 2007). Since it is based on well-known protocols, i.e., RTSP, RTP, and HTTP, it is compatible with legacy clients and servers, i.e., clients require streams to the proxy in a transparent manner as if it were a standard web server. SProxy provides the capability of dividing multimedia content in several segments related to different time intervals: client requests are split in multiple subrequests, in order to retrieve only the multimedia content segments actually required by the client instead of the whole content, thus lowering the traffic imposed on servers. Moreover, SProxy implements prefetching techniques, thus tuning the traffic rate according to the dynamically detected bandwidth between proxy and server. Finally, the retrieved segments are stored at the proxy by adopting a popularity-based replacement policy, in order to exploit caching capabilities in an efficient way.

Park et al. (2006) propose a solution that is based on frame relevance discrimination and that drops frames in a selective way. It evaluates connectivity quality based on MAC/PHY parameters, thus tightly coupling its evaluation metric to the IEEE 802.11 protocol. Home gateways distribute multimedia content gathered from the Internet to local clients via RTP-MPEG2. The proposed solution differently manages RTP/MPEG2 packets in relation to their content, considering that MPEG2 video is composed of self-contained intra-coded (I) video frames including any information required for rendering them and predictive-coded/bidirectionally predictive-coded (P/B) frames relying on data included in other previous/following frames. In particular, it gives maximum priority to RTP control packets, then to audio RTP packets, to I frame packets, and finally to P/B frame packets. Note that RTP packets are possibly split at the proxy in order to contain only one type of frame at a time, e.g., to avoid transmitting TS packets (188B packets representing MPEG2 basic data unit; typically each RTP packet contains 7 TS packets) related to I and P frames together within the same RTP packet.

In the work by Burza et al. (2007), proxies assess bandwidth saturation based on the number of buffered frames that still wait to be sent to the client. In case of saturation, they lower stream bitrate by selectively dropping B and P frames in this order, thus according to their importance for the rendering process on the client. Instead, since I frames are crucial for the correct rendering of the multimedia stream, the proposed solution does not drop them but, in case of saturation, briefly delay their transmission. In fact, even if delayed, I frames can be very useful to decode the following P/B frames. Similar to the work by Park et al. (2006), the proposed solution actually changes how TS packets are encapsulated into RTP packets: It forces any RTP packet to contain at most one frame in order to simplify content-type determination and differentiated packet management. However, RTP packets frequently tend to have a payload of limited size (even less than the traditional 7 TS packets) and, therefore, the relative overhead due to RTP/UDP/IP headers is not negligible.

8.3.3 Collaborative Solutions

In the recent literature, novel streaming adaptation solutions that also support collaborative management of multimedia content, eventually based on peer-to-peer cooperation, are emerging. The goal is to enable content adaptation (quality reduction) only if and where actually needed, by keeping maximum quality otherwise. For instance, Jurca et al. (2007) provide a general overview of peer-to-peer systems applied to self-organized networks, with the main purpose of delivering multimedia content in a scalable way. In particular, Jurca et al. (2007) point out that intermediary nodes should actively cooperate not only to dispatch the provided streams, but also to gather context information related to their locality, to properly dispatch content along the most suitable path (with finer-grained and greater flexibility if compared with traditional end-to-end solutions), and eventually to adapt/filter the multimedia content.

Narayanan et al. (2007) exploit the availability of multiple relays close to the node that is providing multimedia content in a peer-to-peer way. Every device can play the role of relay; therefore, there is the need to support the dynamic discovery of relays in such an opportunistic network. Discovery should be scalable, because the number of available relays could be high, and should be based on proximity considerations, because closer relays generally provide multimedia content with higher quality, e.g., with lower latency.

Uhm et al. (2010) provide nearby devices with content adaptation features in a user-centric way. Even if the proposed solution is also proxy based and exploits special-purpose equipment for multimedia delivery, we position (Uhm et al. 2010) in the category of collaborative solutions because of the prevalence of its collaborative and opportunistic contributions. In fact, Uhm et al. (2010) propose a portable gateway that dynamically retrieves devices and sensors close to the user, interacts with them to gather their characteristics and profiles, and adapts/transcodes the provided multimedia content to fit user/device requirements. In particular, the portable gateway offers the notable advantages of, on the one hand, adapting its behavior in relation to the current location, and, on the other hand, having full knowledge of user profile/requirements, thus enabling the periodic reassessment of connectivity quality and dynamic readaptation of multimedia content to fit changing conditions. For instance, if the user moves from home to her car, the portable gateway actively interacts with dynamically discovered nearby devices to keep active sessions alive, by switching them from the home IEEE 802.11 network to the cellular-based HSDPA one.

Mukherjee et al. (2005) propose a tailoring mechanism suitable for arbitrary formats, based on a universal model fitting all scalable bit streams. The general solution is based on metadata, associated to the stream, that specify how it is possible to tailor the stream and which are the network/terminal constraints to support the real-time process of quality tailoring selection. Adaptation can be performed also at intermediary nodes; since network/terminal constraints/information are sent to the preceding adaptation entity, it is possible to deploy chains of adaptors, each one collecting feedback from the successive entity, either a client or another adaptation entity. In this manner, it is possible to optimize the overall end-to-end transmission efficiency, by adapting the stream only where actually required.

Ip et al. (2007) merge a proxy-based architecture with a peer-to-peer one, exploiting an inherently collaborative approach. In fact, several proxies, colocated near their mobile clients, collaboratively host multimedia streams in a partitioned and partially replicated way. The main purpose is to minimize bandwidth consumption by properly deploying middleware components close to either final users or popular stream sources. Thus, Ip et al. (2007) achieve the twofold objective of lowering transmission costs while providing greater reliability if compared with pure peer-to-peer solutions without any proxy support.

Hutter et al. (2005) couple SVC-based tailoring with a collaborative approach that distributes the burden of content adaptation among intermediate nodes. In particular, each adaptation node consists of three modules. The context aggregation module gathers context information provided by clients. The adaptation engine module exploits information provided by each client to select the proper quality level (local consideration), possibly different for each client. The context merging module considers together all the information provided by its clients (global consideration) and sends to the server (or another cooperating node) the quality level suitable for the client capable of receiving the best quality at the moment.

Mastronarde et al. (2007) exploit the multipath nature of mesh networks, together with the layered structure of SVC streams, to take full advantage of available connectivity resources. In particular, different layers of the same multimedia stream are sent as subflows toward the same destination along different paths, thus achieving a considerable quality gain. Nodes collaborate to monitor and assess wireless link conditions with the twofold goal of deciding which subflows should be admitted and of selecting which are the paths where the accepted subflows should be sent through.

Finally, the work by Hsien-Po Shiang and van der Schaar (2007) introduces the concept of "network horizon" in a wireless multi-hop path. Its purpose is to efficiently disseminate quality-related information within a given network scope, identified in terms of hop-distance. Intermediary nodes provide transmitting nodes with lightweight monitoring feedback, in order to limit the number of hops that monitoring data should go backward along the stream path. On the one hand, larger horizons allow deeper knowledge of global network conditions at the cost of imposing greater overhead; on the other hand, tighter horizons limit the monitoring awareness of the adaptation mechanism but permit to receive more up-to-date information, since control packets traverse only a limited set of intermediate nodes.

Just to summarize some crucial points analyzed in this related work section, it is evident that traditional end-to-end solutions cannot properly and fully take into consideration the specific characteristics of mobile wireless streaming. Proxy-based solutions certainly represent a notable improvement and evolution step in the related architectures of solution, since they permit to lower the load on servers while supporting fine-grained and per-client content adaptation. However, proxy solutions do not apply well to collaborative environments stemming from the impromptu interaction of mobile devices getting and providing connectivity in a peer-to-peer way. For this reason, collaborative solutions have recently emerged as the most promising approach to fully exploit resource sharing among nearby devices carried by users socially interacting and willing to collaborate. However, while first proposals have already delineated interesting solution guidelines and notable high-level models of general applicability, they lack in providing actual prototypes supporting the dynamic discovery and exploitation of shared resources in collaborative wireless environments suitable for the proper distribution and adaptation of multimedia content.

8.4 The RAMP Case Study

The previous section has clearly pointed out that node cooperation can greatly improve multimedia content delivery in multi-hop wireless networks stemming from the impromptu interaction of peer nodes. On the one hand, the overhead imposed on intermediary nodes can be properly lowered down by redistributing the same content toward multiple clients. On the other hand, final user satisfaction can be maximized by tailoring streams dynamically depending on specific requirements and capabilities of client devices. However, the reported related work has mostly privileged the proposal of high-level guidelines and analytical solutions, sometimes applied to simulation-based deployment environments (Hsien-Po Shiang and van der Schaar 2007, Hutter et al. 2005, Jurca et al. 2007, Mastronarde et al. 2007); few prototypes have been proposed in the related literature, and they tend to be specialized for some limited and technology-specific deployment environments (Narayanan et al. 2007, Uhm et al. 2010).

We claim that the RAMP middleware is a notable example of solution for collaborative recasting of live multimedia in multi-hop collaborative mobile networks, which overcomes the aforementioned limitations (Bellavista et al. 2010, Bellavista et al. 2012, Bellavista and Giannelli 2010). In particular, this section focuses on how node cooperation in RAMP can enable novel mechanisms for dynamic quality assessment and adaptation, for the support of multimedia tailoring only when and where actually needed, and for the distribution of tailoring load on multiple nodes in a fully distributed way. In particular, as better detailed in the following, the RAMP middleware exploits node collaboration to achieve multiple goals at different layers. RAMP collaborative nodes are as follows:

- Dispatch packets among nodes by exploiting middleware-layer routing facilities
- Recast multimedia flows at the application layer in order to avoid traffic duplication
- Assess stream quality and tailor multimedia content by considering both transport-layer features, e.g., packet delay and jitter, and application-layer characteristics, e.g., the dynamic bitrate variation of H.264 video streams

RAMP adopts many of the guidelines and principles proposed in Section 8.3.3, with the nonnegligible additional proof offering a prototype solution that is easily deployable on most widespread devices. Compared with the work presented in Section 8.3.3, our proposal aims to demonstrate that it is possible to support cooperative multimedia dispatching, monitoring, and tailoring in a lightweight way, also by exploiting off-the-shelf streaming components (i.e., the multiplatform VLC streamer/player for audio/video encoding) and by adopting middleware-based solutions that does not force to modify packet routing at the operating system level. In fact, we have already implemented and extensively validated the RAMP middleware for live streaming in collaborative mobile wireless networks on most common operating systems, i.e., Android, MS Windows XP/Vista/7, Linux Ubuntu/ Debian, and MacOSX Snow Leopard. The prototype code is available at our website (http://lia.deis.unibo.it/Research/RAMP/). The goal is also to provide the researchers' community with a practical contribution to facilitate collaborative networking experimentation and to leverage an emerging critical mass of application developers in the field. In the following, the chapter devotes some space to the presentation of the design guidelines of our original RAMP-based collaborative recasting solution for multimedia delivery and adaptation; this should work as a clear practical example of the general solution guidelines described earlier.

8.4.1 RAMP-Based Collaborative Multimedia Recasting

RAMP supports multimedia content dispatching and tailoring in mobile environments at the middleware layer, enabling to easily share resources and provide services in a collaborative manner. In particular, RAMP collaborative multimedia recasting works based on two primary ideas:

1. Dynamic exploitation of collaborating nodes to split end-to-end multimedia paths in different RAMP-managed segments (Figure 8.4)
2. Dynamic reconfiguration of clients to get their requested stream from their closest source, by autonomously switching among the available path segments (Figure 8.5)

In this way, RAMP frees the multimedia source from the burden of directly managing service provisioning to every client, by lowering the traffic load on it and on its neighbors. At the same time, RAMP has the goal of imposing minimum overhead on collaborating nodes that offer themselves to recast live multimedia content. Furthermore, RAMP manages path segments as if they were the original and complete stream path, e.g., by possibly tuning the multimedia quality of each segment differently, via temporary downscaling of video quality only in correspondence of low-performance links.

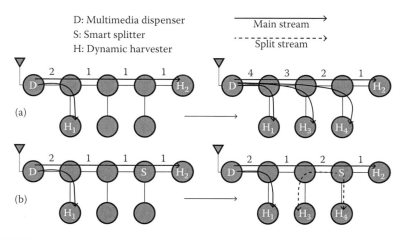

FIGURE 8.4

Active streams in case of two additional harvesters (a) without and (b) with splitter. Values over links indicate the number of per-hop active streams.

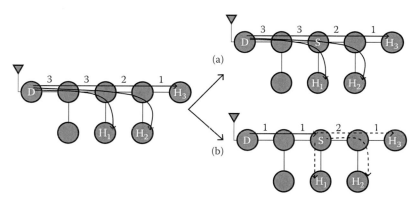

FIGURE 8.5

Active streams in the case of splitter activation (a) without and (b) with dynamic reconfiguration.

To clarify how RAMP nodes collaborate to effectively provide live multimedia to multiple receivers, let us consider again the case of DVB-T program recasting. In this scenario, we identify three different roles:

- *Multimedia Dispenser*, a RAMP node acting as streaming server because equipped with a DVB-T receiver, e.g., node D in Figure 8.4
- *Dynamic Harvester*, a RAMP client receiving the requested stream and rendering it to its user, e.g., node H
- *Smart Splitter*, an intermediary RAMP node that is aware of traversing streams and possibly offers them to clients as if it were a Multimedia Dispenser, e.g., node S

The users of the Multimedia Dispenser nodes can specify which multimedia flows they are willing to redistribute, e.g., a soccer match. Dynamic Harvesters look for the closest Multimedia Dispenser (or Smart Splitter) providing a specific match and possibly request it. The Smart Splitter nodes passively monitor their traversing traffic; thus, Smart Splitters can decide to split and offer traversing programs to other new clients via secondary stream segments, as if they were the real source (origin stream splitting into different segments).

If one adopts a trivial solution with no content recasting on intermediary nodes, the same multimedia flow must be replicated, even in the common case of delivery of the same multimedia content over shared links toward different destinations. Instead, the exploitation of splitters on intermediary nodes can relevantly reduce content delivery redundancy. For instance, in the example of Figure 8.4, the splitter reduces the impact of two new harvesters (H_3 and H_4) by dividing end-to-end streams into two segments (from D to S and from S to $H_{3/4}$), thus lowering the total amount of per-hop active streams from 10 to 6 (Figure 8.4a-right and 8.4b-right, respectively). In particular, this relevantly reduces the traffic load on both the dispenser and its neighbors.

Let us clarify that traditional IP multicast and broadcast mechanisms are unsuitable for these collaborative mobile scenarios. They would require the proactive configuration of operating system routing tables on intermediary nodes, updated and modified whenever the exploited network topology varies, with the consequent frequent and resource-consuming transmissions of both monitoring data and operating system-level reconfigurations. In addition, the RAMP middleware approach can enable multimedia management with per-segment granularity, with the possibility of fine-grained quality tailoring, especially on the intermediaries close to low-performance links. Moreover, traditional IP multicast/broadcast cannot be easily used because, in our targeted environments, collaborative nodes typically reside in different and uncoordinated IP subnets.

In the previous lines, for the sake of clarity, we have described only the "static" behavior of our solution, exploited at receiver start-up. In addition to that, our proposal optimizes resource exploitation also by dynamically reconfiguring multimedia streams at runtime. For instance, if our approach were only static, in the case of a new Smart Splitter becoming active on an intermediary node, already active receivers would keep using their "old" Multimedia Dispenser and could not benefit from new Smart Splitter availability (Figure 8.5a); only new receivers could be able to take advantage of it (such as H_3 and H_4 in Figure 8.4b). On the contrary, to fully exploit dynamic node cooperation, RAMP enables Dynamic Harvesters and Smart Splitters to autonomously reconfigure/manage paths during provisioning. In particular, Dynamic Harvesters periodically look for other sources of requested multimedia programs: whenever they find a closer source along the activated path (either a Multimedia Dispenser or a Smart Splitter), they update their path to the new source. For instance, harvesters in Figure 8.5b can use S as their source.

8.4.2 Distributed and Collaborative Quality Tailoring

Cooperating nodes actively work to maximize the stream quality perceived by final users, by mimicking the behavior of components in a feedback-based self-regulating system (Figure 8.6):

- The input is the original stream
- The output is the tailored stream (via proper selective video frame dropping) sent by the sender to the receiver
- The monitor evaluates the received stream error (defined on the basis of percentage of lost packets and jitter) and tries to achieve a good trade-off between stream error and quality (e.g., video frame rate) by dynamically modifying the number of dropped frames

Before going into deeper details, let us note that there is a different tailor/monitor pair of middleware components for each path segment. In fact, while Multimedia Dispenser and Dynamic Harvester provide only tailoring and monitoring features, respectively, Smart Splitter includes both, the former on the sending side and the latter on the receiving one. Thus, we can evaluate the performance of each segment and tailor each stream separately. In addition, by increasing the number of splitters, the stream is divided into a larger set of shorter segments, thus enabling the desired granularity in estimating where bandwidth bottlenecks are and allowing quality downscaling only where required.

8.4.2.1 Per-Segment Quality Metric

We have experimentally found that user-received quality is affected mainly by two factors: the number of lost/unordered packets and the regularity of

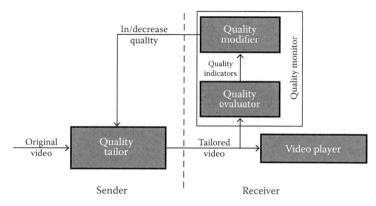

FIGURE 8.6
Per-segment quality tailoring.

packet reception (Luan et al. 2010). While the former can be easily computed by monitoring the sequence number of received RTP headers, the correct determination of the latter deserves more attention because packet arrival regularity depends not only on path conditions (e.g., link saturation largely affects packet dispatching time) but also on packet creation time.

In the simple case of constant bit rate (CBR) coding techniques, such as MPEG2, the generated amount of data per second is always the same. To estimate reception regularity, we have decided to consider the RAMP packet delay variation (rampPDV), defined as the variation of inter-packet interval times (similar to Demichelis and Chimento [2002], but considering RAMP RTP packets instead of IP ones). In particular, we define the normalized rampPDV as the ratio between rampPDV standard deviation and average RAMP inter-packet interval. However, in the more challenging case of variable bitrate coding (VBR) techniques, such as H.264, the desirable throughput varies at runtime, e.g., RTP packets have higher rates in scenes with more moving objects. Moreover, during transitions between motion/ nonmotion (and vice versa) scenes, inter-RTP packet delivery time changes due to bitrate variations. To correctly estimate these cases as "regular" bitrate modifications, we define Stream Jitter (SJ) as follows:

$$SJ = \text{Normalized rampPDV} - \text{Normalized timestampPDV} \qquad (8.1)$$

i.e., the jitter computed for CBR coding minus the ratio among the standard deviation of RTP header timestampPDV and the average inter-packet RTP timestamps. In other words, we estimate RAMP arrival regularity by considering that sometimes RTP packets are deliberately created in a nonperiodic way. In case of CBR coding, normalized timestampPDV value is usually about 1/20 of normalized rampPDV, thus not influencing SJ (it is not zero because of small differences in packet creation times due to multithreading on collaborating nodes).

On the one hand, our experience with in-the-field prototyping and validation has suggested us to consider stream quality still fine only if the percentage of lost/unordered RTP packets (LP) is very low, e.g., less than 2%, nonnegligibly bad when LP does not guarantee stream continuity, e.g., LP greater than 8%, acceptable otherwise. On the other hand, we have experimentally verified that SJ is lower than 0.2 in lightly loaded network conditions, by ensuring very regular packet delivery and thus fine-quality perception. Instead, SJ quickly rises to values greater than 0.8 when the communication channel is heavily loaded, with packet traffic close to saturation bandwidth. Note that the adoption of the SJ indicator allows us to be independent of specific audio/video bitrates, codec techniques, and inter-packet interval, thus facilitating the general applicability of this metric to different collaborative mobile scenarios. In addition, it is worth noting that both LP and SJ are computed by exploiting only data that are already available in RTP packets, without any additional monitoring overhead and negative effect on the size of transmitted packets.

Our receiver (either Dynamic Harvester or Smart Splitter) monitors received packets in a very lightweight way in order to evaluate LP and SJ of the path segment between it and the source (either Smart Splitter or Multimedia Dispenser). In case of need and with a default period of 1.0 s, it notifies its streamer to increase/decrease the quality of the provided stream. In particular, it requests increasing/decreasing quality value of

$$\Delta \text{Quality} = (\Delta \text{ LP} * \Delta \text{ SJ} * 1.5) - 1.0 \tag{8.2}$$

where

$$\Delta \text{LP} = \frac{\left(1 - \tanh\left((\text{LP} - 5) * 0.5\right)\right)}{2} \tag{8.3}$$

$$\Delta \text{SJ} = \frac{\left(1 - \tanh\left((\text{SJ} - 0.5) * 5\right)\right)}{2} \tag{8.4}$$

In other words, if LP is lower than 2% and SJ is lower than 0.2, quality is increased by ≈ 0.5 (the available bandwidth permits to sustain higher video bitrates); if LP is greater than 8% or SJ is greater than 0.8, quality is decreased by ≈ 1.0 (network conditions call for decreasing video bitrate); and by adopting intermediate values otherwise (see Figure 8.7).

8.4.2.2 Per-Segment Quality Tailoring

Our multimedia streamer receives quality feedback by receivers and tailors the stream accordingly, with fine granularity. The solution guideline is to

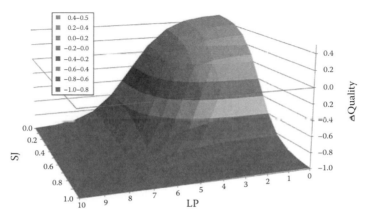

FIGURE 8.7
ΔQuality, depending on LP and SJ.

lower stream throughput selectively by dropping video frames (possibly losing perfect continuity when the supported collaborative path offers very low bandwidth), while preserving TS packets containing audio content. In other words, our solution does not drop RTP packets, but reduces their size discarding part of their payload, thus lowering the total bitrate. In particular, the streamer assigns each client with a quality value (QV) in the [0, 10] range (10 maximum quality, 0 minimum, default value at session startup = 10). Whenever a Dynamic Harvester sends a ΔQuality request, the sender increments/decrements QV accordingly (within the admitted value range). Thus, with the period of 1 s, our solution takes 20/10 s to move from worst/best quality to best/worst quality. In addition, QV is decreased by 2.0 if the stream source does not receive any ACK message from a given receiver for 2 s.

We classify video frames in two different groups, in relation to their role in the rendering process: self-determining (SD) and non-self-determining (NSD). SD frames are self-contained and include any information required for rendering (similar to JPEG images). On the contrary, NSD frames rely on data included in other frames (either SD or NSD). For instance, MPEG2 I frames are classified as SD ones, while MPEG2 P/B are NSD frames. As Figure 8.8 shows, depending on QV, the streamer drops video frames in a differentiated way and with different probabilities (the adopted strategy can be improved and extended with more sophisticated tailoring algorithms, outside of the scope of this chapter):

- If QV = 10.0 (maximum quality), no dropping.
- If QV ≥ 5.0 (mid quality), it drops only NSD frames, with probability $1 - (QV - mid)/(max - mid)$.

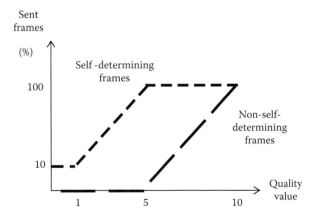

FIGURE 8.8
The simple strategy adopted for quality tailoring in RAMP.

- If QV < 5.0, it drops every NSD frame and SD frames with probability $1 - (mid - QV)/(mid - min)$.
- Exceptional handling—in the unusual situation of very low QV (lower than 1.0), it drops every frame and 90% of SD frames.

In other words, QV variations modify the dropping probability of SD/NSD video frames. Bounding QV variations to a maximum of two units per second permits to modify the video quality slowly, thus gracefully decreasing/increasing the quality perceived by final users without any abrupt disruption. At the same time, stream throughput varies gradually, permitting to the receiver quality monitor to assess whether the current bitrate fits the current channel capabilities. This quality adaptation strategy (with gradually varying quality modifications) smoothes bouncing effects and reduces useless quality adaptation efforts. In addition, as widely recognized, decreasing quality faster than increasing it (fast-decrease, slow-increase) permits to quickly react to bandwidth degradation and to properly tune bandwidth allocation in the case of multiple clients that perceive bandwidth growth concurrently (Allman et al. 2009). Finally, note that quality tailoring is performed with per-segment granularity: one sender can provide the same multimedia stream along different segments with different quality levels based on differentiated quality feedback from receivers.

8.4.3 Performance Evaluation

To present a rapid evaluation of our middleware performance, in the following we consider the simple basic case of one subpath composed of two segments, where a Multimedia Dispenser on node D provides a remote harvester (node H) with multimedia streams, while an intermediate node I dispatches packets between node D and node H. Node D–node I link is based on ad hoc IEEE 802.11g, while node I–node H exploits infrastructured IEEE 802.11b. This scenario represents a typical segmentation case of end-to-end multimedia streaming in spontaneous networking scenarios and permits to have a quantitative but simple overview of the management operations (and related performance) possibly performed in any path segment.

The employed video bitrate is 768 kbps (VBR/H.264 video codec, GOP size of 4), and the audio bitrate is 64 kbps (MPEG-1 Layer 1 audio codec), transmitted via RTP/MPEG-TS; RTP maximum size is 5 KB and, thus, each RTP packet includes at most 27 TS packets. The overall maximum throughput requested for each stream (audio/video bitrate plus RAMP/RTP/MPEG-TS overhead) is about 1050 kbps. Additional experimental results, e.g., based on other multimedia samples and exploiting the less challenging CBR/MPEG2 video codec, are available on our companion website http://lia.deis.unibo.it/Research/RAMP.

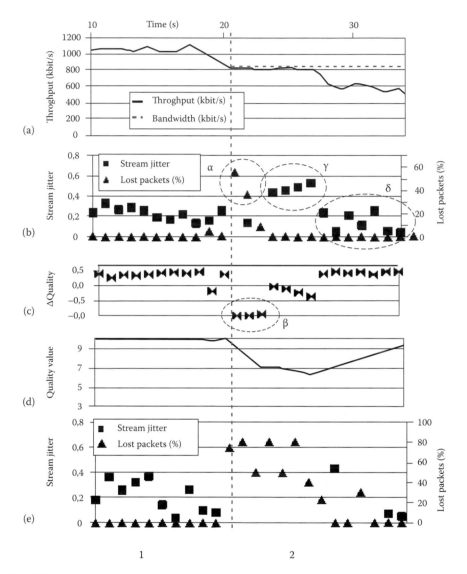

FIGURE 8.9
Node D outgoing throughput (a), received quality on node H with (b) and without (e) monitoring/tailoring, messages sent by harvester to dispenser (c), and quality value assigned by dispenser to harvester (d).

Figure 8.9 shows the significant performance improvements that our middleware can achieve in the case of limited bandwidth availability and with quality monitoring/tailoring activated on node H/node D, respectively (always an active harvester on node H). Figure 8.9a shows node D outgoing throughput without (interval 1) and with (interval 2) bandwidth limitation at 850 kbps. Figure 8.9b and e report the quality received by node H harvester in terms of SJ and LP, exploiting and not exploiting our monitoring/tailoring mechanism, respectively. Figure 8.9c illustrates the ΔQuality messages sent by receiver to sender. Figure 8.9d shows the QV assigned by dispenser to harvester. During interval 1, the RTP stream reaches node H steadily, as shown by SJ and LP values, which are rather low and stable. When bandwidth is limited to 850 kbps (interval 2), node H is notified of a quality degradation event (α, high LP value) and requires node D to greatly lower QV (β); usable SJ values are not available due to insufficient RTP packets received in this interval. After a brief interruption (31 nonconsecutive RTP packets lost in about 2.5 s), the harvester starts receiving the stream again.

Let us stress that now the exchanged RTP packets include only a small subset of TS packets, i.e., every audio and video SD frames and only part of video NSD frames because some of the latter have been discarded by our tailoring middleware (experimented latency of less than 0.03 ms per RTP packet in case of RTP 5088B size). Therefore, on the one hand, throughput is lowered enough to permit to send RTP packets without losing them; on the other hand, final users experience only a very short streaming interruption. After sending few negative ΔQuality messages, LP value gets better again, i.e., the stream bitrate goes below the bandwidth limitation. However, even if LP is zero, QV does not increase because it is limited by the relatively high value of SJ (γ); only when the video bitrate decreases under the bandwidth limitation, the SJ value goes down considerably (δ), by triggering positive ΔQuality messages.

Instead, without our middleware quality monitoring/tailoring, most RTP packets fail to reach node H, in particular in correspondence of H.264 higher bitrates, by producing severe service interruptions (see Figure 8.9e, high LP values for a long period). Let us stress that our bandwidth-shaping mechanism (based on Linux tc) drops packets with layer 2 MTU granularity (1500B): if at least one layer 2 fragment in an RTP packet is lost, the whole RTP packet is lost; for this reason, the percentage of received RTP packets is lower than expected. Then, when the bitrate of the VBR/H.264-based multimedia stream goes below the bandwidth limitation, RTP packets are correctly received (at t = 28.5).

To go into finer details about the performance of the parsing/tailoring mechanisms we have implemented, Table 8.1 and Figure 8.10 show their overhead depending on RTP packet size (set by Multimedia Dispenser prior to VLC transcoder activation at session startup). The reported experimental results have been measured by filtering out the 5% worst data, thus removing spurious measurements, which are generally due to local multithreading management and which made the performance trend less easily understandable.

Moreover, our packet parsing/tailoring mechanisms have demonstrated to be lightweight, with very limited overhead and good scalability while growing the number of simultaneous harvesters. In particular, parsing is applied only once for each program and only on Multimedia Dispensers;

TABLE 8.1

Performance of Parsing/Tailoring Mechanisms

Maximum RTP Size (Byte)	TS Packets	RTP Size (Byte)	Inter-RTP Interval (ms)		Parsing (ms)		Tailoring (ms)	
			Mean	Standard Deviation	Mean	Standard Deviation	Mean	Standard Deviation
1,500	7	1,328	10.429	7.726	0.324	0.351	0.012	0.005
3,072	16	3,020	23.922	12.016	0.746	0.68	0.02	0.007
5,120	27	5,088	43.316	16.899	1.559	1.302	0.028	0.008
7,168	38	7,156	56.204	25.982	2.024	1.809	0.036	0.008
10,240	54	10,164	81.338	29.082	3.205	2.708	0.046	0.011
12,288	65	12,326	96.776	36.694	4.458	3.446	0.053	0.012
15,360	81	15,240	124.639	45.641	5.807	4.451	0.065	0.015

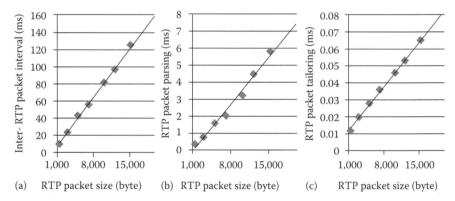

(a)   RTP packet size (byte)   (b)   RTP packet size (byte)   (c)   RTP packet size (byte)

FIGURE 8.10
Inter-RTP interval (a), RTP parsing (b), and RTP tailoring (c) depending on RTP size.

Smart Splitters and Dynamic Harvesters do not perform it. Instead, tailoring is performed on per-segment basis (additive overhead), on both Multimedia Dispensers and Smart Splitters. In particular, to correctly deliver an RTP packet, Multimedia Dispenser has the following constraint:

$$interRTP > parsing + (tailoring * \#segments)$$

In other words, the interval between successive RTP packets must be greater than parsing + tailoring overhead (the latter linearly depends on the number of served segments). Therefore, the maximum number of segments that our multimedia recasting application can serve is as follows:

$$\#segments < \frac{(interRTP - parsing)}{tailoring}$$

By neglecting bandwidth limitation and considering a maximum RTP size of 5 kB (RAMP default value) for 5% filtering, in the worst case scenario of lower standard deviation for inter-RTP (26.417 = 43.316 − 16.899) and upper standard deviation for parsing (2.861 = 1.559 + 1.302) and tailoring (0.036 = 0.028 + 0.08), each Multimedia Dispenser can serve at most 654 segments concurrently. Similar considerations apply to the Smart Splitter case, but with the notable advantage of having to take into account only tailoring overhead (no RTP packet parsing): the result is that Smart Splitter can manage at most 733 concurrent segments. These performance figures have been obtained for Intel Core2 Duo PCs with 2.00 GHz processor and 3 GB RAM; Smart Splitter role is expected to be played by nodes with less resource capabilities but, anyway, the figures provided earlier are definitely larger than what required in any realistic collaborative scenario. In addition, let us note that splitting/tailoring operations have demonstrated to be scarcely CPU-intensive, thus further ensuring that they can be effectively performed also on very resource-limited portable devices.

8.5 Conclusive Remarks

The widespread diffusion of wireless devices is pushing toward the pervasive adoption of mobile services accessed from everywhere and anytime. Multimedia streaming is both the key application in users' demands and the most challenging service to support. On the one hand, users are willing to get on-demand multimedia content from "traditional" servers, such as web-based ones, and to generate live multimedia streams, e.g., audio/video peer-to-peer communications over IP. On the other hand, due to the inherently unreliability and limited performance of wireless links, it is very challenging to support high-quality multimedia streaming in mobile environments.

As pointed out in the chapter, the envisioned mobile deployment environments of the near future call for the adoption of powerful adaptation mechanisms coping with limited connectivity performance and ever-increasing heterogeneity of client devices. To this purpose, end-to-end solutions involve clients and servers to negotiate and properly select the stream quality that best fit the requirements of entities involved in content distribution, e.g., server load, client capabilities, and network available bandwidth. Quality negotiation can be performed either at session start-up or dynamically at provisioning time: in the former case, static constraints are taken into account, e.g., client display size/resolution; in the latter case, the increased complexity due to periodic monitoring and dynamic tailoring is largely counterbalanced by the highly valuable capability of finely tuning stream quality in a much more flexible manner. To simplify multimedia management in mobile environments, proxy-based solutions split paths between servers and clients in two segments, explicitly considering that wired and wireless segments have highly differentiated performance. The adoption of proxies permits to increase performance of legacy servers and clients, e.g., by transparently pre-fetching and caching multimedia content close to mobile devices in order to minimize delivery latency. Collaborative solutions go a step forward, distributing the burden of dynamic quality adaptation among multiple nodes; however, to limit the imposed overhead on collaborating nodes, only low-cost adaptation techniques should be exploited, e.g., avoiding to use CPU-intensive quality tailoring solutions for on-the-fly multimedia transcoding.

We envision future mobile scenarios characterized by increasing heterogeneity where the ever-growing availability of powerful mobile devices will be coupled with the spread of low-end devices. In fact, while most expensive smartphones and tablets will exhibit increasing computing capabilities, the forthcoming mobile ecosystem will probably comprise very heterogeneous devices, also including low-end market devices with limited resources, differing in terms of hardware (e.g., small/medium displays and low-consumption CPUs with reduced computing power), of software (e.g., different mobile platforms supporting different video codecs), and of connectivity, ranging from IEEE 802.11 interfaces to Wi-MAX/HSDPA ones. Moreover, mobile devices will be exploited more massively for audio/video communications, as a valuable alternative to traditional cellular-based ones. In addition to that, the increasing user expectations for multimedia contents, which have to be seamlessly delivered everywhere, will push the adoption of content adaptation techniques for dynamic stream tailoring. In fact, we claim that more traditional solutions based on the static definition and creation of a small set of versions, with differentiated quality, of the same multimedia contents will not fit the ever-increasing demand of different quality levels and the capability of dynamically switching among them in a fine-grained way.

Finally, we believe that user and node cooperation will play a relevant role in the near future. In fact, the widespread diffusion of mobile devices equipped with multiple wireless interfaces, coupled with the willingness of users to share their generated content, is paving the way to a growing market of pervasive and participatory applications. Many related technical challenges still need proper and effective solutions. Among them, there are some crucial technical aspects calling for further investigation: (a) the support of cross-layer-optimized cooperation for multimedia management in order to fulfill quality requirements and constraints while minimizing overhead on content generators, clients, and intermediary nodes; (b) the full exploitation of node cooperation not only to dispatch/adapt multimedia content but also to enable simultaneous multipath techniques and content pre-fetching/caching; and (c) the capability to assess user levels of collaboration in a completely decentralized way, with the main purpose of awarding most collaborative users and punishing too selfish ones, also via incentive-based mechanisms to stimulate node cooperation.

References

Adobe, HTTP dynamic streaming datasheet, http://wwwimages.adobe.com/www.adobe.com/content/dam/Adobe/en/products/hds-dynamic-streaming/pdfs/hds_datasheet.pdf, 2010

Allman, M., V. Paxson, E. Blanton, TCP congestion control, http://tools.ietf.org/html/rfc5681, September 2009.

Bellavista, P., A. Corradi, C. Giannelli, Multi-hop multi-path cooperative connectivity guided by mobility, throughput, and energy awareness: A middleware approach, *The Academy Publisher Journal of Software* (SEUS 2008 special issue), 4(7), 644–653, September 2009.

Bellavista, P., A. Corradi, C. Giannelli, The real ad-hoc multi-hop peer-to-peer (RAMP) middleware: An easy-to-use support for spontaneous networking, *IEEE Symposium on Computers and Communications (ISCC'10)*, Riccione-Rimini, Italy, June 2010.

Bellavista, P., A. Corradi, C. Giannelli, Middleware for differentiated quality in spontaneous networks, *IEEE Pervasive Computing*, 11(3), 64–75, March 2012.

Bellavista, P., C. Giannelli, Internet connectivity sharing in multi-path spontaneous networks: Comparing and integrating network- and application-layer approaches, *Conference on Mobile Wireless Middle Ware, Operating Systems, and Applications (Mobilware 2010)*, Chicago, IL, pp. 84–99, June–July 2010.

Burza, M., J. Kang, P. Van Der Stok, Adaptive streaming of MPEG-based audio/video content over wireless networks, *Journal of Multimedia (JMM)*, 2(2), 17–27, April 2007.

Camp, J., E. Knightly, The IEEE 802.11s extended service set mesh networking standard, *IEEE Communications Magazine*, 46(8), 120–126, August 2008.

Chen, S., B. Shen, S. Wee, X. Zhang, SProxy: A caching infrastructure to support internet streaming, *IEEE Transactions on Multimedia*, 9(5), 1062–1072, August 2007.

Cheung, G., W. Tan, T. Yoshimura, Real-time video transport optimization using streaming agent over 3G wireless networks, *IEEE Transactions on Multimedia*, 7(4), 777–785, August 2005.

Demichelis, C., P. Chimento, IP packet delay variation metric for IP performance metrics (IPPM), IETF, RFC 3393, November 2002, http://www.ietf.org/rfc/rfc3393.txt

Feeney, L.M., B. Ahlgren, A. Westerlund, Spontaneous networking: An application oriented approach to ad hoc networking, *IEEE Communications Magazine*, 39(6), 176–181, June 2001.

Ferreira, L.S., M.D. De Amorim, L. Iannone, L. Berlemann, L.M. Correia, Opportunistic management of spontaneous and heterogeneous wireless mesh networks, *IEEE Wireless Communications*, 17(2), 41–46, April 2010.

Frojdh, P., U. Horn, M. Kampmann, A. Nohlgren, M. Westerlund, Adaptive streaming within the 3GPP packet-switched streaming service, *IEEE Network*, 20(2), 34–40, March–April 2006.

Guenkova-Luy, T., A.J. Kassler, D. Mandato, End-to-end quality-of-service coordination for mobile multimedia applications, *IEEE Journal on Selected Areas in Communications*, 22(5), 889–903, June 2004.

Hutter, A., P. Amon, G. Panis, E. Delfosse, M. Ransburg, H. Hellwagner, Automatic adaptation of streaming multimedia content in a dynamic and distributed environment, *IEEE Conference on Image Processing (ICIP 2005)*, 3, 716–719, September 2005.

Ip, A.T.S., J. Liu, J.C.-S. Lui, COPACC: An architecture of cooperative proxy-client caching system for on-demand media streaming, *IEEE Transactions on Parallel and Distributed Systems*, 18(1), 70–83, January 2007.

Jeong, H.M., M.J. Lee, D.K. Lee, S.J. Kang, Design of home network gateway for real-time A/V streaming between IEEE1394 and ethernet, *IEEE Transactions on Consumer Electronics*, 53(2), 390–396, May 2007.

Jurca, D., J. Chakareski, J.-P. Wagner, P. Frossard, Enabling adaptive video streaming in P2P systems, *IEEE Communications Magazine*, 45(6), 108–114, June 2007.

Liu, Z., Z. Wu, P. Liu, H. Liu, Y. Wang, Layer bargaining: Multicast layered video over wireless networks, *IEEE Journal on Selected Areas in Communications*, 28(3), 445–455, April 2010.

Luan, T.H., L.X. Cai, X. Shen, Impact of network dynamics on user's video quality: Analytical framework and QoS provision, *IEEE Transactions on Multimedia*, 12(1), 64–78, January 2010.

Mastronarde, N., D.S. Turaga, M. Van Der Schaar, Collaborative resource exchanges for peer-to-peer video streaming over wireless mesh networks, *IEEE Journal on Selected Areas in Communications*, 25(1), 108–118, January 2007.

Mukherjee, D., A. Said, S. Liu, A framework for fully format-independent adaptation of scalable bit streams, *IEEE Transactions on Circuits and Systems for Video Technology*, 15(10), 1280–1290, October 2005.

Narayanan, S.R., D. Braun, J. Buford, R.S. Fish, A.D. Gelman, A. Kaplan, R. Khandelwal, E. Shim, H. Yu, Peer-to-peer streaming for networked consumer electronics, *IEEE Communications Magazine*, 45(6), 124–131, June 2007.

Park, S., H. Yoon, J. Kim, A cross-layered network-adaptive HD video streaming in digital A/V home network: Channel monitoring and video rate adaptation, *IEEE Transactions on Consumer Electronics*, 52(4), 1245–1252, November 2006.

Salameh, H.B., M. Krunz, Channel access protocols for multihop opportunistic networks: Challenges and recent developments, *IEEE Network*, 23(4), 14–19, July–August 2009.

Schwarz, H., D. Marpe, T. Wiegand, Overview of the scalable video coding extension of the H.264/AVC standard, *IEEE Transactions on Circuits and Systems for Video Technology*, 17(9), 1103–1120, September 2007.

Shen, J., B. Han, M.-C. Yuen, W. Jia, End-to-end wireless multimedia transmission system, *IEEE 60th Vehicular Technology Conference (VTC2004-Fall)*, Los Angeles, CA, 4, 2616–2620, September 2004.

Shiang, H.-P., M. van der Schaar, Informationally decentralized video streaming over multihop wireless networks, *IEEE Transactions on Multimedia*, 9(6), 1299–1313, October 2007.

Uhm, Y., M. Lee, J. Byun, Y. Kim, S. Park, Development of portable intelligent gateway system for ubiquitous entertainment and location-aware push services, *IEEE Transactions on Consumer Electronics*, 56(1), 70–78, February 2010.

Wang, Z.-J., S.-P. Chan, C.-W. Kok, Receiver-buffer-driven layered quality adaptation for multimedia streaming, *Conference on Signals, Systems and Computers*, Pacific Grove, CA, pp. 1235–1239, October–November 2005.

Wi-Fi Alliance announces groundbreaking specification to support direct Wi-Fi connections between devices, Wi-Fi Alliance press release, Austin, TX, October 2009.

Wu, H., Y. Liu, Q. Zhang, Z.L. Zhang, SoftMAC: Layer 2.5 collaborative MAC for multimedia support in multihop wireless networks, *IEEE Transactions on Mobile Computing*, 6(1), 12–25, January 2007.

Yuan, W., K. Nahrstedt, S.V. Adve, D.L. Jones, R.H. Kravets, GRACE-1: Cross-layer adaptation for multimedia quality and battery energy, *IEEE Transactions on Mobile Computing*, 5(7), 799–815, July 2006.

9

Multilayer Iterative FEC Decoding for Video Transmission over Wireless Networks

Bo Rong, Yiyan Wu, and Gilles Gagnon

CONTENTS

Forward error correction (FEC) plays an important role in modern wireless networks to protect video transmission from errors. For example, low-density parity-check (LDPC) codes, a class of popular FEC option, have been widely adopted by the most advanced wireless communication systems, including IEEE 802.16e, 802.11n, and DVB-S2/T2 [1–3]. To improve the bit error rate (BER) performance, this chapter develops a novel multilayer LDPC iterative decoding scheme using deterministic bits for multimedia communication. These bits serve as deterministic information in the LDPC decoding process to reduce the redundancy during video transmission. Unlike the existing work, our proposed scheme addresses the deterministic bits that can be repositioned, such as moving picture experts group (MPEG) null packets and service information (SI) bits, rather than the widely investigated protocol headers. Simulation results show that our proposed scheme can achieve considerable gain in today's most popular broadband wireless multimedia networks, such as WiMAX and WiFi.

The structure of this chapter is organized as follows. Section 9.1 introduces the state of the art of video transmission, with further explanation on information redundancy of packetized multimedia transmission in Section 9.2. Section 9.3 gives an overview on wireless channel coding, as well as LDPC decoding and shortening. Section 9.4 reviews the existing work on joint protocol-channel decoding (JPCD) and develops our multilayer iterative decoding scheme for LDPC codes using deterministic bits. Section 9.5 presents numerical results to justify the performance of the proposed scheme, followed by Section 9.6 to conclude the chapter.

9.1 Basics of Video Transmission

MPEG-4 Advanced Video Codec and Advanced Audio Codec (AVC/AAC) have been accepted as an audio–visual encoding format by a number of standards. Despite the agreement on the encoding format, there exist two major competing specifications for the transmission of MPEG-4, i.e., MPEG-4 over IP (real-time protocol) and MPEG-4 over MPEG-2 transport stream (TS).

- *Real-time protocol (RTP)*: The RTP was developed to deliver real time data over IP networks [4]. RTP is a native internet protocol and thus works harmonically with the general IP suite. RTP runs in conjunction with the RTP control protocol (RTCP). While RTP carries the media streams, RTCP is responsible for monitoring transmission statistics and quality of service (QoS) [4]. RTP and RTCP run on top of transport layer protocols, such as UDP. RTP stream has to rely on UDP to provide multiplexing over an IP network.

- *MPEG-2 transport stream*: TS is a format specified in MPEG-2, with the design goal of multiplexing digital video and audio and synchronizing the output [5]. MPEG-2 TSs consist of a number of 188 byte TS Packets, each of which has a 4 byte header and a payload. The payload of TS packets may contain program information or packetized elementary streams (PES). PES packets are typically video/audio streams broken into 184 byte chunks to fit into the TS packet payload. TSs are designed such that they can be used in the environment where errors are likely. TSs are commonly used in broadcast applications such as DVB and advanced television systems committee (ATSC). DVB-T/C/S uses 204 bytes and ATSC 8-VSB uses 208 bytes as the size of emission packets (TS packet + FEC data).

9.2 Redundancy in Packetized Multimedia Transmission

There exists tremendous information redundancy in the packetized multimedia transmission over today's wireless communication systems. For example, some bits are constant or predictable during the transmission and, as such, defined as deterministic bits in our work. Those bits can be incrementally identified and recursively fed back in the iterative decoding process to improve the decoding performance.

As shown in the following, deterministic bits may come from three sources, including (1) packet headers, (2) null packets, and (3) SI.

- *Packet headers:* An advanced wireless communication system usually resorts to the RTP/UDP/IP protocol stack to transmit multimedia data. Figure 9.1 illustrates an example of packetized transmission in the IEEE 802.11 standard (WiFi). It has been found that packet headers in each protocol are highly repetitive and thus predicable, due to the intra-layer and inter-layer redundancy [6].

- *Null packets:* Video encoders usually output variable bit rate data depending on the video content, whereas most communication and broadcasting channels are using fixed rate. A data buffer is always implemented at the video encoder output to smooth out the data rate. When there is a buffer underflow, the transport layer at the multiplexer will fill the output data with deterministic null packets.

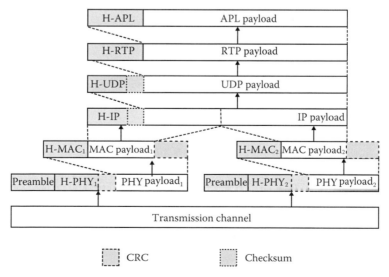

FIGURE 9.1
Protocol stack for multimedia transmission over WiFi.

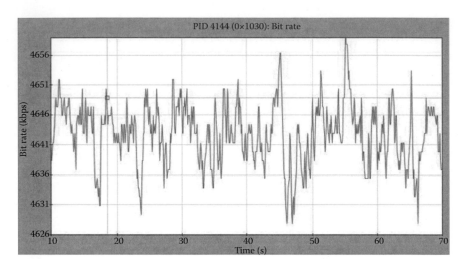

FIGURE 9.2
H.264 video encoder output data rate variation over 1 min period (encoder output rate is set at 5 Mbps).

Our study reveals that there are a considerable number of null packets in today's MPEG TS as well as other CBR video transmission. Figure 9.2 shows an H.264 or MPEG AVC encoder output data rate fluctuation, with an output setting at 5 Mbps. Over 1 min observation time, the maximum data rate is 4.66 Mbps, the minimum data rate is 4.60 Mbps, and the average data rate is 4.64 Mbps. That means that up to 7.2% of null packets need to be inserted by the transport layer to keep a 5 Mbps constant data rate.]

- *Service information*: In television or other services of broadcasting nature, one of the most useful parts of a TS is the SI. SI provides information to enable automatic configuration of the receiver to demultiplex and decode the various streams of programs within the TS. SI may go far beyond simply describing the structure of the stream. For instance, schedule information, detailed information about the elements of a service, language information, and network information are all available. SI has to be sent repetitively and periodically to inform receivers of the programs content and thus contains a plenty of information redundancy.

In this work, we mainly address the controllable deterministic bits, such as MPEG null packets and SI, which are easy to relocate in the encoding buffer, on the transmitter side. As demonstrated in the following, controllable deterministic bits are more flexible and efficient than the uncontrollable deterministic bits such as packet headers.

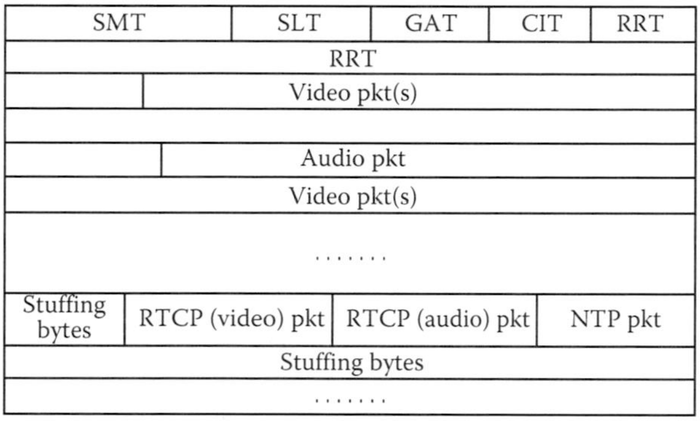

SMT		SLT	GAT	CIT	RRT
RRT					
		Video pkt(s)			
		Audio pkt			
		Video pkt(s)			
		. . . .			
Stuffing bytes	RTCP (video) pkt	RTCP (audio) pkt		NTP pkt	
	Stuffing bytes				
		. . . .			

FIGURE 9.3
Example of ATSC mobile DTV RS frame.

The controllable deterministic bits are as follows:

1. *Easy to be identified by location*: We illustrate the frame structure of ATSC mobile digital television (DTV) in Figure 9.3 to show MPEG null packets and SI at the same time [7]. WiMAX and WiFi will have the null packets if they employ MPEG or other constant bit rate (CBR) standards for video streaming. However, WiMAX and WiFi will not have SI, unless carrying broadcasting service such as IPTV. Figure 9.3 is a possible organization of ATSC mobile DTV RS frame, where *SMT, SLT, GAT, CIT, RRT* are the SI placed at the beginning of the frame and stuffing bytes are the MPEG null packets placed at the end of the frame [7]. As a result, these deterministic bits are easy to be detected and identified by a receiver. It is usually not possible to place packet headers in the beginning and end of the frame, since they are always followed by a payload of variable length.

2. *Easy to be identified by content*: MPEG null packet is an MPEG TS packet of 188 bytes filled mostly with 0x00. In case of errors, say 10% of 0x00s become nonzeros, we can still identify the null packet either with the packet ID in the header (0x1fff) or with the rest of 90% 0x00s by cross-correlation. Packet headers, on the other hand, are not that easy to be identified by their content. For instance, it is difficult to tell an error occurred or not if an IP address like "198.101.10.8" has been corrupted to "198.101.20.3."

3. *High efficiency*: As stated in (2), an MPEG null packet can easily provide a bunch of 188 bytes deterministic bits. It is not possible to obtain so many deterministic bits from a packet header over a short period though. Moreover, each field in an IP header must be compared and analyzed independently, which significantly increases the computational cost.

4. *Stable decoding gain*: As shown in the rest of the chapter, controllable and uncontrollable deterministic bits correspond to the LDPC shortening at fixed and random positions, respectively. As a result, controllable deterministic bits are able to contribute constant decoding gain, whereas the uncontrollable ones produce different gains case by case.

9.3 Channel Coding for Wireless Communications

9.3.1 FEC and LDPC Codes

FEC is a method commonly used in wireless communications to handle losses, enabling a receiver to correct errors/losses without further interaction with the sender. As shown in Figure 9.4, an (n, k) block code converts k source data into a group of n coded data. Usually, the first k data in each group are identical to the original k source data (systematic code); the remaining $(n-k)$ data are referred to as parity data. In coding theory, a parity-check matrix is a basis for a linear code, i.e., a codeword $\mathbf{x}$ belongs to a linear block code C if and only if $H\mathbf{x} = \mathbf{0}$.

Today's wireless networks have to face an increasing demand of high data rate and reliability, which significantly depends on the error correction schemes with near Shannon limit performance. LDPC codes are among the best candidates for this need. For example, quasi-cyclic LDPC (QC-LDPC), a class of LDPC codes, has been widely adopted by the most advanced wireless communication systems, such as IEEE 802.16e, 802.11n, and DVB-S2/T2 [1–3].

LDPC codes have two major features differentiating themselves from other block codes, i.e., the LDPC matrix and the belief propagation (BP) decoding algorithm. An $m \times n$ parity-check matrix H can be associated to a bipartite graph with each column (each component of a codeword) corresponding to a bit node and each row (a parity-check constraint) corresponding to a check

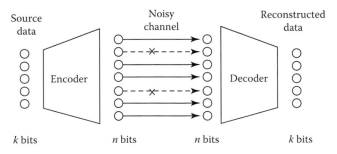

FIGURE 9.4
FEC encoding/decoding.

node. An edge connects the ith bit node and the jth check node, if the ith codeword component participates in the jth constraint equation, i.e., $H_{ji} = 1$. In this way, the BP decoding process can be interpreted to be the exchange of information iteratively between the two kinds of nodes over edges.

Unlike the maximum a posteriori (MAP) decoding algorithm that seeks for the global optimization over the whole codeword space, BP algorithm seeks only for the local optimization, which means a bit node in the bipartite graph can only make use of the information that flow into it without the knowledge of other bit nodes. This local optimization of BP lowers the decoding complexity at the expense of the decoding performance. In recent years, many modified BP decoding algorithms have been proposed to alleviate this problem, such as joint row and column algorithm (JRC) BP algorithm [8,9], oscillation based algorithm (OSC) [10], and so on. These improvements in decoding algorithms refine the updating process of the extrinsic information between the bit nodes and check nodes to some extent.

9.3.2 Deterministic Bits for LDPC Decoding

Using deterministic bits in decoding is also known as "shortening" from the perspective of channel coding theory. This topic can be traced back to the shortened Bose-Chaudhuri-Hocquenghem (BCH) and RS codes, especially the latter that can be shortened almost at will. For LDPC codes, only a few studies were published recently [11–13]. These works show that decoding with deterministic bits is equal to the shortening of information bits, and the extrinsic information transfer (EXIT) chart can be applied for performance evaluation.

Take the (8, 4) LDPC code defined by Equation 9.1 as an example, where the code has a length of 8 with 4 information bits and 4 parity bits:

$$H = \begin{pmatrix} 1 & 1 & 1 & 1 & 0 & 0 & 0 & 0 \\ 1 & 0 & 0 & 1 & 1 & 0 & 0 & 1 \\ 0 & 0 & 1 & 0 & 1 & 1 & 1 & 0 \\ 0 & 1 & 0 & 0 & 0 & 1 & 1 & 1 \end{pmatrix}. \tag{9.1}$$

According to H, the bits $(v_1, \ldots, v_8)$ satisfy the following four constraints ("+" denotes modulo 2 operation in this chapter):

$$\begin{cases} v_1 + v_2 + v_3 + v_4 = 0 \\ v_1 + v_4 + v_5 + v_8 = 0 \\ v_2 + v_5 + v_6 + v_7 = 0 \\ v_2 + v_6 + v_7 + v_8 = 0 \end{cases} \tag{9.2}$$

The corresponding bipartite graph is shown in Figure 9.5a.

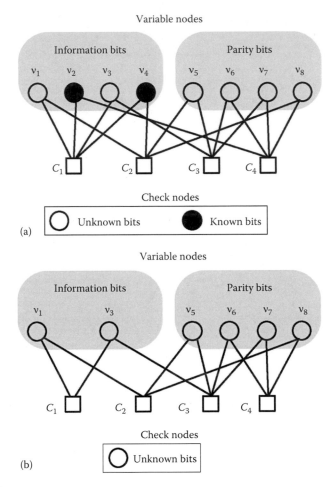

FIGURE 9.5
Bipartite graphs: (a) original graph and (b) residual graph.

Suppose the second and fourth information bits, (v_2, v_4), are set as known. Thus, there remain 2 unknown bits for the receiver. From the perspective of bipartite graph, variable nodes and codeword bits have a one to one correspondence. As v_2 and v_4 are known to the decoder, then in the decoding iterations these 2 nodes are always sending the correct message to the check nodes. In this way, the constraints reduce to

$$\begin{cases} v_1 + v_3 = v_2 + v_4 \\ v_1 + v_5 + v_8 = v_4 \\ v_3 + v_5 + v_6 + v_7 = 0 \\ v_6 + v_7 + v_8 = v_2 \end{cases}. \tag{9.3}$$

Without loss of generality, we set the known bits as zeros, then the constraints further reduce to

$$\begin{cases} v_1 + v_3 = 0 \\ v_1 + v_5 + v_8 = 0 \\ v_3 + v_5 + v_6 + v_7 = 0 \\ v_6 + v_7 + v_8 = 0 \end{cases}' \qquad (9.4)$$

and the parity-check matrix representation is given by

$$H = \begin{pmatrix} 1 & 1 & 0 & 0 & 0 & 0 \\ 1 & 0 & 1 & 0 & 0 & 1 \\ 0 & 1 & 1 & 1 & 1 & 0 \\ 0 & 0 & 0 & 1 & 1 & 1 \end{pmatrix}. \qquad (9.5)$$

From this step, it is clear that deterministic or known bits are equivalent to eliminating the correspondent columns in parity-check matrix or eliminating the correspondent variable nodes with the edges incident to them in bipartite graph as shown in Figure 9.5b. Another fact of this example is that the elimination of columns in H alters the code rate from $1 - 4/8 = 1/2$ to $1 - 4/6 = 1/3$ and the code length from 8 to 6. As a result, the original (2, 4) regular code regenerates to a column-regular, row-irregular one with column degree 2 and row degree (2, 3, 4).

From the aforementioned example, we can reach the following conclusions on "deterministic bits":

1. Deterministic information is equivalent to deleting columns in the information part of the parity-check matrix.
2. Deterministic bits result in the decrease of code rate and length. That is, having m bits known in a (n, k) code will achieve a $(n-m, k-m)$ code.
3. Decoding will be performed on the residual bipartite graph, which is a correspondent column-reduced matrix.
4. Deterministic bits will change the degree distribution of variable and check node.

Therefore, given the number of deterministic bits, the shortening pattern has to be optimized with regard to the performance of residual code. In other words, the degree distributions of shortened code need to be optimized for a low decoding threshold.

Deterministic information not only changes the structure of parity-check matrix but also causes the encoding matrix to mutate. The systematic encoding matrix, denoted by G, is of size k by n and has the information part of a k by k square matrix and the parity-check part of a k by $(n-k)$ matrix. The rows

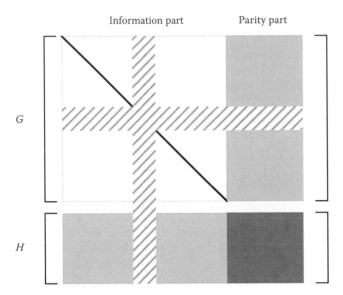

FIGURE 9.6
Correspondence of G and H with m deterministic bits removed.

of G and H span the n-dimensional vector space; the subspaces spanned by G and H are orthogonal to each other, i.e., $GH^T = 0$. We demonstrate the correspondence of G and H in Figure 9.6. As shown in the figure, having m known bits in the codeword is equivalent to deleting m columns in H and both m columns and m rows in G. Suppose the deleted columns are indexed by $G_{del} = \{ci, i = 1, \ldots, m\}$, the columns and rows deleted from G are indexed by the same set. The aforementioned is a general case of encoding. In case of (I) RA codes, there is no need for the G, and the encoding can be done using the column deleted H matrix.

9.4 Multilayer Iterative LDPC Decoding

9.4.1 Architecture Design

We develop in the following a novel multilayer iterative LDPC decoding scheme to improve the BER performance in a multimedia wireless communication system. Our scheme aims at achieving extra decoding gain by using deterministic bits, which may be exploited at any layers above the physical layer. These deterministic bits may involve the widely investigated packet headers [6,14] as well the proposed null packets and SI bits. We especially address the null packets and SI, since they are more flexible and efficient than the packet headers.

The deterministic bits can be gradually identified and used in the iterative decoding process to improve the decoding performance. Our scheme explores as many deterministic bits as possible from all the layers according to the unique feature of a given communication system. Then, these deterministic bits are used to substitute the corresponding received bits iteratively to help the decoding. In terms of coding theory, our scheme is equivalent to achieving a series of subcodes, whose code rates are lower than the original.

The proposed multilayer iterative decoding is different from the conventional joint source-channel coding (JSCC) [15]. JSCC addresses the information redundancy in source coding process, whereas our scheme addresses the transmission and packetization redundancy. JSCC has never been widely deployed in practical systems for two reasons: (1) it requires a combination of special source coding and channel coding algorithms and (2) standard protocol stacks do not allow damaged packets to reach the application layer. Our scheme, on the other hand, is applicable to most existing communication systems with a variety of source and channel coding configurations.

The proposed multilayer iterative decoding is also different from the newly emerging JPCD [16]. JPCD exploits the redundancy present in the protocol stack to facilitate packet synchronization and header recovery at the receiver side. This redundancy is due to the presence of cyclic redundancy checks (CRCs) and checksums at various layers of the protocol stack, as well as from the structure of these headers. Our scheme, however, aims to improve physical layer error correction by applying deterministic bits to LDPC codes. Deterministic bits involve much more than packet headers and could be manipulated at both transmitter and receiver sides.

Figure 9.7 illustrates our proposed approach that can improve the performance of FEC iterative decoder with deterministic information from all possible layers. During each decoding iteration, the decoded symbol from the physical layer is passed to upper layer packet analyzers to find the deterministic bits. Then, these deterministic bits are fed back for the next round of decoding iteration.

Figure 9.8 illustrates where the gain of our multilayer iterative decoding comes from. Modern channel coding technology makes it possible to closely approach the Shannon limit with long code length. However, there does exist information redundancy inside the transmitted symbols before channel coding, i.e., sync, pilot, training sequence, upper layer headers, signaling bits, null packets, SI, etc. Using our approach of multilayer iterative decoding, one can make use of the information redundancy to reduce the receiver C/N threshold and improve coverage or signal robustness.

The scheme of multilayer iterative decoding can be applied to a number of wireless communication systems with different protocol stacks. For example, Figures 9.9 and 9.10 present the architecture and flowchart of a DVB-T2 DTV multilayer iterative decoder. In particular, the smart controller in Figure 9.9 detects a variety of source packets, stores a list of historical deterministic

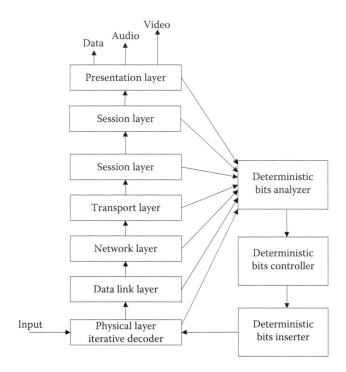

FIGURE 9.7
Multilayer iterative decoding.

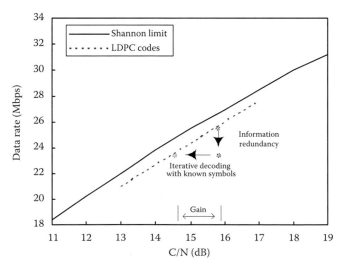

FIGURE 9.8
C/N gain of multilayer iterative decoding.

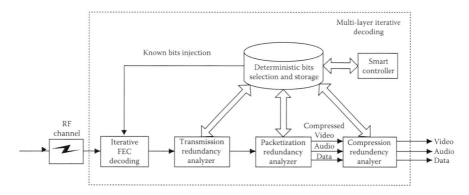

FIGURE 9.9
Architecture of DVB-T2 multilayer iterative decoder.

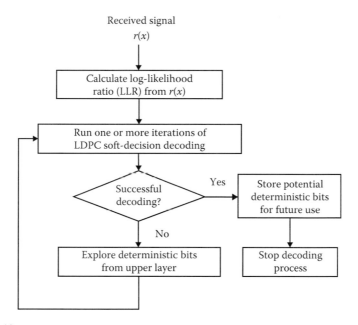

FIGURE 9.10
Flowchart of DVB-T2 multilayer iterative decoder.

bits from all layers, analyzes the correlation between past and future packets, and intelligently obtains deterministic bits to be used in each decoding iteration.

One may argue that the proposed deterministic bits feedback algorithm makes the receiver too complicated. Actually, all those deterministic bits do exist and are used in today's regular receivers for upper layer purpose. They only need to be identified, stored, and fed back to the physical layer to improve the FEC performance.

9.4.2 Arranging Deterministic Bits at Transmitter Side

A study showed that the shortening pattern or the position of deterministic bits in an LDPC block can be optimized to achieve maximum decoding gain [13]. Accordingly, we propose to reposition the controllable deterministic bits in the transmitter buffers before performing LDPC encoding. Figure 9.11 illustrates that the repositioning of controllable deterministic bits involves two steps: (1) repositioning in the frame buffer and (2) repositioning in the FEC encoding buffer.

There are two types of LDPC codes, i.e., regular and irregular. An LDPC code is regular if the rows and columns of parity-check matrix, H, have uniform weight; otherwise, it is irregular. For regular LDPC codes, only Step 1 is necessary; for irregular LDPC codes, both Step 1 and Step 2 are critical. Most existing standards, such as DVB-T2, adopt irregular codes.

Step 1: In practice, there are several ways to transmit the deterministic bits for an easy detection at receiver side. It's worth noting that some deterministic bits, such as SI, have fixed total length, whereas others, such as null packets, have fixed packet size, but variable number of packets, resulting in a variable total length eventually.

1. *By time interval*: For example, SI can be transmitted periodically, like once a second, so that receiver can easily detect them based on fixed time.

2. *By location*: Figure 9.12 shows a data frame with location-sensitive structure for wireless transmission. It specifies that all fixed length deterministic bits, like SI, are inserted at the beginning of a frame. A pointer could also be added as part of the SI to indicate the end of the SI in a frame. Moreover, the deterministic bits of variable

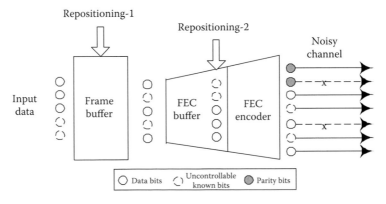

FIGURE 9.11
Repositioning the deterministic bits at transmitter side.

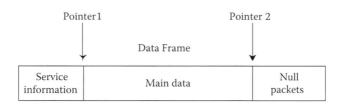

FIGURE 9.12
Example of deterministic bits in frame buffer.

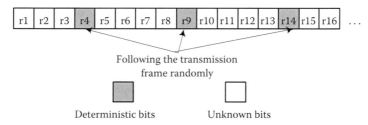

FIGURE 9.13
Random insertion of the deterministic bits in LDPC encoding block.

total length, like null packets, should be appended at the end of a frame as frame paddings. An optional pointer can also be added to indicate the start of them.

3. Aforementioned (1) and (2) work concurrently.

Step 2: Conventionally, the FEC encoder follows the transmission frame to organize the encoding block and generate parity bits. As shown in Figure 9.13, the controllable deterministic bits are inserted randomly and may not be in the best position from the perspective of decoding gain.

We propose to place the controllable deterministic bits to specific positions in encoding block (FEC buffer) before generating parity bits. The specific positions mentioned earlier are related to the shortening property of the error correction code used and are able to be located by certain algorithms. This step, as shown in Figure 9.14, aims at achieving as much decoding gain as possible by manipulating the positions of deterministic bits during the FEC encoding but not the data transmission.

With respect to Figure 9.14, the performance of LDPC shortening is mainly determined by the decoding threshold. In case of irregular codes, the decoding threshold is closely related to degree distribution pair and can be estimated by an EXIT chart [17]. The aforementioned argument leads to a straight forward solution of brute-force searching, i.e., looking for the shortening pattern with the lowest decoding threshold using EXIT chart. Brute-force searching can land on an optimal solution with the computational

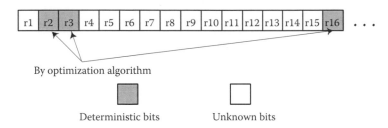

By optimization algorithm

Deterministic bits   Unknown bits

FIGURE 9.14
Smart insertion of the controllable deterministic bits in LDPC encoding block (FEC buffer).

cost varying with the size of parity-check matrix H. For example, it has a computational complexity of $O(n!/p!)$, if p ($<k$) information bits are removed from a (n,k) LDPC codes.

9.5 Numerical Results

In this section, numerical results are presented to demonstrate the performance of our proposed iterative decoding scheme using deterministic bits. There are a number of different irregular QC-LDPC codes listed in 802.16e and 802.11n standards. For example, 802.16e has the code rates of 1/2, 2/3A, 2/3B, 3/4A, 3/4B, 5/6 with possible lengths of 576, 1440, 2304 bit; 802.11n has the code rates of 1/2, 2/3, 3/4, 5/6 with possible lengths of 648, 1296, 1944 bit. Without loss of generality, we take 802.16e and 802.11n LDPC codes as examples to compare different combinations of deterministic bit arrangement.

Let R be the code rate, then the parity-check matrices of 802.16e and 802.11n QC-LDPC are defined by a $24(1 - R) \times 24$ base or mother matrix M with each entry representing the shift of a circulant permutation submatrices (see Figure 9.15).

The base matrices of 1/2 and 2/3 codes have a size of 12×24 and 8×24, respectively, with 12 information columns for 1/2 code and 16 information columns for 2/3 code. In the following, we shorten the QC LDPC codes at the level of base matrix, instead of parity-check matrix, to keep the quasi-cyclic property.

We evaluate the asymptotic performance of different LDPC shortening algorithms by EXIT chart [17]. We remove 4 columns out of all the possible 12 or 16 information columns using different algorithms. Table 9.1 (802.16e) and Table 9.2 (802.11n) list the index of eliminated columns in base matrix M, the weight of deleted column (in parentheses) and the decoding threshold in binary-input additive white Gaussian noise (BAWGN) channel.

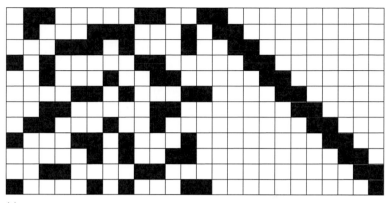

(a)

57				50		11		50		79		1	0										
3		28		0				55	7				0	0									
30				24	37			56	14					0	0								
62	53			53			3	35							0	0							
40			20	66				22	28							0	0						
0				8		42		50				8					0	0					
69	79	79				56		52				0						0	0				
65				38	57			72		27									0	0			
64				14	52			30				32								0	0		
	45		70	0				77	9												0	0	
2	56		57	35							12											0	0
24		61		60				27	51			16	1										0

(b)

FIGURE 9.15
Base matrices in 802.16e and 802.11n standards: (a) 802.16e and (b) 802.11n.

TABLE 9.1

Shortening for the Codes in 802.16e

802.16e		Origin	Brute Force Searching	Average of Random Shortening
Shortening 4 columns 1/2 code	Pattern		5(3),6(6),7(3),9(3)	
	threshold	0.5960 dB	0.1390 dB	0.2260 dB
Shortening 4 columns 2/3A code	Pattern		1(3),2(3),3(6),4(3)	
	threshold	1.4700 dB	1.1050 dB	1.1285 dB

TABLE 9.2

Shortening for the Codes in 802.16N

802.16n		Origin	Brute Force Searching	Average of Random Shortening
Shortening 4 columns 1/2 code	Pattern		4(3),6(3),7(3),10(3)	
	threshold	0.4950 dB	−0.0180 dB	0.2923 dB
Shortening 4 columns 2/3 code	Pattern		1(8),5(6),6(3),8(3)	
	threshold	1.3880 dB	0.9850 dB	1.0359 dB

9.6 Conclusion

This chapter presents a new scheme of multilayer iterative decoding with deterministic bits to improve the performance of LDPC coded wireless communication systems. In our scheme, we address the controllable deterministic bits, such as MPEG null packets, with both transmitter-side and receiver-side mechanisms. We identify the flexible length of deterministic bits as a key problem and further develop a recursive LDPC shortening scheme to overcome it. Numerical results demonstrate that our proposed scheme has a good performance in terms of flexibility, efficiency, and complexity.

References

1. IEEE standard for local and metropolitan area networks Part 16: Air interface for fixed and mobile broadband wireless access systems, IEEE Std 802.16TM-2005.
2. IEEE standard for information technology-telecommunications and information exchange between systems-local and metropolitan area networks-specific requirements Part 11: Wireless LAN medium access control (MAC) and physical layer specifications, IEEE Std 802.11n TM-2009.
3. S.-M. Kim, C.-S. Park, S.-Y. Hwang, A novel partially parallel architecture for high-throughput LDPC Decoder for DVB-S2, *IEEE Transactions on Consumer Electronics*, 56 (2), 820–825, May 2009.
4. Schulzrinne, H., Casner, S., Frederick, R., and V. Jacobson, RTP: A transport protocol for real-time applications, STD 64, RFC 3550, July 2003.
5. ISO/IEC 13818-1:2000 (ITU-T Recommendation H.222.0), Generic coding of moving pictures and associated audio information: Systems, October 2000.
6. C. Marin, Y. Leprovost, M. Kieffer, and P. Duhamel, Robust header recovery based enhanced permeable protocol layer mechanism, *SPAWC 2008*, Recife, Brazil, pp. 91–95, July 6–9, 2008.

7. Advanced Television System Committee, ATSC mobile DTV standard, Part2—RF/transmission system characteristics, Doc. A/153, October 15, 2009.
8. D. E. Hocevar, A reduced complexity decoder architecture via layered decoding of LDPC codes, *Proceedings of IEEE Workshop on Signal Processing Systems*, Austin, TX, pp. 107–112, October, 2004.
9. Z. He, S. Roy, and P. Fortier, Lowering error floor of LDPC codes using a joint row-column decoding algorithm, *Proceedings of IEEE International Conference on Communications*, Glasgow, U.K., pp. 920–925, June 2007.
10. S. Gounai, T. Ohtsuki, and T. Kaneko, Modified belief propagation decoding algorithm for low-density parity check code based on oscillation, *Proceedings of 63rd IEEE Vehicular Technology Conference—VTC 2006-Spring*, Melbourne, Australia, Vol. 3, pp. 1467–1471, May 2006.
11. O. Milenkovic, N. Kashyap, and D. Leyba, Shortened array codes of large girth, *IEEE Transactions on Information Theory*, 52 (8), 3707–3722, August 2006.
12. T. Okamura, A hybrid ARQ scheme based on shortened low-density parity-check codes, *IEEE WCNC 2008*, Las Vegas, NV, pp. 82–87, April 2008.
13. X. Liu, X. Wu, and C. Zhao, Shortening for irregular QC-LDPC codes, *IEEE Communication Letters*, 13 (8), 612–614, August 2009.
14. C. Marin, Y. Leprovost, M. Kieffer, and P. Duhamel, Robust mac-lite and soft header recovery for packetized multimedia transmission, *IEEE Transactions on Communications*, 2010, 58 (3), 775–784, March 2010.
15. M. Fresia, F. Perez-Cruz, H. V. Poor, and S. Verdu, Joint source and channel coding, *IEEE Signal Processing Magazine*, 27 (6), 104–113, November 2010.
16. M. Kieffer and P. Duhamel, Joint protocol and channel decoding: An overview, *Future Network and Mobile Summit 2010*, Florence, Italy, pp. 1–16, June 2010.
17. S. ten Brink, G. Kramer, and A. Ashikhmin, Design of low-density parity-check codes for modulation and detection, *IEEE Transactions on Communications*, 52 (4), 670–678, April 2004.

10

Network-Adaptive Rate and Error Controls for WiFi Video Streaming

JongWon Kim and Sang-Hoon Park

CONTENTS

10.1 Introduction

With the rapid growth of mobile devices, such as smartphones and tablets, huge amount of data traffic transmitted by end-users as well as content providers severely affects the overall performance of 3G/4G wireless networks. Especially, it is reported that Internet-based personal video broadcasting services, such as YouTube and AfreecaTV,* generate the most of data traffic and saturate the provisioned capacity as soon as they are provided. For this reason, telecommunication service providers induce end-users to leverage WiFi wireless networks for bandwidth-savvy video applications. Due to the recent advances of IEEE 802.11 technology, WiFi networks can now enable maximum achievable throughput higher than 100 Mbps. Recently, the WiFi Alliance is also trying to make the WiFi hotspot program that provides seamless WiFi connections to mobile devices in urban areas. Furthermore, for direct multimedia data transmission between mobile devices, the WiFi display (WFD) has been standardized by the WiFi Alliance (Wi-Fi 2011).

Figure 10.1 illustrates a typical video streaming environment over a WiFi wireless network. In this environment, typically diverse video contents from a video server are transmitted to mobile nodes via the WiFi wireless network. For high-quality video streaming, however, there have been a few challenges, detailed later, due to the time-varying nature of underlying WiFi wireless networks and multiple competing video flows from many mobile users, which may result in severe network congestion (Zhang et al. 2005). First, the available network bandwidth for multiple flows of video streams is scarce and limited when compared to the wired networks. For example,

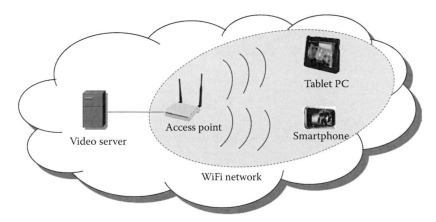

FIGURE 10.1
Typical video streaming environment over a WiFi wireless network.

---

* AfreecaTV is a P2P-based video streaming service in South Korea, which mainly retransmits TV channels, but also allows users to upload their own videos and shows.

the maximum achievable throughput of most IEEE 802.11n-based WiFi products is less than 150 Mbps (Pefkianakis et al. 2010). The packet delay and jitter is also fluctuating due to the time-varying wireless network condition, which is affected by competing flows, channel fading and interferences, and so on. Additionally, both random and burst losses of packets are happening due to the unstable and congested wireless channels. To the received compressed video streams, the packet losses result in the error propagation, since the dependency within/between video frames is heavily used for the video compression. Mobility-induced variations, including the handoff between WiFi networks, result in the transient loss (i.e., disruption) of packets. As a result, the playout quality degrades with the uncontrollable pause and skip of playout at the mobile nodes.

To address the aforementioned challenges, in this chapter, we introduce *network-adaptive streaming* that aims to improve the quality of video playout at the mobile nodes while utilizing given limited resources in the WiFi wireless network. A key idea behind the network-adaptive video streaming is that video quality could be gracefully adjusted to match the time-varying conditions of underlying wireless network. That is, we basically apply the *monitoring and adaptation* principle to handle the quality challenge of wireless video streaming. To be more specific, there are two main modules, which are required to cover the job of monitoring and adaptation, respectively:

1. *Monitoring*: We need to monitor current status of underlying network (i.e., wireless network monitoring) as well as streaming applications at both ends (i.e., application monitoring). First, the monitoring module for wireless network receives various types of information, which reflects the underlying wireless network conditions measured by the joint effort of the server and mobile nodes. By utilizing measured information about estimated available bandwidth, observed packet loss/delay, and others, the monitoring module for wireless network can provide the status information about selected parameters of network conditions. Next, the status information about streaming applications can be collected by the monitoring module for streaming applications. The collected status information of mobile nodes is informed to the server through the feedback channel.

2. *Adaptation*: Depending upon the wireless network conditions, we should carry out network adaptations* at several places in the wireless video streaming environment to absorb the impact of temporal network variations. For example, we may adopt the network adaptation module that adjusts several parameters of video rate and error protection. As the network adaptation schemes, in this chapter, we introduce two network adaptation schemes such as video rate control and selective error control. The network-adaptive

---

* In this chapter, the term, *network adaptation*, actually means the adaptation *to* the network.

control of video rate adjusts the transmission rates of video streams according to the given estimation about wireless network status. On the other hand, the network-adaptive selective error control improves the video playout quality by selectively protecting against the packet losses of error-prone WiFi wireless network.

Thus, in this chapter, after reviewing the network-adaptive video streaming framework, we will extensively review two representative network adaptation schemes: network-adaptive rate control and error control. The rest of this chapter is organized as follows. In Section 10.2, several fundamentals of network-adaptive video streaming in the WiFi networks are introduced. After that, two case studies of network-adaptive rate control and selective error control in the WiFi networks will be introduced in Sections 10.3 and 10.4, respectively.

10.2 Fundamentals of Network-Adaptive Wireless Video Streaming

10.2.1 Network-Adaptive Video Streaming Example

Figure 10.2 shows an example of feedback-based network adaptation over a typical wireless channel (Kim and Shin 2002). The network adaptation for video packets are mainly executed with network-adaptive error control, i.e., forward error correction (FEC)-based unequal error protection (UEP). A product FEC, composed of both Reed—Solomon (RS) and rate-compatible punctured convolutional (RCPC) error correcting codes, is utilized to protect the video packets of varying priorities and sizes. This same product FEC is also utilized to cope with the instantaneously time-varying wireless channel.

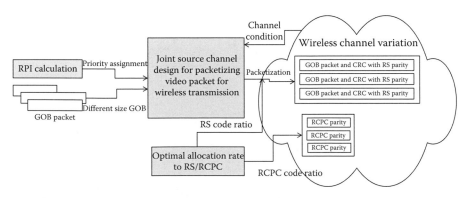

FIGURE 10.2
Feedback-based network-adaptive error control for wireless video.

Based on the layered-RPI (relative priority index) and the size of each video packet along with the feedback channel state information, the joint network adaptation of UEP is conducted with the RS/RCPC product code. The resulting channel packet is modulated and transmitted over the underlying wireless channel. At the mobile node, the received signal is decoded by the maximum likelihood-based scheme, and the long-term fading parameter is estimated. The calculated fading parameter is then fed back via a reliable channel to the server. The level of protection is chosen to give the maximum protection to the video packet stream with both bandwidth and packetization constraints.

10.2.2 Network-Adaptive Video Streaming Framework

In real practice, the aforementioned network adaptation for wireless channel is facing severe difficulty in the monitoring aspect. It is very difficult to timely capture the dynamic variation of wireless channel. Also, the feedback channel back to the server is available in a limited scope, only enabling delayed and intermittent feedback. Compared to this, the WiFi wireless network is exhibiting slightly better environment for network adaptation, since the bidirectional nature of networking enables the consistent and relatively speedy end-to-end feedback to help the dissemination of monitored information. However, the uncoordinated nature of WiFi network environment can sometimes cause the heavy uncontrollable competition among all WiFi traffic. That is, since there is no central entity to govern the flows under the WiFi wireless network, flow unfairness problem arises. This flow fairness issue should be taken care of via a coordination scheme, such as FlowNet by employing source rate control (Yoo and Kim 2010) for applications and backpressure scheduling for WiFi network switches. Note that, in this chapter, we are *assuming that an effective flow coordination scheme for targeted WiFi wireless network is being executed* to ensure appropriate operation conditions for wireless video streaming applications and others.

Figure 10.3 illustrates a network-adaptive video streaming framework for WiFi wireless networks,* with the aforementioned flow coordination assumption. In this framework, the input video stream is compressed, encoded either by scalable or nonscalable video encoders. The compressed video stream is then passed to the real-time parsing and prioritized packetization module for network-based packetized delivery. In this module, the video stream is packetized with priority before going through the network adaptations at the server. The prioritization of video stream is performed by following the layered -RPI concept (Kim and Shin 2002), which reflects the relative influence of each stream to the end-to-end quality. All video packets will be marked by the content-aware streaming application in the granularity of a session, flows, layers, and/or packets. In the case of nonscalable

---

* This framework follows the dynamic network adaptation framework introduced in "Kim and Shin 2002" and provides a simplified framework by focusing on the WiFi video streaming.

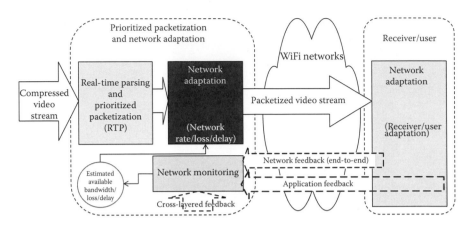

FIGURE 10.3
Network-adaptive video streaming framework in WiFi networks.

MPEG-2 video stream (International Organization for Standardization 1996), therefore, the real-time parsing process is additionally required to prioritize video stream. Based on the associated priority and the time-varying network condition, we match the transport of video packets according to estimated network status (i.e., in terms of available bandwidth, packet loss, and packet delay) of the underlying network. That is, the selected video packets may be forwarded with priority, protected with differentiation, or discarded.

Note that various types of feedbacks are available to guide the required adaptation. Guided by the end-to-end network/application-level feedback (e.g., playout status) from the mobile nodes as well as by the cross-layered feedbacks directly obtained at the server, the video application at the server may control the network adaptations of video. Also, at the mobile nodes, application-level feedback could be used to adapt by speeding up and slowing down for synchronized playback. Finally, the delivered video packets are adaptively processed at the mobile nodes to match its capability and user preference.

10.2.3 Wireless Network Monitoring

In Figure 10.3, the network monitoring module measures underlying wireless network conditions, albeit in semantic forms—available bandwidth, loss and delay, for network-based adaptation of video. In general, the network monitoring could be categorized into receiver-driven and sender-driven approaches.

First, the receiver-driven approach, adopted by most of network monitoring schemes, utilizes *end-to-end* feedback in upper (i.e., transport and application) layers, as shown in Figure 10.3. Main advantage of this approach is that it can use the actual receiving or playback status of mobile nodes for wireless network monitoring. In a severely fluctuating wireless network, however, this approach exhibits serious performance degradation due to the reliability problem (e.g., uncontrollable loss and delay) of the end-to-end feedback packets. That is, the end-to-end feedback packets from mobile nodes could be lost (or delayed) in the severely congested wireless network condition. It could also suffer from the complications of distinguishing packet losses due to buffer overflow and those due to physical-layer errors (Chen and Zakhor 2006).

Next, in comparison, the sender-driven approach aims to mitigate the problem raised in the aforementioned receiver-driven one. Without the help of end-to-end feedback, the sender-driven approach captures the underlying wireless network condition by directly interacting with lower medium access control (MAC) layer (Liu et al. 2003; Park et al. 2006). That is, the *cross-layer* monitoring is adopted to realize the direct monitoring interaction between upper and lower layers (Van Der Schaar and Shankar 2005).

10.2.4 Prioritized Packetization

As discussed already, the role of adaptation is to adaptively match the quality demand of video stream to the underlying network condition by adjusting the sending rate and/or protection. For successful adaptations, however, it is highly important to leverage the awareness on the video stream. The prioritized packetization module packetizes the source video stream into multiple video packets with different priorities that reflect the influence of the video packets to the end-to-end quality. For the prioritization, the dependency relations between the video packets, such as semantic and prediction dependency, can be utilized. In the traditional video encoding schemes, there exists a semantic packet-level dependency between a packet that includes video-encoding header parameters and other dependent packets that need them for decoding. Also, linked by spatial or temporal prediction, the video corruption caused by a packet loss can affect the decoding of subsequent packets. Figure 10.4 illustrates an example of error propagation behavior where transmission errors inserted in $k+2$nd video frame propagates to the next frames (from $k+3$rd to $k+5$th) due to the dependency relations (Tsekeridou and Pitas 2000).

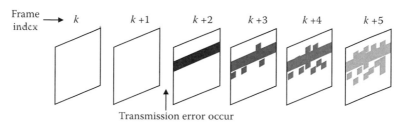

FIGURE 10.4
Example of error propagation behavior.

10.2.5 Network Adaptations with Rate and Error Controls

According to the obtained network monitoring information, the network adaptation module (located at the transport end of server) tries to match the prioritized video packets to the underlying network conditions by focusing on the sending rate and/or protection. In this adaptation process, the adaptation granularity has to be manipulated intelligibly. Its effectiveness is dependent on the accurate association of layered-RPI to each packet, which is one of the key investigation issues. In general, the issue of solving all these adaptations for rate/loss/delay at the same time is too challenging to be solved simultaneously. Thus, depending on the situation, one may focus on the error (or delay) control, leaving the rate control issue behind.

The rate control–based adaptation focuses on maximizing the video quality by adjusting transmission rate under the given estimated available bandwidth. For the rate adjustment, video packets are selectively discarded depending upon the layered-RPI. For example, video frame rate control can be adopted for temporally layered video packets. Figure 10.5 shows a basic temporal scalability structure used in MPEG standards where the video packets containing B frame and P frame have lower priority index compared with the video packet containing I frame due to the dependency relation between frames (International Organization for Standardization 1996). Therefore, the video packets with low priority can be discarded for limited network resources (i.e., bandwidth). In general, spatial and quality scalability can be utilized in addition to the temporal scalability for the priority-based rate control. Note that the granularity of rate control is dependent on the granularity of prioritization.

On the other hand, the error control–based adaptation addresses the reliable video transmission over the fading WiFi channels, which relies heavily on the coordinated protection effort in response to time-varying channel and source variations. In general, most of the existing error control schemes is are based on automatic repeat request (ARQ) and FEC (Sachs et al. 2001; Chan and Zheng 2006). The ARQ scheme represents a reactive error control that retransmits lost/corrupted packets depending upon explicit

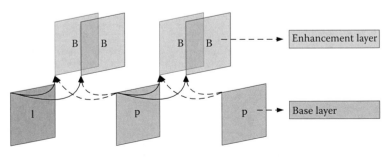

FIGURE 10.5
Basic-level temporal scalability of encoded video.

request(s) by the mobile nodes. However, ARQ should consider the feedback delay of request packets in the real-time video streaming. In comparison, the FEC scheme can be classified to several different proactive error controls that put some redundant information to video packets for error correction. Therefore, the FEC scheme requires additional overhead in the sense of data allocation and computational complexity, while it does not require additional feedback and retransmission.

10.3 Network-Adaptive Rate Control with Sender-Driven Network Monitoring for WiFi Video Streaming

10.3.1 Introduction and Related Work

Now we describe a case study of priority-based network-adaptive video streaming in IEEE 802.11a-based WiFi (Park et al. 2007). In this case, a sender-driven, cross-layered wireless network monitoring is utilized. It intends to speed up the network adaptation and thus can mitigate the problem of receiver-driven wireless network monitoring in a severely congested WiFi network. Furthermore, an intelligent sender interacting with an embedded wireless access point (AP) is adopted, which is capable of performing both video rate adaptation and wireless network monitoring. In this setup, the wireless channel monitoring tightly interacts with underlying MAC and physical (PHY) layers. Without the help of end-to-end feedback, the cross-layered wireless channel monitoring scheme estimates the available bandwidth in terms of the *idle time* of the WiFi channel for a given time period using MAC and PHY transmission statistics. The estimated available bandwidth information is then utilized to adapt the temporally scalable video stream.

Regarding the monitoring aspect, most existing techniques to estimate available bandwidth in wired/wireless networks are receiver-driven and use end-to-end statistics such as packet loss rate and round-trip time (e.g., TCP friendly rate control [TFRC]) (Floyd et al. 2000; Chen and Zakhor 2006). In wireless network, however, these end-to-end schemes require a complicated methodology to distinguish between packet loss due to buffer overflow and that due to physical layer errors. In addition, the estimation schemes for available bandwidth based on end-to-end statistics are not sufficiently accurate, and exhibit serious performance degradation over unstable wireless channel condition due to reliability problem of end-to-end feedback packets (e.g., delay or loss) (Chen and Zakhor 2006). On the other hand, existing active queue management (AQM) algorithms (e.g., RIO), which drop packets based on queue occupancy at the sender, have similar problems as receiver-driven schemes since queue occupancy cannot be an accurate guideline for wireless channel condition (Clark and Fang 1998). To cope with this problem, sender-driven monitoring schemes, which estimates the available bandwidth at sender (e.g., on top of base station), have been proposed. Recently, some work proposed network-adaptive video streaming schemes in wireless local area network (WLAN) based on the sender-driven channel monitoring approach (Schulzrinne et al. 1996; International Organization for Standardization 1999). However, they are limited to heuristic threshold-based estimation of channel conditions.

10.3.2 System Overview

Figure 10.6 shows the system-level overview of the proposed WiFi wireless video streaming, where an intelligent sender with rate adaptation and wireless network monitoring capability is adopted. For the video source, a scalable video stream consists of a base layer (l_1), and $n-1$ enhancement layers $(l_2, l_3, \ldots, l_n)$. It is assumed that a rate profile of the scalable video source, a bitrate set of sublayer streams, is already known as $\{r_1, r_2, \ldots, r_n\}$ to the sender. This rate profile is then converted into a time profile which that is equivalent to the expected transmission time of the video stream over the WiFi network.

For the wireless channel monitoring, a time-based available bandwidth estimation scheme is utilized. Note that the monitoring procedure is triggered for every predefined estimation period, T_{ep}. First, both transmission and retransmission statistics from MAC and PHY layers are observed. Then, these statistics are converted into the form of utilized time resource for transmission and retransmission of MAC data frames (i.e., MAC service data unit [MSDU]), namely, T_{tx} and T_{ro}, respectively. From these parameters, estimated idle time resource, namely, T_{idle}, is calculated as

$$T_{idle} = T_{ep} - T_{tx} - T_{ro}, \quad 0 \le T_{idle} \le T_{ep}. \tag{10.1}$$

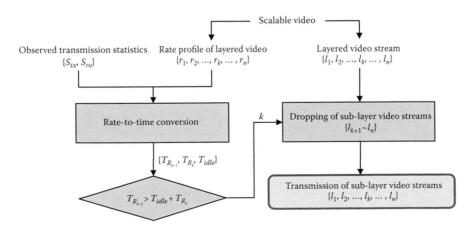

FIGURE 10.6
System-level overview of sender-driven network-adaptive WiFi video streaming.

Note that detailed estimation methods on T_{tx} and T_{ro} are described in Section 10.3.3.

10.3.3 Cross-Layered Video Rate Adaptation

The obtained T_{idle} is compared with the time profile of scalable video to determine the number of maximally allowed video layers, namely, up to k, for transmission.

10.3.3.1 Rate-to-Time Conversion

First, the following two functions based on the MSDU structure of IEEE 802.11a WiFi can be defined.

$$F_{tx}(S,i,\bar{i}) = \left(\frac{L_h + S}{BpS[i]} + \frac{L_h + S}{BpS[i]} \right) \times tSymbol + cSIFS + (tPreamble + tSignal) \times 2 + T_{wait},$$

(10.2)

$$F_{ro}(S,i) = \left(\frac{28 + L_h + S}{BpS[i]} \right) \times tSymbol + tPreamble + tSignal + T_{wait} + ACKtimeout,$$

(10.3)

where $F_{tx}(S, i, i)$ and $F_{ro}(S, i)$ calculate the expected time for transmission and retransmission of data with size S under given PHY mode i for data frame and mode i for ACK frame, respectively. Note that IEEE 802.11a supports eight PHY transmission modes for data frame and three for ACK frame, respectively. IEEE 802.11a specific parameters such as L_h, BpS, $tSymbol$, $B\_ACKsize$, $cSIFS$, $tPreamble$, $tSignal$, and $ACKtimeout$ are stated in the IEEE 802.11a specification (International Organization for Standardization 1999). Also, T_wait, which denotes an average value of backoff duration, can be estimated by using observed transmission failure probability.

10.3.3.2 Observation of MAC and PHY Transmission Statistics

The wireless channel monitoring observes an average size of MAC protocol data unit (MPDU) (i.e., data and ACK frames) from statistics of successfully transmitted and retransmitted MSDUs, $S_{tx}[i]$ and $S_{ro}[i]$, respectively. They are calculated as

$$S_{tx}[i] = \frac{S_{acc_{tx}}[i]}{N_{tx}[i]} \quad \text{and} \quad S_{ro}[i] = \frac{S_{acc_{ro}}[i]}{N_{ro}[i]},$$

(10.4)

where $N_{tx}[i]$ and $S_{acc\_tx}[i]$ denote the number and accumulated size of successfully transmitted MSDUs, respectively. Note that they can be counted for each PHY mode i during T_{ep}. Similarly, $N_{ro}[i]$ and $S_{acc\_ro}[i]$ denote the number and accumulated size of retransmitted MSDUs, respectively.

10.3.3.3 Estimation and Video Rate Adaptation

Now two utilized time resource parameters, T_{tx} and T_{ro}, can be estimated using Equations 10.2 and 10.3 as

$$T_{tx} = \sum_{i=1}^{8} \left\{ F_{tx}\left(S_{tx}[i], i, \overline{i}\right) \times N_{tx}[i] \right\}, \tag{10.5}$$

$$T_{ro} = \sum_{i=1}^{8} \left\{ F_{ro}\left(S_{ro}[i], i\right) \times N_{ro}[i] \right\}, \tag{10.6}$$

where $\overline{i}$ denotes PHY mode to transmit ACK frame. Note that $\overline{i}$ is automatically determined according to i of the data frame in Equation 10.4 (International Organization for Standardization 1999). Also, the rate profile of scalable video can be converted into a time profile using Equations 10.2 and 10.3 as

$$T_{R_k} = \left\{ F_{tx}\left(S_k, \xi, \overline{\xi}\right) + F_{ro}\left(S_k, \overline{\xi}\right) \times p \right\} \times \frac{R_k}{S_k} \times T_{ep}, \tag{10.7}$$

where T_{R_k} is equivalent to the expected time to transmit R_k, which denotes aggregated bit rate of up to kth layers out of n video layers $\left(\text{i.e.,} \sum_{i=1}^{k} r_i \right)$. Also, S_k and p denote the average size of transmitted MPDUs and observed transmission failure probability, respectively. Note that ξ and $\overline{\xi}$ are the respective average of PHY modes during T_{ep} for data and ACK frames, calculated by flooring operation (i.e., $[x]$). Finally, k is determined as

$$k = \max\left\{ \left(h \mid T_{R_h} < T_{idle} + T_{R_\tau}\right) \right\}, \quad h \in 1, 2, \ldots, n, \tag{10.8}$$

where T_{R_τ} is the expected transmission time. Here τ denotes the selected transmission video layer during previous T_{ep}. In Equation 10.6, k is decreased only when T_{idle} becomes zero. Note that in real implementation a damping factor is applied on Equation 10.6 to mitigate this strict condition.

10.4 WiFi Video Streaming with Network-Adaptive Transport Error Control

10.4.1 Introduction and Related Work

Wireless video streaming over the WiFi network is inherently vulnerable to burst packet losses caused by dynamic wireless channel variations with time-varying fading and interference. To alleviate this limitation, especially in the transport layer, the error control schemes based on FEC, ARQ, interleaving, and their hybrid are essential. However, each error control mode shows different performance according to the target application requirement and the channel status. In this section, a case study of network-adaptive selection of transport-layer error (*NASTE*) control is introduced in IEEE 802.11g-based WiFi network, where transport-layer error control modes are dynamically selected and applied (Moon and Kim 2007).

If we briefly review related work, in order to improve the performance of error recovery while matching the given WiFi, numerous hybrid schemes, which combine several error control modes, have been proposed (Hartanto and Sirisena 1999; Sachs et al. 2001; Aramvith et al. 2002). In Moon and Kim (2007), a receiver-initiated, hybrid ARQ scheme is proposed for error recovery of packets in wireless video streaming. It combines the reliability advantage of FEC with the bandwidth-saving properties of ARQ. However, it does not consider multiple retransmissions even when the receiver can tolerate the corresponding delay. In Hartanto and Sirisena (1999), a switching scheme among ARQ, FEC, and hybrid ARQ modes is proposed to match the channel status by considering both playback buffer status and packet loss rate. However, since the burst packet losses are not considered, it cannot avoid sudden video quality degradation. In Aramvith et al. (2002), a low-delay interleaving scheme that uses the playback buffer as a part of interleaving memory is proposed. In this work, the interleaving scheme does not increase the delay significantly and the memory requirement is limited by the amount that is imposed by the video encoder. However, this interleaving scheme uses static depth, which results in severe computational overhead. Regarding the depth of interleaving, in Liang et al. (2002), it is shown that matching the interleaving depth to current burst length is important to convert burst packet losses into isolated ones.

10.4.2 NASTE Framework

The *NASTE* focuses on how to improve the packet loss recovery while utilizing bandwidth efficiently and minimizing delay. For this purpose, the *NASTE* adopts three error control schemes, packet-level FEC, delay-constrained ARQ, and interleaving. Figure 10.7 shows a framework of *NASTE*

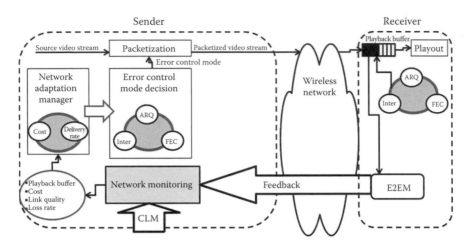

FIGURE 10.7
Framework of *NASTE*.

that the three error control schemes are selectively chosen depending upon monitoring result of underlying wireless network for reliable transmission. For the monitoring, the *NASTE* incorporates the end-to-end monitoring (*E2EM*) and the cross-layered monitoring (*CLM*). The *E2EM* periodically monitors the playback buffer status at the receiver and sends feedback packets to the sender. On the other hand, the *CLM* monitors packet loss rate and link quality in sender-driven manner. Given the monitoring information by *E2EM* and *CLM*, the network-adaptive manager in the sender selects an error control mode that appropriately combines the three error control schemes. Once the error control mode is selected, the packetized video stream is appropriately encoded and transported depending upon the selected error control mode. The video packets transmitted from the sender are stored in the playback buffer and attempts to recover packet errors, if necessary, at the receiver nodes.

10.4.3 Error Control Modes

In the *NASTE*, the dejittering buffer at the client needs to tolerate delay jitters caused by packet interleaving whose maximum depth is 10. Unless specified, the available bandwidth for video streaming is assumed to be sufficient so that the required capacity for error control can be covered. The *NASTE* defines four different error control modes, *AFEC* (adaptive packet–level FEC), *DCARQ* (delay-constrained ARQ), *AFEC & ARQ* (*AFEC* and *DCARQ*), and *IFEC & ARQ* (interleaved *FEC* and *DCARQ*). Each mode has different properties and different performance depending upon the application requirement and underlying wireless channel condition. The error control modes are defined as follows:

- *Mode #1—AFEC$_{n,k}$*: This mode adds *h* redundant packets to *k* original source packets to recover lost packets within bounds (Rizzo 1997). As long as at least *k* packets out of whole *n* packets are received, it can recover all of *k* source packets. However, if the receiver loses *l* (>*h*) packets, it cannot recover all lost packets.
- *Mode #2—DCARQ*: This mode detects lost packets at the client and automatically requests retransmissions from the server after considering delay budget (Shinha and Papadopoulos 2004). It thus can partially reduce unnecessary retransmission trials by avoiding too late retransmissions. Each retransmitted packet, if successfully received, can be used for incremental quality improvement (so-called incremental recovery). Thus, if packet loss rate is high and the delay budget is available, this mode is more efficient than *AFEC* mode.
- *Mode #3—AFEC & ARQ$_{n,k}$*: This mode combines both *AFEC* and *DCARQ* modes. To improve the error control performance further, this mode spends more delay and overhead. Thus, if the client can tolerate resource spending caused by both *AFEC* and *DCARQ*, this mode is selected (Papadopoulos and Parulkar 1996).
- *Mode #4—IFEC & ARQ$_{n,k,d}$*: This mode adds interleaving to *AFEC & ARQ* mode so that it can convert burst packet losses into isolated (random) packet losses. This mode should tolerate delay jitter caused by interleaving as well as retransmission. To fine-tune this, the depth of interleaving should be selected (Liang et al. 2003; Chen et al. 2004).

10.4.4 Mode Switching Algorithm

Figure 10.8 illustrates the mode switching procedure in the *NASTE*. Note that the mode switching procedure is activated whenever there are noticeable changes in monitoring information by *E2EM* and *CLM*. For every periodic interval (P_{ms}), the *E2EM* captures playback buffer level ($D_{playback}$) and *Cost, while CLM captures* packet loss rate (P_l) and link quality (L_Q). After that, these monitoring values are utilized as input parameters of mode selection function (F_{ms}), whose criteria include packet delivery rate (P_{DR}) and *Cost* used for error recovery. The P_{DR} indicates packet delivery rate after applying error

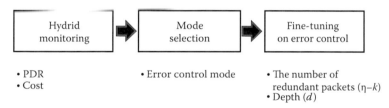

FIGURE 10.8
Procedure of error control mode switching.

recovery procedure while, *Cost* indicates the ratio of redundant packets used for error recovery. Then, by properly balancing P_{DR} and *Cost*, the best error control mode is determined. Once a mode is selected, subsequent fine-tuning of selected mode is applied so that specific parameters (e.g., number of redundant packets and depth of interleaving) for selected mode can be decided to match time-varying WiFi environment.

10.4.4.1 Hybrid Monitoring

To assist mode switching, the *NASTE* measures both the application and channel status by *E2EM* and *CLM*. First, *E2EM* monitors the level of play-back buffer by counting the number of frames stored. Note that, in wireless video streaming, rather long-size playback buffer is essential to absorb the gradual degradation of wireless channel as well as the instantaneous glitches of packet delay (i.e., jitter). The playback buffer level can be converted into time-domain quantity, $D_{playback}$, approximated as

$$D_{playback}(i) = TS\{Q(i-1,n)\} - TS\{Q(i-1,1)\}, \tag{10.9}$$

where
   $TS(\cdot)$ indicates the time stamp function of input packet
   $Q(i,j)$ denotes jth packet in playback queue at ith time instance, respectively

On the other hand, the *CLM* attempts to capture underlying wireless channel condition by monitoring WiFi link-layer statistics. First, to represent time-varying channel quality, such as *SINR* (signal to interference noise ratio), L_Q, which indicates the overall quality of WiFi link, is utilized. After that, P_{LR} is measured by inferring link parameters about the number of received ACKs and transmitted packets. For every P_{ms}, relevant parameters are collected and used to calculate P_{LR}. Then, *NASTE* calculates and updates P_{LR} employing exponentially weighted moving average (EWMA) as

$$P_{LR}(i) = \alpha \times P_{LR}(i) + (1-\alpha) \cdot P_{LR}(i-1), \tag{10.10}$$

where α is set to 0.9.

10.4.4.2 Mode Selection

Based on the monitoring parameters, *NASTE* decides the best error control mode by employing mode selection function (F_{ms}) designed to consider both P_{DR} and *Cost*, approximated as

$$F_{ms}(Mode\#i) = P_{DR}(Mode\#i) + \lambda \times (1 - Cost(Mode\#i)), \qquad (10.11)$$

where i has a value from 1 to 4, while λ is a constant value that determines degree of *Cost*, which indicates the ratio of how many packets are used for error recovery. The Cost can be approximated as

$$Cost(MODE) = \frac{N_{red}}{N_{ori} + N_{red}}, \qquad (10.12)$$

where N_{ori} and N_{red} indicate the number of original packets for video data and the number of redundant packets for error recovery, respectively.

The P_{DR} of each error control mode, denoted as $P_{DR}(Mode\#i)$ in Equation 10.9, varies according to the monitoring information by *E2EM* and *CLM*. To estimate $P_{DR}(Mode\#i)$, a training-based approximation scheme is adopted (Choi 2002; Choi and Shin 2006).

For the fine-tuning of selected error-control mode, *NASTE* first controls the number of redundant packets. Before choosing the number of redundant packets for packet-level FEC, by assuming that packet loss happens independently, the throughput of FEC (T_{FEC}) can be calculated as

$$T_{FEC} = \frac{k}{n} Prob(X \le n - k) = \frac{k}{n} \sum_{j=0}^{n-k} \binom{n}{j} P_{LR}^{j} (1 - P_{LR})^{(n-j)}, \qquad (10.13)$$

where X is the random variable that indicates the number of lost packets in a FEC block.

After that, the interleaving depth (d) is calculated as

$$d = \left\lceil \frac{D_{playback}}{n \cdot L \cdot \left((1/r_{source}) + (1/r_{sending}) + RTT + D_{process}\right)} \right\rceil, \qquad (10.14)$$

where
   n and L indicate the size of FEC block and packet length, respectively
   r_{source}, $r_{sending}$, and $D_{process}$ indicate video source rate, transmission rate, and processing delay due to interleaving, respectively

Once the interleaving scheme is used, the code ratio (i.e., the number of redundant packets) of *AFEC* should be fixed, since overall packet loss rate is spread until interleaving is finished.

References

Aramvith, S., Lin, C.-W., Roy, S. et. al. (2002) Wireless video transport using conditional retransmission and low-delay interleaving. *IEEE Transactions on Circuits and Systems for Video Technology*, 12(6), 558–565.

Chan, S. and Zheng, X. (2006) Video loss recovery with FEC and stream replication. *IEEE Transactions on Multimedia*, 8(2), 370–381.

Chen, L., Sun, T., Sanadidi, M. et al. (2004) Improving wireless link throughput via interleaved FEC. In: *Proceedings of the IEEE Symposium on Computer and Communications*, Alexandria, Egypt, Vol. 1, pp. 539–554.

Chen, M. and Zakhor, A. (2006) Rate control for streaming video over wireless. *IEEE Wireless Communications*, 12(4), 32–41.

Choi, S. (2002) IEEE 802.11e MAC level FEC performance evaluation and enhancement. In: *IEEE GLOBCOM'02*, Taipei, Taiwan, pp. 773–777.

Choi, J. and Shin, J. (2006) A novel design and analysis of cross-layer error control for H.264 video over wireless LAN. In: *Proceedings of the WWIC'06*, Bern, Switzerland, pp. 247–258.

Clark, D. and Fang, W. (1998) Explicit allocation of best-effort packet delivery service. *IEEE/ACM Transactions on Networking*, 6(4), 362–337.

Floyd, S., Handley, M., Padhye, J. et al. (2000) Equation-based congestion control for unicast applications. In: *Proceedings of the ACM SIGCOMM*, Stockholm, Sweden, pp. 43–56.

Hartanto, F. and Sirisena, H. R. (1999) Hybrid error control mechanism for video transmission in the wireless IP networks. In: *Proceedings of the IEEE LANMAN'99*, Sydney, New South Wales, Australia, pp. 126–132.

International Organization for Standardization (1996) Information technology-generic coding of moving pictures and associated audio information, part 2: Video, ISO/IEC 13818-2.

International Organization for Standardization (1999) Wireless LAN medium access control (MAC) and physical layer (PHY) specification: High-speed physical layer in 5 GHz band, IEEE 802.11a WG, Part 11, IEEE 802.11 Standard.

Kim, J. and Shin, J. (2002) Dynamic network adaptation framework employing layered relative priority index for adaptive video delivery. *Lecture Notes in Computer Science, Springer*, 2532, pp. 936–943.

Liang, Y. J., Apostolopoulos, J. G., and Girod, B. (2002) Model-based delay distortion optimization for video streaming using packet interleaving. In: *Proceedings of the Asilomar Conference on Signals, Systems, and Computers*, Pacific Grove, CA, pp. 1315–1319.

Liang, Y. J., Apostolopoulos, J. G., and Girod, B. (2003) Analysis of packet loss for compressed video: Does burst-length matter? In: *Proceedings of the IEEE ICASSP'2003*, Hong Kong, China, vol. 5, pp. 684–687.

Liu, D., Makrakis, D., and Groza, V. (2003) A channel condition dependent QoS enabling scheme for IEEE 802.11 wireless LAN and its Linux based implementation. In: *Proceedings of the IEEE CCECE'03*, Montreal, Quebec, Canada, pp. 1807–1810.

Moon, S. M. and Kim, J. W. (2007) Network-adaptive Selection of Transport Error Control (NASTE) for video streaming over WLAN. *IEEE Transactions on Consumer Electronics*, 53(4), 1440–1448.

Papadopoulos, C. and Parulkar, G. (1996) Retransmission-based error control for continuous media applications. In: *Proceedings of the NOSSDAV'96*, Zushi, Japan.

Park, S., Yoon, H., and Kim, J. (2006) Network-adaptive HD MPEG2 video streaming with cross-layered channel monitoring in WLAN. *Journal of Zhejiang University, Science A*, 7(5), 885–893.

Park, S., Yoon, H., and Kim, J. (2007) A cross-layered network-adaptive video streaming using sender-driven wireless channel monitoring. *IEEE Communications Letters*, 11(7), 619–621.

Pefkianakis, I., Hu, Y., Wong, S. H. Y. et al. (2010) MIMO rate adaptation in 802.11n wireless networks. In: *Proceedings of the ACM MOBICOM'10*, Chicago, IL, pp. 257–268.

Rizzo, L. (1997) On the feasibility of software FEC. DEIT Technical Report LR-970131.

Sachs, D. G., Kozintsev, I., and Yeung, M. (2001) Hybrid ARQ for robust video streaming over wireless LANs. In: *Proceedings of the ITCC 2001*, Las Vegas, NV, pp. 317–321.

Schulzrinne, H., Casner, S., Frederick, V., et al. (1996) RTP: A transport protocol for real-time applications. RFC 1889, Audio-Video Transport Working Group.

Shinha, R. and Papadopoulos, C. (2004) An adaptive multiple retransmission technique for continuous media streams. In: *Proceedings of the NOSSDAV'04*, Cork, Ireland, pp. 16–21.

Tsekeridou, S. and Pitas, I. (2000) MPEG-2 error concealment based on block-matching principles. *IEEE Transactions on Circuits and Systems for Video Technology*, 10(4), 646–658.

Van Der Schaar, M. and Shankar, N. S. (2005) Cross-layer wireless multimedia transmission: Challenges, principles, and new paradigms. *IEEE Wireless Communications*, 12(4), 50–58.

Wi-Fi Alliance Technical Committee Display TG (2011) Wi-Fi display specification, Draft Version1.16.

Yoo, J. and Kim, J. (2010) Centralized flow coordination for proportional fairness in enterprise wireless mesh networks (Poster Paper at *ACM MobiCom 2010*). *ACM SIGMOBILE Mobile Computing and Communications Review (MC2R)*, 14(3), 52–54.

Zhang, Q., Zhu, W., and Zhang, Y. Q. (2005) End-to-end QoS for video delivery over wireless Internet. *Proceedings of the IEEE*, 93(1), 123–134.

11

State of the Art and Challenges for 3D Video Delivery over Mobile Broadband Networks

Omar Abdul-Hameed, Erhan Ekmekcioglu, and Ahmet Kondoz

CONTENTS

11.1 Introduction

Three-dimensional (3D) video is an emerging technology that allows enjoying 3D movies, TV programs, and video games by creating the illusion of depth in the images displayed on the screen. With the recent advances in 3D video acquisition, postprocessing/formatting, coding, transmission, decoding, and display technologies, 3D video is taking the next step toward becoming a reality in the entertainment and telecommunication domains with realistic opportunities. One of the exciting applications that are enabled by 3D video is the extension of traditional TV entertainment services to 3DTV broadcasting that targets the users' homes by large displays capable of 3D rendering and by immersive 3DTV content. For example, a stereoscopic TV is the simplest form of 3DTV where the display is capable of rendering two views, one for each eye, so that the video scene is perceived as 3D. More sophisticated displays are able to render and display multiple views in such a way that the perceived 3D video depends on the user's location with respect to the TV. For example, the user can move his/her head while watching TV to see what is behind a certain object in a video scene. This feature brings a new form of interactivity and an immersive feeling to TV viewers that have never been experienced before.

Besides these advances, the delivery of 3D video services, such as mobile 3DTV over wireless networks, is believed to be the next significant step in visual entertainment experience over wireless, which is expected to appeal to mobile subscribers by bringing 3D content to next-generation handheld devices that are characterized through their portable 3D display technologies (e.g., auto-stereoscopic displays). Globally, the research and development of 3D content delivery in the form of mobile 3DTV over wireless networks has also been actively conducted based on mobile broadcasting technologies, such as terrestrial-digital multimedia broadcasting (T-DMB) and digital video broadcasting-handheld (DVB-H) systems. At the same time, the rapid evolution of mobile broadband technologies into an all-IP network architecture with significant improvements in data rates and reliability would benefit from the rich content leading to new business models for the service provider and where the end user can benefit from the improved network infrastructure for enjoying a 3D video experience on the move. This chapter addresses the state of the art and challenges in delivering 3D video content to wireless subscribers over mobile broadband networks, tackling key issues, such as the effective delivery of 3D content in a system that has limited resources in comparison to wired networks, network design issues, such as the deployment area and the efficient utilization of radio resources, as well as scalability and backward compatibility with conventional two-dimensional (2D) video mobile devices.

11.2 State of the Art

11.2.1 3D Video Representation Formats and Coding

Following the 3D content creation stage, which aims to produce 3D content based on various data sources or data generation devices, the next two stages in the end-to-end 3D video delivery chain are the 3D scene representation formats and the coding approaches, respectively. Several 3D video representation formats and coding approaches for the delivery and storage of 3D video have been established and are foreseen to realize 3D video applications (Smolic et al. 2009). The frame-compatible stereo video format was developed in order to provide 3DTV services over existing digital TV broadcast infrastructures that were originally designed for conventional 2D video coding and transmission. This format is based on a spatial subsampling approach where pixel subsampling is employed in order to keep the resulting frame size and rate same as that of conventional 2D video. The original left and right views are subsampled into half resolution and then they are embedded into a single video frame for compression and transmission. There are several subsampling techniques as illustrated in Figure 11.1 including side by side, top and bottom, line interleaved, and checkerboard.

The side-by-side technique applies horizontal subsampling to the left and right views, thus reducing their horizontal resolution by 50%. The subsampled frames are then put together into one frame side by side. The top-and-bottom technique applies vertical subsampling to the left and right views and the resulting subsampled frames are then put together into one frame top-bottom. In the line-interleaved technique, the left and right views are also subsampled vertically, but the resulting subsampled frames are put together in an interleaved fashion. The checkerboard format subsamples the left and right views in an offset grid pattern and multiplexes them into

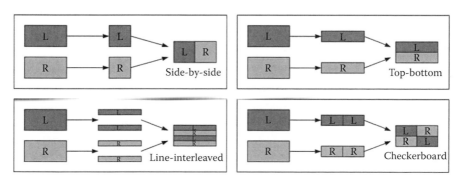

FIGURE 11.1
Illustration of spatial subsampling techniques for frame-compatible formats.

one frame in a checkerboard layout. This format can be compressed using a nonscalable video codec, such as the H.264/MPEG-4 advanced video coding (AVC) video coding standard (ITU-T H.264 2011). At the decoder side, demultiplexing and interpolation are applied in order to reconstruct the left and right views. The spatial subsampling approach is very simple to implement without changing the existing video codec system. However, it does not exploit the correlation between the left and right views for compression efficiency and it shows certain limitations in guaranteeing backward compatibility with legacy 2D receivers.

The two-channel stereo format considers the simplest scenario in the form of left-eye and right-eye views captured from two slightly different viewpoints that roughly correspond to the human eyes' separating distance. This format can be regarded as a special case of multiview video with the number of views equal to two. A straightforward approach to the coding of this format is the independent encoding of each view using existing 2D video coding schemes, such as H.264/MPEG-4 AVC (ITU-T H.264 2011). This approach is termed simulcast encoding, which is the process of encoding each view independently only utilizing the correlation between the adjacent frames in a single view, i.e., only exploiting the temporal correlation.

The advantage of this approach is that it is backward compatible, i.e., it is possible to decode both views using conventional 2D video decoders, thereby eliminating extra decoder complexity. The effect of transmission channel errors on the independently encoded transmitted views is similar to the 2D video case, where, in this case, the overall quality at the receiver side is related to how the human visual system responds to artifacts in either of the views. However, the disadvantage of this approach is that it does not exploit the correlation between the two views, i.e., the inter-view correlation. Since the two nearby views have nearly similar content, an alternative coding approach can be adopted to improve the compression efficiency. In this case, the two views can be encoded jointly using the multiview video coding (MVC) extension of H.264/MPEG-4 AVC (ISO/IEC MVC 2008). Multiview codecs achieve higher coding efficiency than the simulcast encoding approach by exploiting the inter-view correlations between the two adjacent views in addition to the traditional temporal correlation along the frames of each view as illustrated in Figure 11.2.

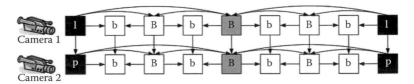

FIGURE 11.2
Illustration of H.264 MVC in case of two views.

An alternative representation format is the color-plus-depth, which consists of a conventional 2D video and a per-sample depth map that determines the position of the associated color data in the 3D space, i.e., each frame of the depth map conveys the distance of the corresponding video pixels from the camera. Depth maps are composed of a single luminance plane without the chrominance components, i.e., gray-level images. This format can be encoded using H.264/MPEG-4 AVC by the independent encoding of the color and depth map. In this case, the desired left and right views are reconstructed at the receiver side using depth-image-based rendering (DIBR) (Fehn 2004) utilizing both the 2D video and the depth map. The color-plus-depth 3D video format allows customized and flexible stereoscopic video rendering with adjustable stereobaseline, depending on the screen size and the viewing conditions (Fehn 2004). The cost of encoding and transmitting this format is usually less in comparison to the two-channel stereo 3D video format. This is mainly due to the texture characteristics of depth maps that comprise large and uniform areas, which can be compressed efficiently using state-of-the-art codecs. The encoding of a depth map presents a small overhead compared to the encoding of textual image. In general, a single depth map requires an extra 10%–20% of the bit rate necessary to encode the original video to encode the depth information such that the decoder has sufficient precision to generate proper stereoscopic images (Smolic et al. 2007). An example deployment of 3D video in the color-plus-depth format was attempted by the European project Advanced Three-dimensional Television System Technologies (ATTEST) (IST FP5 ATTEST) for 3DTV broadcasting over digital video broadcasting-terrestrial (DVB-T). The 2D video is encoded and transmitted using a conventional MPEG-2 approach, while the depth map is encoded independently and transmitted as side information. At the receiver side, the left and right views are reconstructed using DIBR.

The lack of the occlusion information in the color-plus-depth format degrades the scene reconstruction quality to a certain extent. The layered depth video (LDV) format is similar to the color-plus-depth format, except the added occlusion layer information that serves to reconstruct the occlusions on the synthesized images with higher accuracy. Accordingly, better quality views can be synthesized utilizing the additional occlusion layer information. However, like the color-plus-depth format, the quality of the viewing range is still restricted with the LDV format. The addition of another LDV viewpoint results in the so-called depth enhanced stereo (DES) format (Smolic et al. 2009) that enhances the view range and gives premium stereo-video quality, since the base format comprises stereoscopic 3D. However, this comes at the cost of an increased bandwidth requirement.

The multi-view plus depth map (MVD) format consists of multiple (more than 2) viewpoints with associated depth maps to serve limitless viewing range with the ability to synthesize any view point in the range. No occlusion layer information is necessary with this format as any occluded region in any arbitrary synthesized viewpoint can be filled using the texture information of the nearby viewpoints. In this format, N views and N depth maps are used to generate M views at the decoder with $N \leq M$ (Merkle et al. 2007). Nevertheless, the usage of MVD format brings an increased amount of bandwidth requirement depending on the total number of views.

11.2.2 Key Mobile Broadband Technologies

The next stage in the end-to-end 3D video delivery chain is the system used for the delivery of 3D video to the wireless subscribers. Wireless cellular technologies are continuously evolving to meet the increasing demands for high data rate mobile services. Starting with the global system for mobile communications (GSM), it is currently the most widely deployed cellular system with more than three billion subscribers in more than 200 countries (GSMWorld 2009). The proliferation of the Internet and the continuously escalating market demand for mobile data services necessitated the evolution of GSM by developing new technologies. For example, general packet radio service (GPRS) and enhanced data rates for GSM evolution (EDGE) are direct enhancements to GSM networks, capable of supporting packet switched mobile services and providing peak data rates of 128 kbps with GPRS and 384 kbps with EDGE (Holma and Toskala 2007). In order to meet the forecasted growth of cellular subscribers and the need for faster and more reliable data services, the third-generation (3G) Universal Mobile Telecommunications System (UMTS) was developed and standardized in 1999 based on a code division multiple access (CDMA) air interface.

The wideband CDMA (WCDMA) mode of UMTS is currently being widely deployed by cellular operators all over the world with more than 400 million subscribers in more than 130 countries (UMTS Forum 2009). WCDMA UMTS provides higher spectral efficiency than GSM, GPRS, and EDGE with peak data rates of 2 Mbps. High-speed packet access (HSPA) and HSPA+ are direct enhancements to UMTS networks that can provide an uplink peak data rate of 5.7 Mbps and a downlink peak data rate of 14.4 Mbps for HSPA and peak theoretical downlink data rates of 21 Mbps for HSPA+. The International Telecommunication Union (ITU) has been working on a new international standard called International Mobile Telecommunications-Advanced (IMT-Advanced), which is regarded as the succeeding and evolutionary version of IMT-2000, the international standard on 3G technologies and systems. Among the few technologies that are currently contending for a place in the IMT-Advanced standard include the Third-Generation Partnership Project (3GPP) Long-Term Evolution (LTE)/LTE-Advanced, 3GPP2 Ultra Mobile Broadband (UMB), and the Worldwide Interoperability for Microwave Access (WiMAX) based on IEEE 802.16e/m. The next two sections provide an overview on two of them that have wider adoption and deployment, since they are strong candidates for carrying 3D multimedia multicast and broadcast services to mobile users.

11.2.2.1 3GPP LTE

One of the major contenders for a place in the IMT-Advanced standard is 3GPP LTE that was introduced by 3GPP in Release 8 as the next major step on the evolution track. The goals of LTE include higher data rates, better spectrum efficiency, reduced delays, lower cost for operators, and seamless connection to existing networks, such as GSM, CDMA, and HSPA (Holma and Toskala 2007). Commercial deployments of LTE networks have already started around the end of 2009. LTE is based on an orthogonal frequency division multiple access (OFDMA) air interface and it is capable of providing considerably high peak data rates of 100 Mbps in the downlink and 50 Mbps in the uplink in a 20 MHz channel bandwidth. LTE supports a scalable bandwidth ranging from 1.25 to 20 MHz and supports both frequency division duplex (FDD) and time division duplex (TDD) duplexing modes. In order to achieve high throughput and spectral efficiency, LTE uses a multiple input multiple output (MIMO) system (e.g., 2×2, 3×2, and 4×2 MIMO configurations in the downlink direction). The one-way latency target between the base station (BS) and the user equipment (UE) terminal is set to be less than 100 ms for the control plane and less than 5 ms for the user plane. In addition, LTE provides IP-based traffic as well as end-to-end quality of service (QoS) for supporting multimedia services.

11.2.2.2 Mobile WiMAX

WiMAX is a broadband wireless access technology that is based on the IEEE 802.16 standard, which is also called wireless metropolitan area network (WirelessMAN), enabling the delivery of last-mile wireless broadband services as an alternative to cable and digital subscriber line (DSL) wireline access technologies. The name "WiMAX" was created by the WiMAX Forum (WiMAX Forum), which was formed in June 2001 to promote conformance and interoperability of the standard. WiMAX covers a wide range of fixed and mobile applications. The fixed version of the WiMAX standard is termed IEEE 802.16d-2004, which is for fixed or slow position changing devices (e.g., personal computers, laptops). The mobile version is termed IEEE 802.16e-2005, which is an amendment to IEEE 802.16d-2004 to support mobility, targeting mobile devices traveling at speeds of up to 120 km/h (e.g., cellular phones, PDAs). Two modes of sharing the wireless medium are specified in the WiMAX IEEE 802.16 standard as being point-to-multipoint (PMP) and mesh modes. The IEEE 802.16 standard considers the frequency band 2–66 GHz, which is divided into two frequency ranges being 2–11 GHz for non-line-of-sight (NLOS) transmissions and 10–66 GHz for line-of-sight (LOS) transmissions.

WiMAX is capable of offering a peak downlink data rate of up to 63 Mbps and a peak uplink data rate of up to 28 Mbps in a 10 MHz channel bandwidth with MIMO antenna techniques and flexible subchannelization schemes.

Mobile WiMAX is based on an OFDMA air interface and supports a scalable channel bandwidth ranging from 1.25 to 20 MHz by scalable OFDMA. The key feature that makes WiMAX more attractive than cable and DSL is its ease of deployment and the low cost of its installation and maintenance, especially in those places where the latter two technologies cannot be used or their costs are prohibitive (Forsman et al. 2004). In addition, WiMAX can become part of the broadband backbone in countries with a scarce wired infrastructure. Similar to LTE, WiMAX is also competing for a place in the IMT-Advanced standard via the IEEE 802.16m, which is an amendment to IEEE 802.16-2004 and IEEE 802.16e-2005. A theoretical data rate requirement for IEEE 802.16m is a target of 100 Mbps in mobile and 1 Gbps in stationary. In contrast to 802.16e, which uses only one carrier, 802.16m can use two or even more carriers in order to increase the overall data transfer rates.

11.3 Challenge of Effectively Delivering 3D Video

The delivery of 3D video services, such as mobile 3DTV over wireless networks to individual users, poses more challenges than conventional 2D video services due to the large amount of data involved, diverse network characteristics and user terminal requirements, as well as the user's context (e.g., preferences, location). In addition, the demand for mobile data access is intense and will continue to increase exponentially in the foreseeable future as smart phones, personalized video traffic, and other bandwidth intensive applications continue to proliferate in unexpected ways and are straining current wireless networks to a breaking point. There can be several mechanisms that can be adopted for the delivery of 3D video over mobile wireless networks.

11.3.1 Video Delivery Mechanisms

11.3.1.1 Unicast

Recent years have featured a substantial interest in mobile TV, where many operators have already launched mobile TV services on their existing cellular infrastructures, enabling their subscribers to watch live television programs on their mobile devices. Cellular network operators are currently using the streaming option over unicast or point-to-point connections for their mobile TV services. The offering of the service in unicast mode may work well for low to moderate number of subscribers, but it is limited when deploying mass media services in a large-scale market both from cost and technical view points with the inevitable decline in QoS at periods of peak demand. Hence, this would limit the audience and prevent a mass-market deployment. This is due to the fact that the spectrum is a limited and an expensive resource and if numerous recipients that are located in the same cell would try to view the service at the same time, the radio access network, in particular, the wireless link, can easily become a bottleneck and the operator would not be able to offer the service adequately. For example, in a single WCDMA/UMTS cell, it is only possible to support a small number of simultaneous high data rate unicast multimedia sessions, typically in the order of four 256 kbps streams, although this capacity would be somewhat improved in HSPA/HSPA+, LTE, and mobile WiMAX systems (Teleca 2011). Considering the data rates for AVC-compatible stereoscopic video streams at QVGA resolution (a commonly deployed resolution for portable mobile devices), such as 300–350 kbps, —two to three such 3D video services can be realized simultaneously, whereas for MVC-coded full-frame stereoscopic videos (450–500 kbps), even lesser number of services can be realized concurrently. For 3DTV formats other than two-channel stereo video, which usually include extra components, such as depth maps and occlusion regions (if existing), it is unlikely to satisfy a handful of 3DTV subscribers over single unicast connections.

11.3.1.2 Overlay Broadcast

An alternative approach is to use an overlay broadcast access network technology to offer mobile TV and mobile 3DTV services for the mass-market (Alcatel-Lucent 2006). There is a variety of competing technologies that are available today for delivering broadcast content to mobile terminals:

- Terrestrial digital broadcast networks and their extensions, such as DVB-H, based on DVB-T standards, for the transmission of digital TV to handheld receivers
- T-DMB, based on the Eureka-147 digital audio broadcasting (DAB) standard, featuring the reception of mobile video and various data services
- Media forward link only (MediaFLO), a Qualcomm proprietary solution improving DVB-H
- Integrated services digital broadcasting-terrestrial (ISDB-T), a Japanese digital television allowing high definition television (HDTV) as well as reception on handsets
- Hybrid satellite/terrestrial systems, such as the S-DMB in Korea and Japan, and digital video broadcasting-satellite services to handhelds (DVB-SH), a hybrid satellite/terrestrial transmission system standard that is designed to deliver video, audio, and data services to vehicles and handheld devices

However, each of the aforementioned broadcast access network technologies would require an independent network infrastructure to be built, implying additional deployment and operational costs for the service provider.

11.3.1.3 Cellular Broadcast Techniques

The other alternative approach for cellular network operators is to take advantage of cellular broadcast techniques, whereby utilizing such techniques would enable them to take advantage of their already-deployed network infrastructure. Such techniques are supported by the emerging mobile broadband wireless access network technologies, such as WiMAX multicast and broadcast service (MBS) and LTE multimedia broadcast multicast service (MBMS), which are considered as one of the viable wireless networking technologies for meeting the uptake of mobile TV and mobile 3DTV services. For example, some commercial solutions for mobile TV MBS over WiMAX have been provided by Cao and Joy (2008) and WiMAX TV (2008).

11.3.1.3.1 MBS Support in WiMAX

The MBS supported by WiMAX provides an efficient transmission method in the downlink direction for the concurrent transport of data common to a group of mobile stations (MSs) through a shared radio resource using a common connection identifier (CID) (IEEE Std 802.16-2009). The MBS service can be supported by either constructing a separate MBS region in the downlink subframe along with other unicast services, i.e., integrated or embedded MBS, or the whole frame can be dedicated to MBS, i.e., downlink only for a standalone broadcast service. This is illustrated in Figure 11.3 that shows the downlink subframe construction when a mix of unicast and multicast/broadcast services are supported. In addition, it is also possible to construct multiple MBS regions.

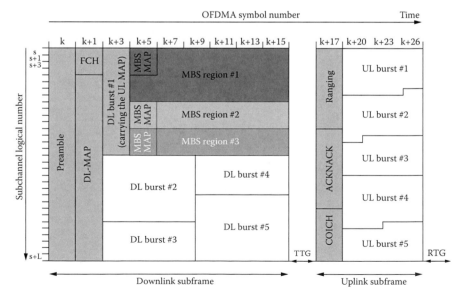

FIGURE 11.3
Illustration of MBS region construction in an OFDMA frame in TDD mode.

In the DL-MAP, there is one MBS MAP information element (IE) descriptor per MBS region that specifies the MBS region PHY configuration and defines the location of each MBS region via the OFDMA symbol offset parameter. For each MBS region, there is one MBS MAP that is located at the first subchannel and first OFDMA symbol of the associated MBS region. The MBS MAP contains multiple MAP data IEs that specify the CID, the location, and the PHY configuration of one MBS burst. The MS accesses the DL-MAP to initially identify the MBS regions and the locations of the associated MBS MAPs in each region. Then the MS can subsequently read the MBS MAPs without reference to the DL-MAP unless synchronization to the MBS MAP is lost.

The area where the MBS service is offered is called an MBS zone, which may consist of one or more than one BS. Each MBS zone is identified by a unique MBS zone ID that is not reused across any two adjacent MBS zones. In the single-BS provisioning of MBS, the BS provides the MSs with MBS using any multicast CID value locally within its coverage area and independently of other BSs. In the multi-BS provisioning of MBS, all BSs within the same MBS zone use the same CIDs and security associations (SAs) for transmitting the content of a certain common MBS. Coordination between these BSs allows an MS to continue to receive MBS transmissions from any BS within the MBS zone, regardless of the MS's operating mode (e.g., normal, idle) and without the need to re-register to the BS. Optionally, MBS transmissions may be synchronized across all BS's within an MBS zone, enabling an MS to receive the MBS transmissions from multiple BSs using macrodiversity, hence improving the reception reliability.

11.3.1.3.2 E-MBMS Support in LTE

One of the key features in LTE Release 9 is the evolved-multimedia broadcast multicast service (E-MBMS) that enables broadcast and multicast services over a cellular network. Although MBMS was initially introduced in 3GPP Release 6 for UMTS, MBMS service trials were only conducted by a small number of operators so far (Teleca 2011). The OFDMA-based LTE air interface now offers a characteristic better suited to these services, including cell-edge spectrum efficiency in an urban or suburban environment that can be achieved by exploiting the special features of E-MBMS. Two scenarios have been identified for E-MBMS operation being the single-cell broadcast and the multimedia broadcast single frequency network (MBSFN). Both the single-cell MBMS and the MBSFN typically use PMP mode of transmission. Therefore, UE feedback such as acknowledgement (ACK)/negative ACK (NACK) and channel quality information (CQI) are not utilized as it is the case in point-to-point mode of transmission. However, 3GPP is currently evaluating the utilization of aggregate statistical ACK/NACK and CQI information for link adaptation and retransmissions.

The idea in MBSFN operation is that the MBMS data is simultaneously transmitted from a set of time-synchronized evolved NodeBs (eNBs) using the same resource block, resulting in the UE receiver observing multiple versions of the same signal with different delays. This enables the UE to combine transmissions from different eNBs, thus greatly enhancing the signal-to-interference-and-noise ratio (SINR) in comparison to the non-SFN operation. This is especially true at the cell edge, where transmissions that would otherwise have constituted inter-cell interference are translated into useful signal energy; hence, the received signal power is increased at the same time as the interference power being largely removed. The cyclic prefix (CP) used for MBSFN is slightly longer, and provided that the transmissions from the eNBs are sufficiently tightly time-synchronized for each to arrive at the UE within the CP at the start of the symbol, there will be no inter-symbol interference (ISI) (3GPP TS 36.300 2011). Figure 11.4 illustrates the overall user-plane architecture for MBSFN operation and MBMS content synchronization.

The eBM-SC is the source of the MBMS traffic and the E-MBMS gateway is responsible for distributing the MBMS traffic to the different eNBs of the MBSFN area. IP multicast may be used for distributing the traffic from the E-MBMS gateway to the different eNBs. In addition, 3GPP has currently assumed that header compression for MBMS services will be performed by the E-MBMS gateway. 3GPP has defined a SYNC protocol between the E-MBMS gateway and the eNBs in order to ensure that the same MBMS content is transmitted from all the eNBs of the MBSFN area. In order to ensure that the same resource block is allocated for a given service across all the eNBs of a given MBSFN area, 3GPP has defined the MBMS coordination entity (MCE), which is a control-plane entity that also ensures that the radio

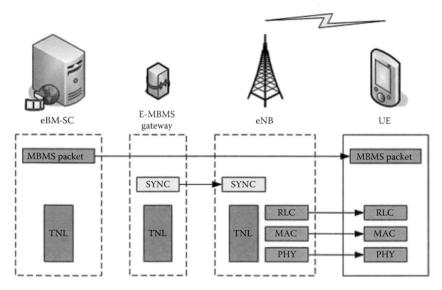

FIGURE 11.4
Overall user-plane architecture for MBSFN operation.

link control (RLC) and medium access control (MAC) sublayers at the eNBs are appropriately configured for the MBSFN operation.

11.3.1.4 Unicast versus Cellular Broadcast Approaches

In comparing the aforementioned unicast and cellular broadcast transmission approaches, it can be observed that, on the one hand, the solutions that are based on unicast mode allow an unlimited number of TV channels to be offered to a limited number of subscribers. On the other hand, the solutions that are based on utilizing cellular broadcast techniques allow a limited number of TV channels to be offered to an unlimited number of subscribers. One possible solution would be the combination of the two transmission approaches into a hybrid approach for meeting the consumer's and the operator's interests (Hong and Lee 2009). For example, the most popular channels can be broadcasted using the cellular broadcast techniques (e.g., WiMAX MBS, LTE MBMS). The secondary channels can be offered using multicast or unicast depending on the demand in each particular cell. Interactive services and niche channels can be selectively offered to the subscribers over unicast links. This way, the radio resources are not wasted by broadcasting channels that only few people watch and yet all users can receive the channels they want to watch.

11.3.1.5 Transport Protocols Used for 2D and 3D TV Services

The transmission of H.264/MPEG-4 AVC encoded streams over the real-time transport protocol (RTP) (Schulzrinne et al. 1996, 2003) is standardized by the IETF in RFC 3984 (Wenger et al. 2005), which describes an RTP payload format for the ITU-T recommendation H.264 video codec. The RTP payload format allows for packetization of one or more network abstraction layer (NAL) units that are produced by an H.264 video encoder in each RTP payload. This RTP payload format is suitable for simulcast coded 3D video representation formats, such as two-channel stereo video encoded independently using H.264/AVC and color-plus-depth. The transmission of MVC-encoded streams such as 3D video streaming, 3DTV, and free-viewpoint video over RTP is described in Wang and Schierl (2009), which is an Internet draft that describes an RTP payload format for the multiview extension of the ITU-T recommendation H.264 video codec. This RTP payload format allows for packetization of one or more NAL units that are produced by the video encoder in each RTP payload. For example, instead of sending the MVC stream as a single H.264/AVC stream over RTP, the stream is divided into two parts where NAL units are transmitted over two different RTP port pairs as if they are separate H.264/AVC streams. Similarly, RTP payload format for scalable video coding (SVC) standard is also defined and the packetizing format allows for splitting the NAL units corresponding to different layers in the 2D/3D video streams to separate RTP sessions.

Currently, AVC-compatible 3DTV services, which offer half resolution pictures to each eye, are most commonly used that utilize RTP packetization and streamed exactly as 2D video services. However, 3D services that offer more flexibility (in terms of comfort-baseline adjustment), as well as higher picture resolution per eye, should use other standards, such as SVC and MVC, or the current most commonly used AVC standard over multiple channels. These formats involve color-plus-depth, LDV, DES, multiview video plus depth, etc.

11.3.2 Radio Resource Efficiency

Since 3D video services, such as mobile 3DTV, demand more radio system resources than conventional 2D video services, one of the key issues to be addressed in the delivery of 3D video over mobile broadband networks is the efficient utilization of the radio resources for providing good quality 3D perception. At the physical layer of OFDMA-based mobile broadband systems, such as WiMAX and LTE, the uplink and downlink transmission properties are described using a set of parameters, such as the modulation type (e.g., QPSK, 16-QAM, 64-QAM), the FEC code type (e.g., CTC, BTC), and the channel-coding rate (e.g., 1/2, 2/3, 3/4, 5/6). These parameters are termed the modulation and coding scheme (MCS) that the base station can use to achieve the best trade-off between the spectrum efficiency and the resulting application level throughput. Table 11.1 lists the OFDMA slot capacity (spectral efficiency) and the theoretical peak data rates in the downlink direction for various MCSs, or in other words, for different number of data bits per slot.

TABLE 11.1

Spectral Efficiency and Peak Downlink Data Rates for Various MCSs

Modulation Type	Number of Bits per Symbol	Channel-Coding Rate	Number of Data Bits per Slot	Peak Data Rate (Mbps)
QPSK	2	1/8	12	0.936
		1/4	24	1.872
		1/2	48	3.744
		3/4	72	5.616
16-QAM	4	1/2	96	7.488
		2/3	128	9.984
		3/4	144	11.232
64-QAM	6	1/2	144	11.232
		2/3	192	14.976
		3/4	216	16.848
		5/6	240	18.72

3D video representation formats, such as two-channel stereo video encoded jointly using H.264/MVC or simulcast AVC, or color-plus-depth and LDV type formats coded using the same codecs with multiple channel streams, all have a multilayered bit stream structure, which may comprise parts that are perceptually more important than others or the decoding of some parts of the bit stream is only possible if the corresponding more important parts are received correctly (e.g., base view or base quality layer). For example, in the case of two-channel stereo video encoded jointly using H.264/MVC, since the right view is predicted from the left view, any error in the left view will affect the quality of the right view directly, thus decreasing the overall perceived 3D video quality. Therefore, the importance and redundancies in 3D video data representation formats and coding can be exploited effectively in order to improve the system performance in terms of the efficient utilization of the radio resources in a system that has limited resources in comparison to wired networks.

There are different transmission approaches that can be adopted for the delivery of a multilayered 3D video bit stream over a mobile broadband network. The first approach is to utilize a robust transmission technique, where the different components of the multilayered 3D video bit stream are transmitted using a robust modulation type and a low channel-coding rate, such as QPSK (CTC) 1/2, i.e., ensuring the service quality in an average RF environment, such as the case in mobile broadcast systems, where the transmission is typically designed to serve the worst-case user. However, this would exhaust the available radio resources and hence would affect the number of offered TV channels and/or other services (VoIP, data, etc.). The second approach is to utilize a radio resource efficient transmission technique, where the different components of the multilayered 3D video bit stream are transmitted using an efficient modulation type and a high channel-coding rate, such as 64-QAM (CTC) 3/4 if allowed by the RF conditions of the target coverage area of the BS, thereby offering the service as efficiently as possible. However, in this case, the area over which the service can be offered would be limited.

The third approach is to utilize an optimized transmission technique, where the first and the second transmission approaches are combined into a unified approach such that the limitations of each are overcome by the primary advantage of the other. In this case, this involves the parallel transmission of the multilayered 3D video bit stream using different spectrally efficient MCSs through different transport connections.

The choice between the aforementioned transmission approaches would depend on the operator's 3D video service business model, where two cases would be mainly considered. In the first case, the business model's aim is to maximize the service quality of the offered 3D video service, such that subscribers at the cell edge would be able to receive the different components of the multilayered 3D video bit stream adequately. This can be realized by utilizing the robust transmission technique. In the second case, the business model's aim is to offer the 3D video service efficiently, thereby consuming fewer resources per TV channel and hence the operator would be able to offer more TV channels (e.g., generate more revenue) or other services (e.g., maximizing the number of subscribers per BS). This can be realized by utilizing the radio resource efficient transmission technique. Considering the two business model cases and the fact that it is essential to ensure that the network is well designed in order to achieve a good compromise between the service quality and the radio resource efficiency in a system that has limited resources in comparison to wired networks, the utilization of the optimized transmission technique would be able to provide such a compromise.

11.3.3 Backward Compatibility

One of the fundamental requirements that should be taken into consideration in the development and deployment of 3D video services, such as mobile 3DTV over mobile broadband networks, is backward compatibility. The aim is to ensure that legacy mobile devices without 3D capabilities are still able to receive and play at least conventional 2D video content. Backward compatibility can be achieved either in the application layer or in the lower layers, such as the data link layer and/or the physical layer, of the mobile broadband technology's radio interface protocol stack.

At the application layer, there are several ways to ensure backward compatibility depending on the utilized 3D video representation format and the coding scheme used. For example, in the case of two-channel stereo format, backward compatibility can be achieved by the independent encoding of the left and right views using existing 2D video encoders, such as H.264/MPEG-4 AVC. Similarly, the color-plus-depth format is also backward compatible since the color and the depth map components are encoded independently using conventional 2D video encoders. Even though other less commonly used coding standards are used, such as SVC and MVC, in either cases, the base layer or the base view that is independently encoded from the other view/layers should be decoded by legacy conforming AVC decoders. For instance, if the MVC extension of H.264/MPEG-4 AVC is used for coding the left and right views of the stereoscopic video, one of the views (e.g., the left view) is encoded according to the AVC standard and the second view (e.g., the right view) is encoded with respect to the left view. In order to differentiate between the AVC-encoded left view and the dependant right view, a new H.264/AVC NAL unit type is used (Type 20), allowing legacy H.264/AVC decoders to ignore this data and extract and decode a 2D version of the content easily. New MVC decoders can gain the full benefits of the additionally coded view and can decode the complete 3D video bit stream including the dependant view.

The advantage of achieving backward compatibility in the application layer is that it is straightforward and easy to implement. However, the disadvantage is that it will be required to receive all the data up to the application layer at the receiver side although the data corresponding to one of the views is not necessary, unless media aware network elements (MANE) are employed at base stations that can truncate/manipulate the RTP streams.

At the data link layer, for example, at the BS's MAC layer in a WiMAX network, the incoming IP packets carrying the left-view and the right-view data in the case of two-channel stereo format or carrying the color and depth map data in the case of color-plus-depth format can be classified into different MAC transport connections. The client software at the MS can include a video format selector that determines how many transport connections can be utilized simultaneously according to the 3D capabilities of the device, i.e., determining in which video format (e.g., 2D, 3D) the content can be viewed. The determination of the transport connections indicates the WiMAX MS MAC/PHY layers to process only those MBS region data that are associated with the video format from the corresponding MBS regions.

11.4 Case Study: Mobile 3DTV over WiMAX Networks

11.4.1 System Overview

An end-to-end system for a 3DTV MBS service over WiMAX mobile broadband networks has been developed within the European project Multimedia Scalable 3D for Europe (MUSCADE) (ICT FP7 MUSCADE). Figure 11.5 illustrates the end-to-end network architecture, which comprises a 3DTV MBS server, a connectivity service network (CSN), an access service network (ASN), and an MBS client. The CSN is owned by the network service provider (NSP) and provides connectivity to the Internet, the application service provider (ASP), or other public or corporate networks. The ASN comprises

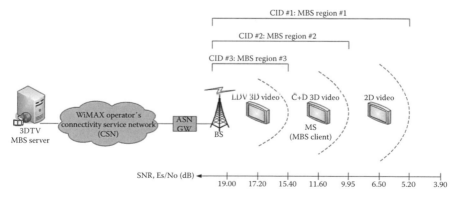

FIGURE 11.5
Network architecture for 3DTV MBS service over WiMAX networks.

an ASN GateWay (ASN GW) and one or more BSs. The MBS client is located in an MS that resides in a WiMAX cell.

The envisaged source 3D video representation format is based on a multilayered 3D video bit stream. The aforementioned LDV format is suitable for this purpose due to its affordable demand on transmission bandwidth and eligibility of better precision stereoscopic video reconstruction at various baselines. The multilayered 3D video bit stream is composed of a first component (color texture video), a second component (per-pixel depth map), and a third component (occlusion information of the color texture video or Δ color) at a specified baseline. Considering the fact that it is essential to ensure that the network is well designed to achieve a good compromise between the service quality and the radio resource efficiency in a system that has limited resources in comparison to wired networks, the coverage area of the BS is divided into three overlapping regions where the BS multicasts/broadcasts the different components of the multilayered 3D video bit stream using different spectrally efficient burst profiles through different transport connections for each TV channel as illustrated in Figure 11.5. The first component is transmitted using a robust MCS through CID #1, allowing the subscribers to receive conventional 2D video over the whole coverage area of the BS. The second component is transmitted using an efficient MCS through CID #2, allowing the subscribers who are receiving a good signal quality to receive CID #2 adequately in addition to CID #1, and thereby allowing them to display color-plus-depth 3D video. The third component is transmitted using an even more efficient MCS through CID #3, allowing the subscribers who are receiving a high signal quality to receive CID #3 adequately in addition to CID #1 and CID #2, and thereby allowing them to display enhanced quality 3D video, i.e., LDV 3D video that allows better modeling of occlusions; hence, enjoying better perceived video quality.

11.4.2 End-to-End System Configuration

This section describes the end-to-end configuration of the system, starting at the 3DTV MBS server and ending at an MS that resides in a WiMAX cell.

11.4.2.1 Multilayered Mobile 3DTV Video Application Characteristics

At the application layer, Figure 11.6 illustrates a sample composition of the LDV multilayered 3D video format, where the three components to be transmitted are shown framed with a dashed line. The first two components, i.e., the left-eye color texture video and the depth map video are used to render the right-eye's view with DIBR (Fehn 2004), with the occlusion areas left blank or filled with background extrapolation for estimation.

In this case, the estimation results in an insufficient quality for the synthesized right-eye video and thus hindering the 3D video perception. Hence, the addition of the third component serves as the quality enhancement layer of the 3D video that estimates the right-eye's view with higher precision.

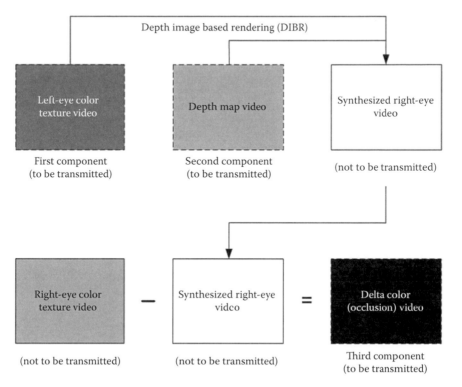

FIGURE 11.6
Envisaged multilayered 3DTV video format.

Using three raw video test sequences of 300 frames each at 30 fps with different content and complexity characteristics, all three components of each video test sequence are encoded separately offline using the H.264/MPEG-4 AVC reference software (Sühring 2011). The characteristics of the utilized video test sequences are summarized in Table 11.2.

TABLE 11.2

Characteristics of Utilized Video Test Sequences

Sequence #1	Sequence #2	Sequence #3
Indoor	Indoor	Outdoor
Simple object motion	Medium complex object motion	Simple object motion
No camera motion	No camera motion	No camera motion
High detail	Medium detail	High detail
Medium complex depth structure	Medium complex depth structure	Complex depth structure
Nine cameras with 5 cm spacing	5 × 3 camera multiview	Twelve cameras with 3.5 cm spacing
1024 × 768 original resolution	640 × 480 original resolution	1024 × 768 original resolution

TABLE 11.3

Main Encoding Parameters' Settings

Parameter	Value
YUV format/image format	YUV 4:2:0/CIF (352 × 288)
Total frames	300
Frame rate (fps)	30
Sequence type	IPPP
Period of I-pictures	30
QP	Color: (I = 27, P = 28)
	Depth: (I = 34, P = 35)
	Δ Color: (I = 36, P = 37)
Entropy coding method	CAVLC
Profile/level IDC	Baseline (66)/20
Motion estimation scheme	SHEX
Slice size (bytes)	498
Bitrate at 30.00 Hz (kbps)	Color: 447.01
	Depth: 198.00
	Δ Color: 213.12

Using unconstrained variable rate coding, i.e., using a constant quantization parameter (QP) for encoding the whole video test sequence and not using any rate control, the inherent rate variation of the video test sequence is not modified and the video encoder in this case generates an encoded video bit stream that has a variable bit rate and variable packet sizes. The main encoding parameters' settings are listed in Table 11.3.

The choice of setting the maximum slice size to about 500 bytes is due to the fact that the video packets should not be large in order to allow better error resilience and to minimize the transmission delay; therefore, the size of each packet is kept smaller than 512 bytes according to the recommendations in 3GPP TS 26.236 (2011).

11.4.2.2 3DTV MBS Server Video Transport and Transmission

The 3DTV MBS server is responsible for video encoding and transmission. RTP/UDP/IP is the typical protocol stack for video transmission. Each compressed component content data, i.e., the color, depth map, and Δ color is encapsulated into RTP packets by encapsulating one slice of a target size per RTP packet. A target RTP packet payload size can be maintained by using the H.264/MPEG-4 AVC error resilience slices feature (ITU-T H.264 2011). At the transport/network layers, the resulting RTP packets are then transmitted over UDP/IPv4 (Postel and Institute 1980, Institute 1981) using variable bit rate packet (VBRP) transmission, where the transmission of a packet solely depends on the timestamp of the video frame the packet belongs to (3GPP TR 26.937 2011). Hence, the video rate variation is directly reflected to the

WiMAX access network. The average IP bandwidth of the transmitted multilayered 3D video bit stream is controlled as follows: for each of the three
components of a video test sequence, the QP is adjusted such that the total
resulting average IP bandwidth would be below 1.5 Mbps. In order to allow
for mapping each IP packetized component into the appropriate transport
connection at the WiMAX BS, the 16 bit destination port field in the UDP
header (Postel and Institute 1980) can be utilized for this purpose. The 3DTV
MBS server is configured to appropriately set the UDP header's destination
port field of each outgoing IP packet based on the component type contained.

11.4.2.3 WiMAX BS and MS

11.4.2.3.1 BS MAC/PHY Layers

At the BS's MAC layer, packet classification is performed at the packet convergence sublayer (CS) by a classification rule, which is a set of matching
criteria applied to each packet entering the BS. A classification rule consists
of some protocol-specific packet matching criteria, a classification rule priority, and a reference to a CID. Protocol-specific packet matching criteria
(e.g., IPv4-based classification) classification rules operate on the fields of
the IP and the transport protocol headers, which include the destination IP
address, IP type of service (ToS), and the transport protocol port number
(IEEE Std 802.16-2009). Based on this specification, the incoming IP packets
can be classified into three different transport connections being CID #1, CID
#2, and CID #3 as illustrated in Figure 11.7 according to the corresponding
transport protocol (UDP) port number value. Once the packets are classified,
they are associated with three real-time variable-rate (RT-VR) service flows
being SFID #1, SFID #2, and SFID #3.

In particular, the IP packets carrying the color, depth map, and the Δ color
components are classified and mapped onto SFID1-CID1, SFID2-CID2, and
SFID3-CID3, respectively, as illustrated in Figure 11.7. At the BS's physical
layer, the incoming MAC protocol data units (PDUs) from the MAC layer are
PHY channel coded using convolutional turbo code (CTC) and the resulting
coded blocks are mapped to the corresponding MBS region for transmission
over the radio channel as follows:

- MAC PDUs corresponding to SFID1-CID1 (contain color component)
  are PHY channel coded and mapped to MBS region #1 for transmission using a robust MCS [QPSK, (CTC) 1/2].
- MAC PDUs corresponding to SFID2-CID2 (contain depth map component) are PHY channel coded and mapped to MBS region #2 for
  transmission using an efficient MCS [16-QAM, (CTC) 1/2].
- MAC PDUs corresponding to SFID3-CID3 (contain Δ color component) are PHY channel coded and mapped to MBS region #3 for
  transmission using an even more efficient MCS [64-QAM, (CTC) 1/2].

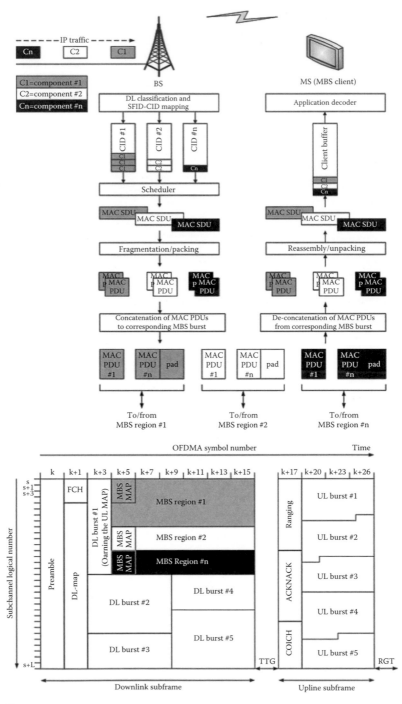

FIGURE 11.7
User-plane data flow over WiMAX BS/MS radio interface protocol stacks.

The idea behind the selection of the lowest order channel-coding rate of 1/2 is to make the three MBS regions' data transmissions as robust as possible and to cover as larger portion of the cell area as possible.

11.4.2.3.2 MS MAC/PHY Layers

As illustrated in Figure 11.7, the reverse operations are performed at the MS's MAC/PHY layers. The MBS client at the MS includes a predecoder buffer, a standard-compliant H.264/MPEG-4 AVC decoder, and a TV channel/video format selector, which can be part of the MBS client software. The TV channel selector determines the CIDs of a certain TV channel according to the selected TV channel ID. The video format selector determines whether one, two, or all the three CIDs can be utilized according to whether the mobile device is 3D capable or not and according to the signal quality measured at the MS. In other words, the video format selector determines in which video format, i.e., 2D video, color-plus-depth 3D video, or LDV 3D video, the TV channel can be viewed. The determination of the video format according to the received signal quality at the MS is facilitated by the fact that the MS monitors the carrier-to-interference-and-noise ratio (CINR) and compares the average value against the allowed range of operation (IEEE Std 802.16-2009). The determination of the CIDs indicates the WiMAX MS MAC/PHY layers to process only those MBS MAC PDUs that are associated with the TV channel/video format from the corresponding MBS regions.

11.4.2.3.3 Deployment and Coverage Considerations

The BS multicasts/broadcasts the different components of the multilayered 3D video bit stream through different transport connections for each 3DTV channel to be offered using different spectrally efficient MCSs. The CID #1 transport connection that carries the first component is mapped to MBS region #1 for transmission using a robust MCS of QPSK (CTC) 1/2. The CID #2 transport connection that carries the second component is mapped to MBS region #2 for transmission using an efficient MCS of 16-QAM (CTC) 1/2. The CID #3 transport connection that carries the third component is mapped to MBS region #3 for transmission using an even more efficient MCS of 64-QAM (CTC) 1/2. Each MBS region's downlink operational MCS has an allowed range of operation in terms of the received signal quality at the MS and this region is bounded by threshold levels (IEEE Std 802.16-2009). Therefore, the coverage area of the BS can be regarded in this case as divided into three overlapping regions that result in three reception cases. In the first case, the subscribers can receive conventional 2D video over the whole coverage area of the BS. In the second case, the subscribers that are receiving a good signal quality can receive CID #2 adequately in addition to CID #1, thereby allowing them to display color-plus-depth 3D video. In the third case, the subscribers that are receiving a high signal quality can receive CID #3 adequately in addition to CID #1 and CID #2, thereby allowing them to display enhanced quality 3D video, i.e., LDV 3D video that allows

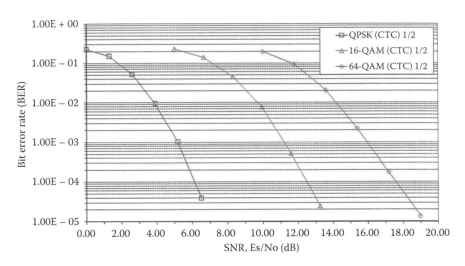

FIGURE 11.8
BER performances over SNR for the three selected MCSs.

better modeling of occlusions; hence, enjoying better perceived video quality. However, since the transmission is in broadcast mode, all the MSs can listen to the three MBS regions' data transmissions in the downlink subframe. The plots in Figure 11.8 show the BER performances of the three selected MCSs over a range of SNR values for an ITU vehicular A test environment and an MS traveling at a speed of 60 km/h.

It can be observed from the plots that in order for the MS to be able to utilize a certain MBS region's MCS while achieving the required target BER, target IP packet loss rate (PLR), and hence the required video quality, the signal quality at the MS needs to be within a certain allowed operating region for the MCS. Table 11.4 lists the OFDMA slot capacity (spectral efficiency) and the required received SNR levels of the three selected MCSs for achieving the minimum required BER of 10^{-4} and IP PLR of 10^{-3} (3GPP TS 26.236 2011, 3GPP TR 26.937 2011) and hence the minimum level of perceived video quality at the MS.

In order to improve the reception reliability while also considering the mobility aspect of the MSs, i.e., avoiding the situation of unstable video decoding result between, for example, 2D and 3D due to a channel quality

TABLE 11.4

Spectral Efficiency and SNR/BER Requirement for Selected MCSs

MBS Region	MCS	Slot Capacity (Data Bits/Slot)	Required SNR, Es/No (dB)	Required BER
1	QPSK (CTC) 1/2	48	SNR > 5.20	<0.0010251
2	16-QAM (CTC) 1/2	96	SNR > 11.60	<0.0005203
3	64-QAM (CTC) 1/2	144	SNR > 17.20	<0.0001762

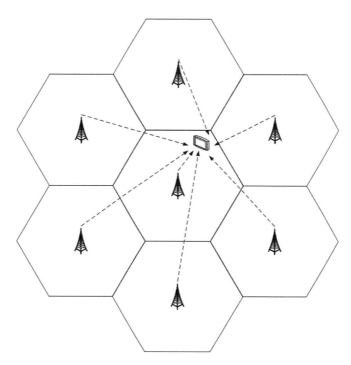

FIGURE 11.9
Illustration of the multi-BS reception for improving the reception reliability.

fluctuation in the reception of the MBS regions, the multi-BS provisioning of MBS concept is utilized as illustrated in Figure 11.9.

In this case, the MBS regions' transmissions are synchronized across all the BSs within an MBS zone, i.e., the same information is transmitted using the same time-frequency resources from multiple synchronized BSs. This enables an MS to receive the MBS transmissions from multiple BSs using macrodiversity, i.e., the resultant signal level at the MS is obtained from the sum of individual signals from all the BSs in the multi-BS MBS zone; hence, improving the reception reliability.

11.4.2.3.4 Radio Resource Allocation

In the mobile WiMAX WirelessMAN-OFDMA PHY, a slot is the minimum possible data allocation unit that requires both a time and a subchannel dimension for completeness. In the considered downlink partial usage of subchannels (PUSC), one slot is defined as one subchannel by two OFDMA symbols (IEEE Std 802.16-2009). Therefore, in order to deliver each component of the multilayered 3D video bit stream according to the required QoS parameters, the radio resources can be allocated as follows: the minimum reserved traffic rate or, equivalently, the number of slots allocated per 5 ms frame QoS parameter of the RT-VR data delivery service is set to a value such

TABLE 11.5

Main WiMAX Network Radio Interface Parameters' Settings

Parameter	Value
PHY layer interface	Wireless MAN-OFDMA
Carrier frequency (GHz)	2.3
Channel bandwidth (MHz)	8.75
Duplexing mode	TDD
DL:UL ratio	2:1
TDD frame length (ms)	5
Subcarrier permutation	PUSC
Total DL slots without preamble	390 slots (13 slots by 30 subchannels)
Total UL slots excluding ranging, CQICH, and ACK/NACK	140 slots (4 slots by 35 subchannels)
Connection mode	non ARQ-enabled
Data delivery service	RT-VR
Fragmentation/packing	ON/OFF
Generic MAC header size (bytes)	6
Fragmentation subheader size (bytes)	2
CRC size (bytes)	4

that any MAC service data unit (SDU) (or an IP packet) is delivered within a maximum latency of 500 ms selected from the extensible look-up table that specifies the maximum latency values in IEEE Std 802.16-2009. Therefore, a predecoder buffer with an initial buffering time of at least 500 ms is required at the MBS client in order to compensate for the difference between the accumulated video encoding or transmission rate and the data delivery rate at the WiMAX radio access network (ITU 2007). To improve link efficiency, the fragmentation optional feature can be enabled and in this case the MAC SDUs and MAC SDU fragments are fragmented according to the available bandwidth. The main WiMAX network radio interface parameters settings are listed in Table 11.5.

11.4.3 Performance Evaluation

11.4.3.1 QoS Performance

The data in Table 11.6 show the QoS performance in terms of the MAC PDU error rate and the IP PLR for the color, depth map, and the Δ color components of sequence #1 video test sequence over different levels of received signal quality at the MS.

The IP PLR is computed above the MS's MAC layer and it can be seen that the IP PLR values are higher (almost double) than the corresponding MAC PDU error rate values. This is due to the fact that each MAC SDU or a remaining MAC SDU fragment in the corresponding BS MAC layer's queue that is scheduled for transmission and did not fit into the available bandwidth is

TABLE 11.6

MAC PDU Error Rate and IP PLR Statistics

SNR (dB)	MAC PDU Error Rate (%)			IP PLR (%)		
	Color	Depth	Δ Color	Color	Depth	Δ Color
2.60	27.002	100	100	50.579	100	100
3.90	1.125	100	100	2.548	100	100
5.20	0.397	99.841	100	0.927	100	100
6.50	0.033	80	100	0.077	98.039	100
8.30	0	23.968	100	0	60.634	100
9.95	0	3.254	98.214	0	11.312	100
11.60	0	0.159	58.418	0	0.603	91.855
13.25	0	0.040	28.852	0	0.151	45.704
13.60	0	0	11.267	0	0	33.183
15.40	0	0	0.850	0	0	3.017
17.20	0	0	0.043	0	0	0.151
19.00	0	0	0	0	0	0

subjected to fragmentation according to the available bandwidth into a number of MAC PDUs. The loss of a MAC PDU that forms part of the parent MAC SDU results in losing the whole MAC SDU.

11.4.3.2 *2D/3D Video Reconstruction Quality*

At the MS, the received video bit stream is decoded using the H.264/MPEG-4 AVC reference software (Sühring 2011). For error concealment, the decoder is configured to conceal the corrupted frame slices if any occur due to packet loss using the motion information of the correctly received previous frame, i.e., motion copy. The video quality is evaluated using the peak signal to noise ratio (PSNR) objective quality metric. The decoded stereoscopic video quality is obtained by calculating the average 2D video reconstruction qualities of the left-eye view and the rendered right-eye view. The following figures show the 2D/3D video reconstruction qualities over a range of SNR values. Figure 11.10 shows the decoded video quality in the 2D video case, i.e., when only the color component is received and decoded. It can be seen that the MS can receive the minimum level of video quality at Point 2 (SNR = 5.20 dB). Below this point, the perceived video quality is not acceptable for all considered video test sequences. The reception quality of the 2D video stabilizes after Point 4 (SNR = 9.95 dB).

Figure 11.11 shows the decoded video quality in the color-plus-depth 3D video format case when only the color and depth map components are received and decoded. In this case, the stereoscopic video pair is generated solely based on the received 2D video and its corresponding depth map, where the occlusions and holes are filled based on the extrapolation of background pixels. At and below Point 3 (SNR = 6.50 dB), the received depth map

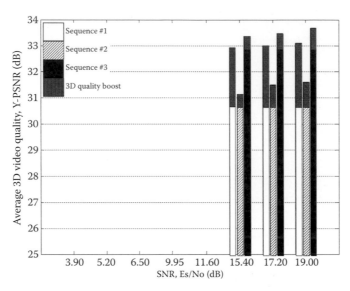

FIGURE 11.10
Average Y-PSNR over SNR for conventional 2D video.

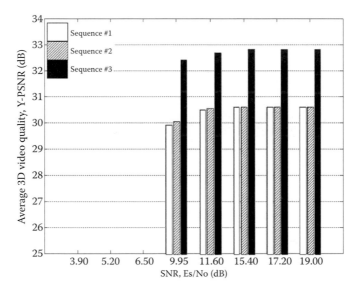

FIGURE 11.11
Average Y-PSNR over SNR for color-plus-depth 3D video.

signal is not decoded due to the heavy corruption in the bit stream. From Point 5 (SNR = 11.60 dB) and above, the perceived 3D video quality becomes acceptable.

Figure 11.12 shows the decoded video quality in the LDV 3D video case when all the three components of the multilayered 3D video bit stream are

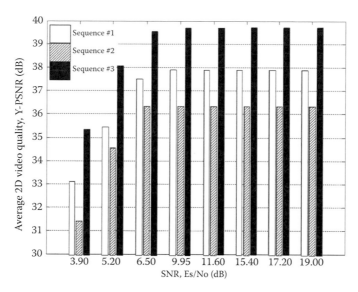

FIGURE 11.12
Average Y-PSNR over SNR for LDV 3D video.

received and decoded. The received Δ color component is utilized in this case for filling the occlusions and holes with much higher precision. It should be noted that when switching from Point 5 (SNR = 11.60 dB) to upper SNR points, the reception and the decoding of the Δ color component improves the perceived 3D stereoscopic video quality significantly by 2.2, 1.0, and 0.9 dB for sequence #1, sequence #2, and sequence #3, respectively. The boost in the decoded 3D video quality in this case is illustrated with dotted textures as illustrated in Figure 11.12. Hence, the reception of the Δ color component on top of the color and the depth map components enables the subscribers who are receiving a high signal quality to watch the 3D content with higher quality. The observed quality difference between the three video test sequences is mainly caused by the different motion and texture characteristics and the fact that they were encoded to generate similar source coded bit rates.

11.4.3.3 Radio Resource Efficiency

In this section, the performance of the optimized transmission technique described earlier, i.e., the parallel transmission of a multilayered 3D video bit stream using different spectrally efficient MCSs through different transport connections is evaluated in terms of the amount of required radio resources. Table 11.7 provides a summary of the radio resource efficiency statistics for the optimized transmission technique and its comparison with the legacy robust transmission technique of utilizing the most robust MCS, such as QPSK (CTC) 1/2 for transmitting all the three components. In addition, it provides an estimation of the possible number of TV channels

TABLE 11.7

Radio Resource Efficiency Statistics

Component	Average IP Bandwidth (kbps)	No. of Slots Allocated (Minimum Reserved Traffic Rate)	
		Proposed	Legacy
Color	489.49	59 (566.4 kbps)	59 (566.4 kbps)
Depth	207.68	19 (364.8 kbps)	38 (364.8 kbps)
Δ Color	248.06	10 (288 kbps)	30 (288 kbps)
Total	945.23	88 (1.22 Mbps)	127 (1.22 Mbps)
No. of TV channels	2D	6–7	6–7
	Color-plus-depth	5	4
	LDV	4–5	3

in 2D, color-plus-depth, and LDV formats that can be provided to the subscribers in the case that all the downlink subframe capacity, i.e., 390 slots is utilized for providing the 3DTV MBS. The presented results correspond to the maximum required radio resources from the three considered video test sequences, since each would require different amount of radio resources depending upon its content complexity and motion characteristics.

It can be seen from the statistics that with the adoption of the optimized transmission technique, 3DTV MBS can be offered using a total of 88 slots per TV channel in comparison to adopting the legacy robust transmission technique that would require 127 slots, thereby saving about 40 slots per TV channel that can be used for offering more TV channels or other services. Furthermore, in comparison to other candidate 3D video formats, two-channel stereo video encoded independently with H.264/MPEG-4/AVC, i.e., simulcast, would require a total of 118 slots per TV channel, i.e., 59 slots would need to be allocated for each of the left-view and right-view channels. Two-channel stereo video encoded jointly using the MVC extension of H.264 (ISO/IEC MVC 2008) would require 1.79 the bit rate required for encoding one of the views using H.264/AVC as in (Akar et al. 2008). Therefore, this would require total of about 106 slots per TV channel and would impose a requirement for the decoder at the MBS client to support MVC.

11.5 Conclusions

3D video delivery over mobile broadband networks is foreseen as one of the main stream research areas in the next few years. One of the major challenges to provide 3D video services, such as mobile 3DTV, over such networks relies on how to efficiently utilize the system resources and the functionalities that are available in the mobile broadband technology's radio interface protocol stack to maximize the subscribers' 3D viewing experience. The case study presented in this chapter demonstrates the importance of the fact that it is essential to ensure that the network is well designed to achieve a good compromise between the service quality and the radio resource efficiency in a system that has limited resources in comparison to wired networks. The utilization of an optimized transmission technique in such systems would be able to provide such a compromise and would allow for the following features: (1) format scalability, for example, ranging from conventional 2D video, to color-plus-depth 3D video, and to enhanced quality 3D video (e.g., LDV 3D video) that allows better modeling of occlusions, hence enjoying better perceived video quality. This enables to make the content displayable with the best possible quality based on (1) the device characteristics (display type, processing power, etc.), the received signal quality at the MS, and the offered package (e.g., free-to-air, pay-per-view, subscription); (2) backward compatibility with legacy 2D video mobile devices; and (3) radio resource efficiency—reducing the amount of required radio resources for transmitting one TV channel in 3D, thereby allowing more TV channels to be offered or more radio resources to be available for other services.

References

Akar Gozde, B., Oguz Bici, M., Aksay, A., Tikanmäki, A., and Gotchev, A. 2008. Mobile stereo video broadcast. Technical report D3.2. ICT FP7 project: Mobile 3DTV content delivery optimization over DVB-H system. http://sp.cs.tut.fi/mobile3dtv/

Alcatel-Lucent. 2006. Unlimited mobile TV for the mass market. Strategy white paper. November 2011. http://www.alcatel-lucent.com/

Cao, J. and Joy, Z. 2008. Mobile TV: A great opportunity for WiMAX. Huawei Technologies solution. Issue 41. http://www.huawei.com/en/static/hw-080436.pdf

Fehn, C. 2004. Depth-Image-Based Rendering (DIBR), compression and transmission for a new approach on 3D-TV. *Proceedings of SPIE Stereoscopic Displays and Virtual Reality Systems XI*, San Jose, CA, January 2004, pp. 93–104.

Forsman, J., Keene, I., Tratz-Ryan, B., and Simpson, R. 2004. Market opportunities for WiMAX take shape. Gartner, Inc, December 24, 2004. http://www.gartner.com

GSMWorld. 2009. Market data summary. December 7, 2009. http://www.gsmworld.com/newsroom/market-data/

Holma, H. and Toskala, A. 2007. *WCDMA for UMTS-HSPA Evolution and LTE*. Chichester, U.K.: Wiley.

Hong, S.-E. and Lee, W.-Y. 2009. Design and analysis of hybrid approach to mobile TV service over WiMAX network. *Wireless Days (WD), 2nd IFIP*, Paris, France, December 15–17, 2009, pp. 1–6.

ICT FP7. Multimedia scalable 3D for Europe (MUSCADE). Integrating project. http://www.muscade.eu/

IEEE Standard for local and metropolitan area networks. Part 16: Air interface for broadband wireless access systems. IEEE Std 802.16-2009 (Revision of IEEE Std 802.16-2004), pp. C1-2004, 2009.

Institute, I.S. 1981. RFC 791: Internet protocol. November 2011. http://www.ietf.org/rfc/rfc791.txt

ISO/IEC JTC1/SC29/WG11. 2008. Text of ISO/IEC 14496-10:200X/FDAM 1 multiview video coding. Doc. N9978. Hannover, Germany.

IST FP5. Advanced Three-dimensional Television System Technologies (ATTEST). http://www.hitech-projects.com/euprojects/attest/

ITU. 2007. Quality of experience requirements for IPTV services. ITU FG-IPTV-DOC-0814.

ITU-T. 2011. H.264: Advanced video coding for generic audiovisual services. November 2011. http://www.itu.int/rec/T-REC-H.264/en

Merkle, P., Smolic, A., Muller, K., and Wiegand, T. 2007. Multi-view video plus depth representation and coding. *IEEE International Conference on Image Processing (ICIP)*, San Antonio, TX, September 16–October 19, 2007, Vol. 1, pp. I-201–I-204.

Postel, J. and Institute, I.S. 1980. RFC 768: User datagram protocol. August 28, 1980. http://www.ietf.org/rfc/rfc768.txt

Schulzrinne, H., Casner, S., Frederick, R., and Jacobson, V. 1996. RFC 1889: RTP: A transport protocol for real-time applications. November 2011. http://www.ietf.org/rfc/rfc1889.txt

Schulzrinne, H., Casner, S., Frederick, R., and Jacobson, V. 2003. RFC 3550: RTP: A transport protocol for real-time applications. November 2011. http://www.ietf.org/rfc/rfc3550.txt

Smolic, A., Mueller, K., Merkle, P., Kauff, P., and Wiegand, T. 2009. An overview of available and emerging 3D video formats and depth enhanced stereo as efficient generic solution. *Picture Coding Symposium (PCS)*, Chicago, IL, May 2009, pp. 1–4.

Smolic, A., Mueller, K., Stefanoski, N., Ostermann, J., Gotchev, A., Akar, G.B., Triantafyllidis, G., and Koz, A. 2007. Coding algorithms for 3DTV-A survey. *IEEE Transactions on Circuits and Systems for Video Technology*, 17(11), 1606–1621.

Sühring, K. 2011. H.264/AVC reference software. JM 18.0 ed. H.264/AVC software coordination. November 2011. http://iphome.hhi.de/suehring/tml/

Teleca. 2011. Increasing broadcast and multicast service capacity and quality using LTE and MBMS. Solution area: e-MBMS in LTE. White paper. November 2011. http://www.teleca.com/

Third Generation Partnership Project (3GPP) TS 36.300. 2011. Evolved universal terrestrial radio access (E-UTRA) and evolved universal terrestrial radio access network (E-UTRAN); Overall description; Stage 2. V10.4.0 (Release 10). November 2011. http://www.3gpp.org/

Third Generation Partnership Project (3GPP) TS 26.236. 2011. Packet switched conversational multimedia applications; Transport protocols. V10.0.0 (Release 10). November 2011. http://www.3gpp.org/

Third Generation Partnership Project (3GPP) TR 26.937. 2011. Transparent end-to-end Packet-switched Streaming Service (PSS); Real-time Transport Protocol (RTP) usage model. V10.0.0 (Release 10). November 2011. http://www.3gpp.org/

UMTS Forum. 2009. Fast facts. December 7, 2009. http://www.umts-forum.org/

Wang, Y.-K. and Schierl, T. 2009. Internet draft: RTP payload format for MVC video. February 18, 2009. http://tools.ietf.org/html/draft-wang-avt-rtp-mvc-03

Wenger, S., Hannuksela, M.M., Stockhammer, T., Westerlund, M., and Singer, D. 2005. RFC 3984: RTP payload format for H.264 video. November 2011. http://www.rfc-editor.org/rfc/rfc3984.txt

WiMAX Forum. http://www.wimaxforum.org/home/

WiMAX TV. 2008. WiMAX TV: Mobile TV broadcasting over WiMAX networks. UDCAST. November 2011. http://www.udcast.com/products/downloads/WiMAX_TV.pdf

12

A New Hierarchical 16-QAM Based UEP Scheme for 3-D Video with Depth Image Based Rendering

Khalid Mohamed Alajel and Wei Xiang

CONTENTS

One of the most important pieces of three-dimensional (3-D) video transmission over wireless channels is the design of error-resilient video transmission. Naturally, many techniques used in two-dimensional (2-D) can be adapted or extended to exploit 3-D video properties. One of those techniques is unequal error protection (UEP) where different parts of 3-D video are protected with different levels of protection. Exploiting the characteristics of 3-D video to provide UEP schemes is not fully investigated.

In this chapter, an UEP scheme based on hierarchical quadrature amplitude modulation (HQAM) for 3-D video transmission is proposed. The proposed scheme exploits the unique characteristics of the color plus depth map stereoscopic video, where the color sequence has a significant impact on the reconstructed video quality. The UEP scheme assigned more protection to the color sequence than the depth map sequence in order to achieve high quality 3-D video. However, the different levels of protection are assigned through 16-quadrature amplitude modulation (16-QAM). The color data with high priority (HP) are mapped onto the most significant bits (MSBs) of QAM constellation points, and depth map with low priority (LP) is mapped onto the less significant bits (LSBs). Simulation results show that the proposed UEP scheme outperforms the classical equal error protection (EEP) by up to 5 dB gain in terms of the received left and right views quality.

12.1 Introduction

With the rapid growth of multimedia communication systems, such as the Internet and wireless networks, many applications that deliver multimedia contents have affected the everyday life of people. Some of these applications include video conference, video telephone, video on demand, and video over mobile networks (Blau 2005, Cherry 2005). Therefore, delivery of multimedia contents over wireless channels is becoming more popular. The quality of services (QoS) required to guarantee multimedia contents transmission will play a key factor in future generation wireless communication systems success. However, the effects of transmission errors on the reconstructed bitstream pose a major problem for QoS guarantee. To this end, more reliable transmission systems require more active research.

According to Shannon's theorem (Shannon 1948), an ideal communication system should adaptively change its information depending on the available channel capacity because the signal-to-noise ratio (SNR) can be highly variable in wireless channels. However, a feedback channel can be used to provide knowledge of channel conditions to the transmitter. Unfortunately, this feedback channel may not be available in many applications such as broadcasting and wireless channels. In wireless video communication, the two major obstacles when transmitting multimedia services are bandwidth limitations and high probability of error. The first problem has been targeted in the last two decades by proposed several video coding standards. In particular, the state-of-the-art video compression standard H.264 advance video coding (H.264/AVC) (AVC 2009) provides better compression with better quality. However, H.264/AVC adopts variable-length codes (VLCs) as entropy codes to achieve a high coding efficiency. The nature of VLC is the root cause of the phenomenon of error propagation and is very sensitive to channel errors. A single bit error could render the entire bitstream undecodable in the worst case, which makes the bit error rate (BER) problem still existent and exacerbated by the high source coding.

A new field in signal processing is representation of 3-D scenes. Interest in 3-D data representation for 3-D video communication has grown rapidly within the last few years. 3-D video may be captured in different ways such as stereoscopic dual-camera and multiview settings. Since 3-D video formats consist of at least two video sequences and possibly additional depth data, many different coding techniques have been proposed (Wang et al. 2004, Yoon and Ho 2007). In any case, compressed 3-D video data such as video plus depth (V + D) or multiview video (MVV) have to be transmitted over error-prone channels, which raise the problem of error protection. However, transmitting 3-D video over error-prone channels poses more challenges than the conventional 2-D video. The existing 2-D video transmission algorithms cannot be applied directly to 3-D video formats. Due to the use of motion-compensated prediction, the error may propagate to subsequent frames, which lead to degraded 3-D video quality at the receiver. However, 3-D video coding techniques utilize temporal and interview prediction to achieve high coding efficiency; thus, an error occurring in one view is propagated not only to the subsequent frames of the same view but also to the other views.

One of the most popular and widely used formats for representing 3-D video is V + D (Merkle et al. 2009). In depth image based rendering (DIBR) technique (Fehn 2004), a depth map is required to generate good quality of 3-D video, of which the quality does not need to be very high to render 3-D scenes as opposed to the color sequence. Color and depth images need to be transmitted over communication channels to the end user for display. However, the color sequence is directly viewed by the user. Therefore, in transmission, if the color sequence is impaired, it will highly degrade the 3-D video quality than the damage of depth sequence. Transmitting 3-D video over networks such as the Internet and wireless networks poses new challenges, and consumer applications will not gain much popularity unless the problem is addressed. Nevertheless, transmission of 3-D video contents is expected to be the next big revolution in multimedia applications.

The aforementioned observation has led to present methods to minimize the effect of transmission errors on the reconstructed video quality over wireless channels, which motivate the research work of this chapter. The loss of synchronization between decoder and encoder caused by error-prone channels could be solved by using retransmission. Error-free transmission could be achieved by retransmitting packets that have been lost or corrupted.

However, the problem with such a scheme is that it increases delays, which could not be acceptable in some applications. An alternative approach is to use effective data protection to create compressed bitstream error resilience to transmission errors. Several standard source coding approaches are available and have been added to H.264 standard. These techniques are used to provide robust source coding for 2-D video, and many of these can also be used for 3-D video. Standardized error resilience techniques may include slice coding, redundant slice, flexible macroblock ordering, and data partitioning. Exploiting the correlation that exists in different 3-D video formats could lead to more error-resilient video encoding schemes for 3-D video transmission.

To address the quality degradation problem caused by channel errors, several error-resilient source and channel coding techniques used in 2-D video can be extended or adapted to consider 3-D video properties of different 3-D video formats. One technique that can be extended or adapted to 3-D video from 2-D video is UEP. UEP can be implemented based on the quality distributed significance of binary bits in the compressed video stream, where the MSBs are assigned higher protection than those with less important bits (Li et al. 2010). UEP is usually used in conjunction with channel coding such as turbo code (Barmada et al. 2006) and low-density parity-check code (Kamolrat et al. 2008). However, in 3-D video, the added overhead will be much higher than in conventional 2-D video. Thus, alternative methods to achieve UEP are required. Another simple and efficient UEP scheme is based on the HQAM in which the data with HP are mapped onto the MSBs of QAM constellation points and the data with LP are mapped onto the LSBs (Lee et al. 2000).

Although UEP using HQAM was proposed for 2-D video transmission (Barmada et al. 2005, 2006, Chang et al. 2006, 2009, Li et al. 2010), UEP based on HQAM for 3-D video transmission has not been considered so far. Moreover, the unequal importance of color and depth map is not considered in formatting the HP and LP data streams. In this chapter, we propose an UEP scheme for color and depth map 3-D video transmission over wireless channels using hierarchical 16-QAM. The proposed system takes into consideration the unequal importance of color and depth map 3-D video contents. A color sequence that is more sensitive to human perception is provided more protection by assigning it to the HP data and the depth map to the LP data of hierarchical 16-QAM.

The rest of the chapter is structured as follows: some related work is briefly discussed in Section 12.2. The fundamentals of the HQAM and DIBR technique are overviewed in Section 12.3. Section 12.4 describes the proposed HQAM-based UEP scheme for 3-D video transmission. In Section 12.5, we show the results obtained with the proposed technique, and Section 12.6 concludes the chapter.

12.2 Overview of 3-D Video Representation and Communication

12.2.1 3-D Video Formats and Coding

The recent interest in 3-D technology is now widespread in different applications including the 3-D cinema (Umble 2008), 3DTV (Morvan et al. 2008), and mobile phones (Flack et al. 2007). Depending on the application, various choices of 3-D video formats are available. According to Merkle et al. (2010), the 3-D video formats can be presented in the following formats: conventional stereo video (CSV), multiview video (MVV), video plus depth (V + D), multiview video plus depth (MVD), and layered depth video (LDV). In this section, these formats are going to be briefly described along with their associated coding methods. The breakdancers 3-D video sequence (Zitnick et al. 2004, Microsoft 2008) will be used to illustrate these formats.

12.2.1.1 CSV Format

CSV is considering the least complex 3-D video format and it is spatial case of multiview (two views only). In CSV, the 3-D video consists of two videos (views) representing the left and right views of the same scene with slightly different angles of view corresponding to the distance of human eyes. Each view forms a normal 2-D video, and the human brain can fuse these two different frames to generate the sensation of depth in the scene being viewed. Figure 12.1 illustrates the CSV formats.

Since both cameras capture the same scene, a straightforward approach is to apply the existing 2-D video coding scheme. Using the 2-D video coding approach, the two separate views can be independently encoded, transmitted, and decoded with a 2-D video codec like H.264/AVC, and this method

Left view   Right view

FIGURE 12.1
CSV formats.

is known as simulcast coding. However, since the two views have similar content and therefore highly redundant, coding efficiency can be increased by combined temporal/interview redundancy. This coding method is called multiview coding (MVC) (Merkle et al. 2006, Flierl and Girod 2007).

12.2.1.2 V+D Format

One of the most popular formats for representing 3-D video is V + D, which consists of a conventional 2-D video with associated per-pixel depth map represented with luma component only. For video and depth information, a stereo pair can be synthesized at the decoder. With this technique, left and right views are generated at the display side by a method known as a DIBR (Fehn et al. 2002, Fehn 2003). The depth map represents the per-pixel distance from the camera and is between $Z_{near} = 255$ and the maximum $Z_{far} = 0$, indicating the distance of the corresponding 3-D point from the camera, where the near objects appear brighter and the far objects appear darker. The V + D format is illustrated in Figure 12.2.

For coding V + D format, both MPEG-2 and H.264/AVC can be used to encode V + D. If MPEG-2 is used, MPEG-C part 3 defines a V + D representation, which allows encoding them as conventional 2-D video (ISO/IEC JTC1/SC29/WG11 2008). The video and depth sequences are encoded independently, where one view is transmitted simultaneously with the depth signal and the other view is synthesized by DIBR techniques at the receiver side. In this case, the transmission of depth map increases the required bandwidth of 2-D video stream by about 20% (Fehn 2003). If the H.264/AVC is used, the H.264 codec is applied to both sequences simultaneously but independently, where the video is the primary coded picture and the depth is the auxiliary coded picture. In this case, the required bandwidth increases by only 8% as mentioned by Fehn (2003).

However, by exploiting the depth data features, higher coding efficiency can be achieved. For instance, the existing correlation between the 2-D video sequence and its corresponding depth map sequence can be exploited to improve the compression ratio as proposed by Grewntsch and Miiller (2004)

 Color video   Depth data

FIGURE 12.2
V + D format.

and Pourazad et al. (2006). Alternative approaches based on so-called plate-lets were also proposed (Merkle et al. 2008).

The V + D concept is highly interesting due to the backward compatibility and to use available video codec. This format is alternative to CSV for mobile 3-D services and is being investigated by Fraunhofer Institute for telecom-munications. However, the advantages of V + D format come at the cost of increased encoder/decoder complexity (Flierl and Girod 2007).

12.2.1.3 MVV Format

For more than two views of CVS, this is easily extended to MVV (Smolic et al. 2007, Vetro et al. 2011). Transmission of a huge amount of data is the major challenge with MVV applications, which requires a high coding effi-ciency scheme. In MVC, N cameras are arranged to capture the same scene from different viewpoints. Therefore, they all share common scene contents. The straightforward method to encode MVV is simulcast coding, where each view is coded independently. Simulcast coding can be done with any video codec including H.264/AVC, where the temporal and spatial correla-tion within one view is exploited. However, MVV contains a large amount of interview statistical dependencies, which can be exploited for combined temporal/interview prediction. The multiple correlation makes MVV coding to have a different structure from single view, where the images are pre-dicted temporally from a neighboring image within the same view and also from corresponding images in adjacent views, as illustrated in Figure 12.3. Significant gain can be achieved by combining temporal/interview predic-tion as proposed by Merkle et al. (2005) and Kaup and Fecker (2006).

12.2.1.4 MVD Format

Transmitting all views requires a high bit rate, where the number of views increases the bit rate linearly. Therefore, MVC is inefficient if the number of views to be transmitted is large. At the same time, the V + D format provides a very limited free viewpoint video (FVV) functionality. The solution to the problem of high bit rate when transmitting all views and the limited FVV is the MVD format. The MVD format contains multiple views and associated depth information for each view as illustrated in Figure 12.4.

The MVV plus depth format is an extension of V + D and is included by MPEG in recent proposals (Kauff et al. 2007, ISO/IEC JTC1/SC29/WG11 2008). In MVD, depth has to be estimated for the N views and then the N colors with N depth videos have to be encoded and transmitted. MVD video sequence can be coded using methods for MVV coding, where the depth image is estimated for each view of the MVVs. Many algorithms have been proposed for coding MVD such as Merkle et al. (2007) and Yoon and Ho (2007). The coding of MVD has been improved by using platelet-based depth coding as shown in Merkle et al. (2008).

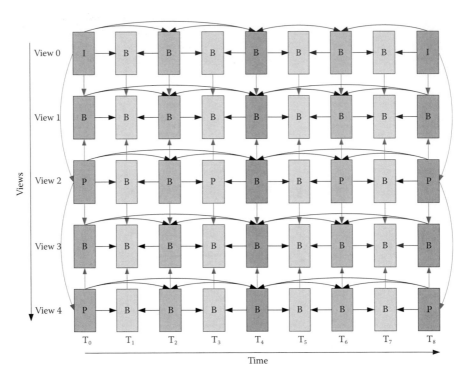

FIGURE 12.3
(See color insert.) Multiview coding structure with temporal/interview prediction.

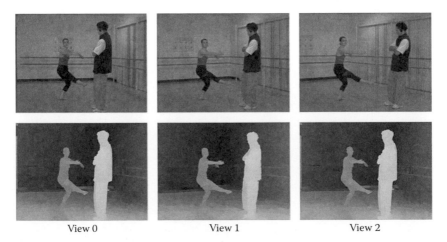

FIGURE 12.4
Multiview video plus depth.

12.2.1.5 LDV Format

Although MVD can reduce the required bandwidth to transmit color and depth data for all views, the overall required bandwidth is still huge. To further reduce the bit rate, LDV is used. LDV (Muller et al. 2008a, Muller 2009) is a derivative and an alternative to MVD, where only one full view with additional residual data is transmitted. One representation of LDV again uses color video with associated depth map (V + D) representation and additional component called the background layer with its associated depth map as illustrated in Figure 12.5.

Another type of LDV consists of a main layer that contains one full or central view and one or more residual layers of color and depth data to represent the side views. One major problem with LDV is disocclusions, where blank spots appear as the distance between the central view and side views increases. Hence, the extra information enables a correct rendering of disoccluded objects. For more details on LDV, the reader is referred to Muller et al. (2008a,b).

12.2.2 Depth Image Based Rendering

Understanding how the human visual system (HVS) (Wandell 1995) works play the key role in understanding how 3-D imaging works. The HVS consists

Color video

Depth data

Background layer

Background layer depth data

FIGURE 12.5
Layered depth video.

of two parts, the eye and the brain. Each eye has a retina that collects information and transfers it to a region called the lateral geniculate body and then to the visual cortex through the optic nerve. The picture produced at each of the retinas is upside down and as the visual information is processed by the visual cortex, it produces one single upright image.

As the human eyes are separated by about 6–8 cm, the 3-D depth perception is realized by two slightly different images being projected to the left and right eye retinas (binocular parallax), and then the brain fuses the two images to give the depth perception. On the receiver side, autostereoscopic or shutter glasses are used to provide each eye with its corresponding video stream. A major drawback of such a scheme is that the camera parameters need to be set at the recording time and cannot be changed later.

DIBR is a more recent technique to overcome this problem. DIBR is a technique of view synthesis and is used to generate a stereoscopic view at the display end in 3-D video applications, where a color image and depth map are used to synthesize two disparate visual views, one for the left eye and the other for the right eye (Fehn 2004). The color image is stored in the same way as normal 2-D video, and the depth map is stored using only the luminance component. The main advantage of DIBR is that it provides high quality 3-D video and reduces storage and bandwidth requirements for stereoscopic video transmission over communication channels compared to traditional left and right views techniques.

Figure 12.6 illustrates an example of color and depth map representation in DIBR. In DIBR, a depth frame is used to generate two virtual views from the original view in the following way: (1) the pixels of the original image are projected into the 3-D domain, utilizing their depth values specified by the depth stream and (2) thereafter, this 3-D model is projected into the image plane of a virtual camera. This process is called 3-D image warping.

Figure 12.7 illustrates the visual view generation process. As shown from the figure, the original image pixels at location (x,y) are moved to the new locations (x_L,y) and (x_R,y) for left and right view, respectively, where x_L and x_R can be calculated as in the following equations:

$$x_L = x + \frac{\alpha_x t_c}{2}\left(\frac{1}{Z} - \frac{1}{Z_c}\right) \tag{12.1}$$

$$x_R = x - \frac{\alpha_x t_c}{2}\left(\frac{1}{Z} - \frac{1}{Z_c}\right) \tag{12.2}$$

where
   α_x is the focal length of the reference camera
   t_c is the distance between left and right camera positions
   Z_c is the convergence distance located at the zero parallax setting plane
      and Z represents the depth value of each pixel in the reference view

(a)

(b)

FIGURE 12.6
Color plus depth representation in DIBR for *Interview* test sequence. (a) Color image and (b) associated depth map.

The quality of rendered views can be determined by calculating the peak signal-to-noise ratio (PSNR) between the rendered view with uncompressed color image and associated depth map and the rendered view with reconstructed color and depth map.

12.2.3 Hierarchical Quadrature Amplitude Modulation

The capacity of a wireless video transmission system could be increased by using a high order modulation scheme such as 16/64-QAM. However, modulation schemes that use uniform signal space constellation result in error decoding even with small errors. To solve this problem, nonuniform signal space constellations are used to provide different classes of error protection, which are known as nonuniform constellation or hierarchical modulation (HM). HM is a physical layer modulation technique that was initially

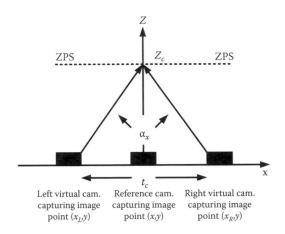

FIGURE 12.7
Virtual view generation in DIBR process.

proposed to provide different classes of data to the end users with different reception conditions (Wei 1993, O'Leary 1997). The main advantage of HM over channel coding–based UEP is that the HP data are protected without an increase in the bitrate.

In HQAM, the data stream is split into two substreams where the most sensitive bits are assigned HP and are known as "high priority (HP)" data, whereas the remaining bits are assigned LP and known as "low priority (LP)" data. In conventional HQAM with signal constellation size M (M-HQAM), the HP data occupy the first two MSBs of each point where all points in the same quadrant have the same HP bits. LP data occupy the rest of the bits and the number of LP bits in each symbol is given by $\log_2(M - 2)$. HM consists of two constellations, which are a basic constellation that includes common bits or HP bits and a secondary constellation that carries enhancement bits or LP bits (Jiang and Wilford 2005).

Figure 12.8 shows the constellation diagram of hierarchical 16-QAM modulation. The two MSBs (in bold face) represent basic information bits, while the two remaining bits represent the secondary bits. The parameters d_1 and d_2 represent the minimum distance between constellation points in different quarters and the minimum distance between constellation points in the same quarter, respectively. Parameters d_1 and d_2 are adjusted to control the degree of protection and to achieve UEP for the HP and LP bits.

Let β be the hierarchical QAM modulation parameter defined as

$$\beta = \frac{d_1}{d_2} \tag{12.3}$$

Ratio β is an important parameter and controls the achievable error rate of the system. When $\beta=1$, the result is a conventional rectangular 16-QAM

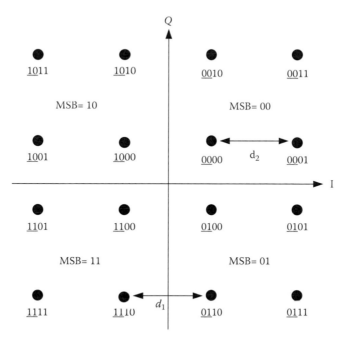

FIGURE 12.8
Hierarchical 16-QAM constellation diagram.

constellation where each layer has the same reliability. On the other hand, when $\beta > 1$, the HP stream is more protected than the LP stream.

A general recursive algorithm of BER expression for uniform M-QAM over Additive White Gaussian Noise (AWGN) channel was derived and the BER performance of an arbitrary square M-QAM constellation has been estimated by Yang and Hanzo (2000). The exact general BER expression of M-ray square QAM in AWGN channel has also been derived and analyzed by Yoon et al. (2000). Vitthaladevuni and Alouini (2001) have obtained the exact and generic expressions in M for the BER of the 4/M-QAM constellations over AWGN and fading channels. Moreover, exact and generic expression in M for the BER of the generalized hierarchical M-PAM constellations over AWGN and fading channels are derived by Vitthaladevuni and Alouini (2003).

12.3 Related Works

Hierarchical QAM has been widely used to provide UEP for wireless video transmission due to its simplicity and efficiency. UEP protects data according to the system requirements. If the channel is good, the receiver recovers the two classes of data (known as basic and enhancement data), while if the channel is poor, the receiver only receives the more important classes. Implementing UEP schemes by exploiting the characteristics of 2-D video is common in research literature (Barmada et al. 2005, 2006, Chang et al. 2006, 2009, Li et al. 2010).

One major disadvantage of conventional HQAM is the fixed capacities for HP and LP data. Therefore, conventional HQAM is not well suited with layered video. To solve this problem, Barmada et al. (2005) first proposed hierarchical QAM to provide UEP for layered H.264/AVC coded video. In this scheme, the HP data include network abstraction layer (NAL)-A, whereas the LP contains NAL-B and NAL-C. The results show that the proposed scheme provides more efficient video transmission compared with nonhierarchical methods. Better protection for HP data than for LP data could be achieved through channel coding such as turbo coding. The drawback of such a scheme is the overhead introduced by the channel coding, which results in low video quality. Therefore, channel coding and hierarchical QAM can be combined together to solve this dilemma as reported by Barmada et al. (2006) and Chung et al. (2010).

Barmada et al. (2006) studied the UEP transmission of scalable H.264 with two layers. The base layer of H.264 bitstream is classified as HP data and is assigned better protection than enhancement layer with LP. Turbo coding and HQAM are combined together to solve the problem of the high overhead introduced by channel coding. Obtained results show that combined HQAM and turbo coding outperforms UEP with turbo coding alone. Chung et al. (2010) proposed a joint design of rate compatible punctured codes (RCPC) and hierarchical PAM/QAM to implement UEP for prioritized H.264 bitstream. BERs of the combined RCPC and HQAM have been driven as a function of system parameters, which have been optimized by minimizing the expected loss of important information at the source.

Another UEP scheme based on the nonuniform importance of intra-frame (I-frame), predictive frame (P-frame), and macroblock (MB) position in each frame has been proposed by Chang et al. (2006, 2009). The idea introduced by Chang et al. (2006) is that frames in groups of pictures (GOP) have different importance. Therefore, the earlier an error occurs in a GOP which is organized as IPPPP sequence, the more frames are affected. In such a scheme, the I-frame which is located at first frame of GOP is classified as HP data while the last two frames are classified as LP data. Additionally, in frame 2 and frame 3, the MBs at the beginning of each frame are treated as HP data, while others are treated as LP data. Similar work with three levels of priority, namely, HP, medium priority (MP), and LP has been reported by Chang et al. (2009). Furthermore, a generic solution for optimally allocating the symbol bits of HQAM to the three classes of data has also been proposed and the performance was evaluated over the AWGN channel. Li et al. (2010) proposed an UEP scheme based on HQAM for H.264/AVC video over frequency elective fading channel. In this scheme, video stream is divided into substreams with different priority using data partitioning. However, orthogonal frequency division multiplexing subcarrier classification is used to avoid HP data to be mapped into subcarriers in deep fading.

One of the most important pieces of 3-D video transmission over wireless channels is the design of error-resilient video transmission. Naturally, many techniques used in 2-D can be adapted or extended to exploit 3-D video properties. One of those techniques is UEP, where different parts of 3-D video are protected with different levels of protection. Exploiting the characteristics of 3-D video to provide UEP schemes is not fully investigated.

Hewage et al. (2009) described a 3-D video transmission scheme based on UEP, where the UEP method assigned more protection levels to the color sequence than the depth map. Different levels of protection have been achieved by allocating unequal transmission power to 3-D video components. The motion correlation between color and depth map is used to provide UEP as shown by Hewage et al. (2008), where the redundant motion information is used to conceal errors at the decoder. In this scheme, more protection for the motion information of the coded data is applied.

Even though FEC can provide error-resilient transmission for 3-D video in wireless channels, it may fail when data are equally protected and channel conditions are very erroneous. Aksay et al. (2009) evaluated the effect of forward error correction (FEC) on 3-D video delivery over digital video broadcasting—handheld under different channel conditions. In this work, UEP is implemented through FEC where HP (left) video is well protected and LP (right) video is less protected.

To further improve the performance of 3-D video transmission, joint source channel coding (JSCC) is an effective method. JSCC scheme for color and depth map 3-D video was proposed by Kamolrat et al. (2008). In this method, different channel coding rates have been implemented to protect color and depth sequences. The obtained results show that the quality of depth image does not significantly affect the reconstructed quality. Therefore, low protection can be used for depth sequence compared to the color one. However, the drawback of this scheme is that it introduces more overhead data to 3-D video compared to conventional 2-D video.

Another novel technique that recently became popular for error residence in wireless channels is rateless code (Byers et al. 1998). The idea of rateless code is different from FEC codes; in rateless code, many parity packets are generated as needed, whereas in FEC, the channel encoding rate is fixed. Tan et al. (2009) proposed an UEP scheme for stereoscopic video, where MVV and rateless code are combined together for error-resilient stereoscopic video transmission. In this approach, the video stream is divided into three layers of importance, where intracoded left view frames are classified as the most important ones, left view predictive coding frames as a medium important, and right view frames are considered as less important. Rate distortion optimization of stereoscopic video system with rateless code is also proposed. The UEP scheme proposed in this chapter uses an alternative concept for achieving UEP, where HQAM is used to assign different levels of protection for 3-D video data.

12.4 UEP Scheme for 3-D Video Transmission

12.4.1 Problem Statement

In color and depth map 3-D video representation, color and depth map sequences have different error sensitivities to the overall quality. In this technique, DIBR is used to project color sequence to 3-D space based on the depth pixel values. The color sequence is the only texture information that is directly viewed by the users. In DIBR, a good detail of depth map is required to generate a high quality 3-D video and provide the end users with a sense of depth but the quality does not need to be very high to render 3-D scenes as opposed to the color sequence. Therefore, when transmission errors occur at the color sequence, the quality of the reconstructed signal will be highly degraded compared to the effect of depth map transmission errors.

By taking this into consideration, the proposed system takes into consideration the unequally important color and depth map 3-D video contents. The protection levels are determined based on hierarchical 16-QAM. A color sequence that is more sensitive to human perception is provided more protection by assigning it to the MSBs and the depth map to the LSBs of hierarchical 16-QAM.

12.4.2 System Model

The system block diagram of the proposed UEP scheme is shown in Figure 12.9. In this scheme, the color and depth map are encoded using H.264/AVC encoders separately. The output data of each encoder are mapped to 16-QAM, where the HP data (color) are mapped to MSBs of 16-QAM and the LP data (depth) are mapped to LSBs, since the two MSBs of the constellation points in 16-QAM have lower BER than the two LSBs. The output of 16-QAM is transmitted over a Rayleigh fading channel and then the color and depth map are reconstructed.

Color and depth map data are used to generate left and right views using the DIBR technique. The overall quality of the proposed system is evaluated using the average PSNR of reconstructed left and right views comparing the original left and right views according to the following equations:

$$PSNR_{left/right} = 10\log_{10}\left(\frac{255^2}{MSE_{l/r}}\right) \tag{12.4}$$

$$PSNR_{joint} = 10\log_{10}\left(\frac{255^2}{(MSE_l + MSE_r)/2}\right) \tag{12.5}$$

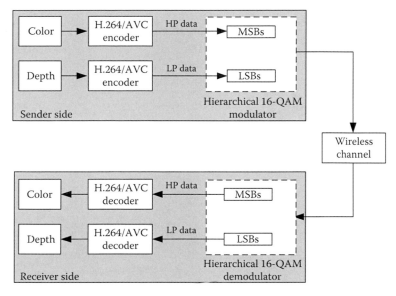

FIGURE 12.9
System model of the proposed UEP scheme.

where MSE_l and MSE_r represent the mean squared error in left and right views, respectively, and is given by

$$MSE(ori, rec) = \frac{1}{W \times H} SSD(ori, rec) \qquad (12.6)$$

where SSD is the sum of squared differences and is given by

$$SSD(ori, rec) = \sum_{i=1}^{H} \sum_{j=1}^{W} \left[ori(i, j) - rec(i, j) \right]^2 \qquad (12.7)$$

where W and H represent the width and the height of the original and reconstructed pictures, respectively.

12.4.3 BER Performance of 16-QAM

As shown in Figure 12.9, the two bit streams (color and depth) are separately fed into H.264 encoders, which are then gray coded and modulated onto 16-QAM constellation. The color and depth streams are combined into one symbol and then transmitted through an wireless channel. In this section, the bit error probability is conducted to determine the performance of 16-QAM.

The BER of QAM can be determined from the BER of PAM as described by Proakis (2001) and Vitthaladevuni and Alouini (2003). Thus, the BER of M-ray QAM is

$$P_M = \frac{1}{2}\left[\left(P_{i,\sqrt{M}} + P_{q,\sqrt{M}}\right)\right] \tag{12.8}$$

where $P_{i,\sqrt{M}}$ and $P_{q,\sqrt{M}}$ are the error probabilities of $\sqrt{M}$-ray PAMs

We assume that each symbol has the same probability and the error probability is given by

$$P = \frac{1}{2}P\left[|r - s_m| > d\right] \tag{12.9}$$

$$P = Q\left(\sqrt{\frac{2d^2}{N_0}}\right) \tag{12.10}$$

where

N_0 is the power spectral density of AWGN channel

$Q(x)$ is the error function and is given by

$$Q(x) = \frac{1}{\sqrt{2\pi}} \int_x^\infty \exp\left(-\frac{t^2}{2}\right) dt \tag{12.11}$$

The average bit energy (ε_b) is

$$\varepsilon_b = d_1^2\left[1 + \frac{1}{(\beta+1)^2}\right] = d_2^2\left[(\beta+1)^2 + 1\right] \tag{12.12}$$

from 8 to 12 and by assuming that

$$P_{i,\sqrt{M}} = P_{q,\sqrt{M}} \tag{12.13}$$

The BER of the HP and LP bits of QAM is given by

$$P_{HP} = \frac{1}{2}\left[Q\left(\sqrt{(2(\beta+2)^2\gamma)/B}\right) + Q\left(\sqrt{2\beta^2\gamma/B}\right)\right] \tag{12.14}$$

$$P_{LP} = Q\left(\sqrt{2\gamma/B}\right) + \frac{1}{2}Q\left(\sqrt{4\gamma(2\beta^2 + 5\beta + 4)/B}\right) - \frac{1}{2}Q\left(\sqrt{4\gamma(2\beta^2 - 5\beta + 4)/A}\right) \tag{12.15}$$

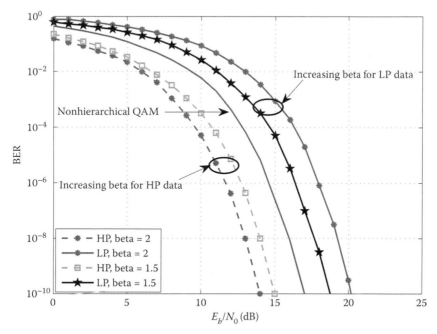

FIGURE 12.10
BER performance of hierarchical 16-QAM over an AWGN channel with different values of β.

where $A = (1 + \beta)^2$, $B = A + 1$, and γ is the average SNR per bit and is given by

$$\gamma = \frac{\varepsilon_b}{N_0} \tag{12.16}$$

Figure 12.10 is drawn according to (12.14) and (12.15), which depicts the BER curves of hierarchical 16-QAM over the AWGN channel as a function of the channel SNR. It can be seen that by increasing the value of β, the BER curve of the HP bits moves backward (improved) while the BER of the LP bits moves forward (degraded) over the SNR axis. Thus, by adjusting the value of β, we can control the BERs of the HP data and LP data, respectively.

12.5 Simulation Results and Discussions

12.5.1 Experimental Setup

To evaluate the proposed UEP scheme, extensive simulations are carried out. The proposed UEP scheme is compared with EEP. In the simulation, two different color and depth map sequences, namely, *Interview* sequence and *Orbi* sequence that represent different motion profiles, are used. *Interview* is a very slow motion sequence and *Orbi* is a very complex sequence with high motion.

The two sequences are encoded at a spatial resolution of 720×576. The H.264/AVC reference software JM version 16.1 is used to encode the video sequences. For each sequence, 30 frames are encoded with the coding structure of IPPP ... sequences, content adaptive binary arithmetic coding is used as entropy coding, and the search range is set to 32×32. Rate distortion optimization is switched off. Simulation results are repeated 30 times for each channel SNR condition to obtain more accurate simulation results and the average of the PSNRs of the left and right views is used as the performance evaluation matrix.

12.5.2 Discussion of Results

Table 12.1 presents the performance of the proposed UEP scheme compared to EEP of *Orbi* sequence with different values of SNRs and with different values of β for color and depth sequences. The average PSNR improvements in comparison with EEP are calculated. The average quality of the reconstructed color sequence achieves improved quality for all SNRs. This is due to assigning more protection to the color sequence by mapping it to MSBs of 16-QAM.

The quality of the depth map degrades from 13.98 dB with EEP to 11.49 and 10.59 dB with β = 1.5 and 2, respectively, at SNR = 10 dB. On the other hand, the color quality increases from 14.09 dB with EEP to 18.79 and 20.81 dB with β = 1.5 and 2 at the same value of SNR for *Orbi* sequence. Consequently, the color sequence will have a significant impact on the received 3-D video, where the DIBR is used to generate the left and right views as will be discussed in the following.

Figures 12.11 and 12.12 illustrate the PSNR performance of left, right, and the average left and right views using hierarchical 16-QAM with different

TABLE 12.1

Average PSNR with and without UEP for *Orbi* Sequence

Channel SNR (dB)	PSNR of EEP (dB)		PSNR of UEP (dB) β = 1.5		PSNR of UEP (dB) β = 2	
	Color	Depth	Color	Depth	Color	Depth
6	11.27	11.13	14.31	8.99	14.98	8.68
8	12.43	12.31	16.14	9.99	17.39	9.51
10	14.09	13.98	18.79	11.49	20.81	10.59
12	16.62	16.43	23.16	13.71	26.24	12.10
14	20.36	20.24	30.02	16.98	32.95	14.45
16	25.94	26.28	38.06	21.81	40.24	17.91
18	35.62	35.28	50.24	28.74	47.10	23.16
20	45.29	42.63	50.24	39.25	47.10	40.29
22	45.33	42.63	50.24	39.25	47.10	40.41

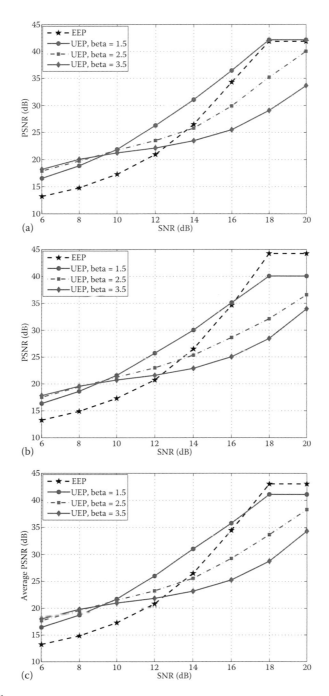

FIGURE 12.11
PSNR performance of the reconstructed 3-D video for *Orbi* sequence. (a) Left view, (b) right view, and (c) average left and right view.

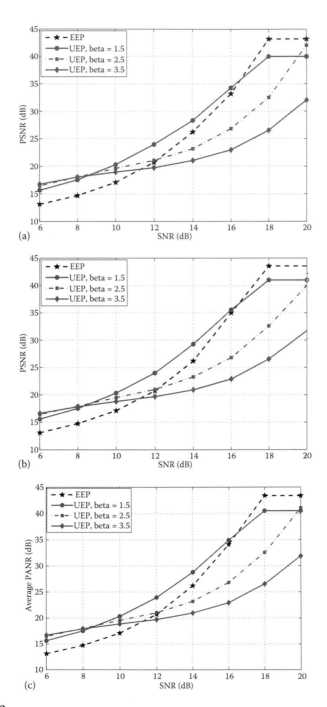

FIGURE 12.12

PSNR performance of the reconstructed 3-D video for *Interview* sequence. (a) Left view, (b) right view, and (c) average left and right view.

β values for the *Orbi* and *Interview* sequences, respectively. As can be seen from the figures, the average PSNR has been improved and outperforms EEP by up to 5 dB at lower SNRs (*SNR* = 6–16 dB) when β = 1.5. By increasing the value of β from 1.5 to 2.5, the UEP performance will increase at small SNRs (i.e., *SNR* = 6–13 dB). However, the performance will decrease compared to EEP when the channel is in a good state.

The obtained results show that the quality of the depth map is not a significant factor in the reconstructed 3-D video quality but still needs to be acceptable to warp the 2-D color sequence to generate the left and right views. However, the average PSNR of the reconstructed left and right views which are used to generate 3-D scene is dominated by the quality of the color sequence. For instance, when *SNR* = 14 dB, the quality of the reconstructed left and right views has been improved by up to 4.5 dB compared to EEP even though the quality of the depth sequence is degraded by about 6 dB at β = 1.5 (Table 12.1).

As a result, UEP performs better at low channel SNRs and performs worse in good channel conditions, especially when β is increasing as shown in Figures 12.12 and 12.13. The figures show that the proposed UEP achieves significant PSNR gains especially in low channel SNR cases. This can be explained as follows. When channel SNR is low, the probability of errors occurring in the color sequence is better protected against channel noise than depth sequence. However, when the channel quality is good, the proposed UEP scheme may result in degraded quality due to the high BER of hierarchical 16-QAM.

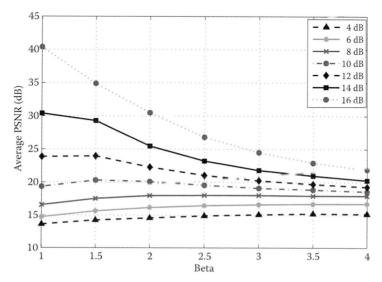

FIGURE 12.13
Average PSNR of a range of β values for *Orbi* sequence.

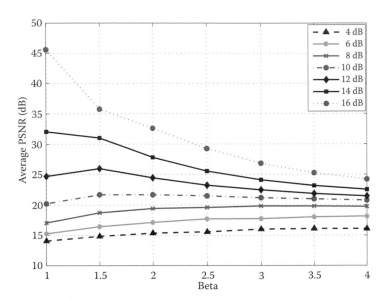

FIGURE 12.14
Average PSNR of a range of β values for *Interview* sequence.

The average PSNR performance versus β values of hierarchical 16-QAM with different SNR for left and right views is also investigated as shown in Figures 12.13 and 12.14. As can be seen from Figure 12.13, when SNR is low ($SNR=4$–10 dB), the average PSNR will increase as the value of β increases (β=1–2.5). For example, when $SNR=8$ dB, the average PSNR increases from 17 to 20 dB at β=1 and 2.5, respectively. On the other hand, when SNR is high ($SNR=12$–16 dB), the average PSNR will decrease by increasing the value of β. That is, the average PSNR decreases from 32 dB at β=1 to 25.5 dB at β=2.5 when $SNR=14$ dB.

For small β values (β=1 and 1.5), the highest average PSNR is achieved when the channel SNR is high. For example, when β=1, the average *PSNR* = 45.6 dB at $SNR=16$ dB and the video quality is approximately 13 dB higher than that in the case of $SNR=14$ dB and 20.8 dB higher than in the case of $SNR=12$ dB with the same value of β. To conclude, it is clear from the figures that at low channel SNRs, a large β value (β=4) has the best performance in the average PSNR but when the channel SNR is high, a large β value has the worse PSNR performance.

12.5.3 Visual Examples for the Proposed Method

Figures 12.15 through 12.18 show the *Orbi* test sequence used for the subjective quality evaluation of the proposed system for the 20th frame of color, depth map, left view, and right view with and without UEP as well as with different β values. These figures can be viewed by the end user by using a red and blue glass with red glass on the right eye. Figure 12.15 illustrates the

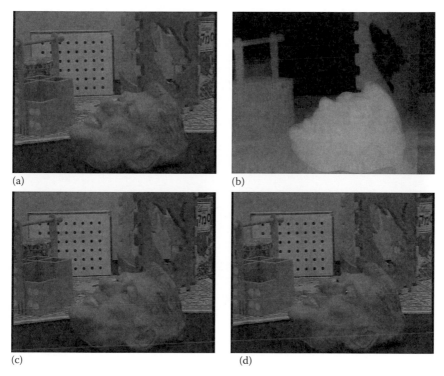

FIGURE 12.15
Original frames of *Orbi* sequence. (a) Color sequence, (b) depth sequence, (c) left view, and (d) right view.

20th frame of the original color, depth sequences, and the generated left and right views. In the case of EEP at $SNR = 12$ dB, where the color quality is low and depth quality is good, the quality of reconstructed left and right views is low due to the low quality of color sequences, which has a significant impact on the overall quality as shown in Figure 12.16.

Figure 12.17 illustrates the decoded frames at $SNR = 12$ dB and $\beta = 1.5$. As shown in the figure, the depth quality is heavily distorted when UEP is used compared to the color quality but the visual quality of left and right views is still acceptable. In the case of UEP with $\beta = 2$ at $SNR = 12$ dB, the quality of color is more increased at the cost of depth quality but the quality of reconstructed left and right views is good as shown in Figure 12.18. In these cases, when displaying left and right views on 3-D display, even though the depth quality is low, users can perceive some sense of depth, but if the depth is completely lost, then the user cannot perceive any sense of depth. This is attributed to the fact that, although the depth map information is lost or highly distorted, which means the sense of depth from binocular disparity is not available, the user can still sense depth from other depth cues and overlapping of objects from the highly protected color sequence.

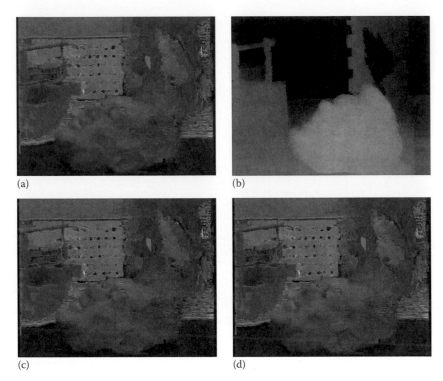

FIGURE 12.16
Reconstructed frames at $SNR = 12$ dB and EEP for *Orbi* sequence. (a) Color sequence, (b) depth sequence, (c) left view, and (d) right view.

Our previous results show that the overall quality highly depends on the color sequence. Using UEP to provide more protection to the color data results in increased left and right views' quality, although the quality of the depth map is degraded significantly with the increase of β. At high SNRs (e.g., $SNR = 18$ dB) and when β value is increased (i.e., $\beta = 2$), EEP outperforms UEP and the visual quality highly degrades especially with high β values. This is due to the high damage of depth map quality. For example, when $SNR = 18$ dB and $\beta = 2$, the depth quality is reduced by about 12 dB compared to EEP as demonstrated in Table 12.1 for the *Orbi* sequence.

12.6 Conclusion

In this chapter, we propose an UEP scheme employing hierarchical 16-QAM by exploiting the differences in the importance of the color and depth map

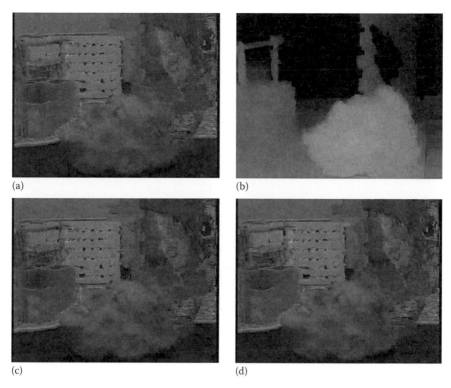

(a)   (b)

(c)   (d)

FIGURE 12.17
Reconstructed frames at $SNR = 12$ dB and $\beta = 1.5$ for *Orbi* sequence. (a) Color sequence, (b) depth
sequence, (c) left view, and (d) right view.

3-D video in terms of the received quality. In this scheme, color sequences
are prioritized then the depth map sequences due to the importance of
the color sequence on the reconstructed visual quality, since even with an
impaired depth map, the user can still perceive a good visual quality and
sense of depth. The MSBs are mapped to the HP data (color), while the LSBs
are mapped to LP data (depth map). The obtained results show that the average PSNR quality of the reconstructed left and right views is dominated
by the quality of the color sequence. Results also indicate that increasing
β will increase the average PSNR performance at low channel SNRs and
decrease the performance at high SNRs. A high β value has the best performance when the SNR is low, whereas it has the worse one when the SNR is
high. It is shown through simulation results that the proposed UEP scheme
significantly outperforms EEP by up to 5 dB in the average left and right
views PSNR gain at $\beta = 1.5$ and $SNR = 12$ dB. Therefore, the UEP scheme proposed in this chapter is effective in delivering better 3-D video over wireless
channels.

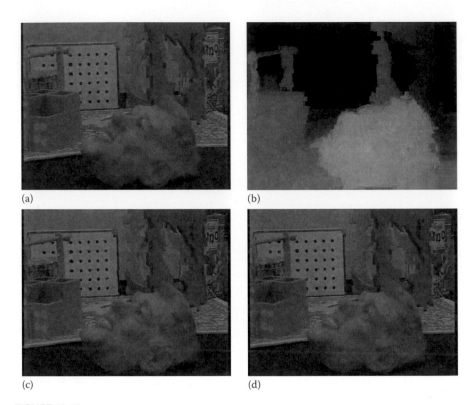

(a)   (b)

(c)   (d)

FIGURE 12.18
Reconstructed frames at $SNR = 12$ dB and $\beta = 2$ for *Orbi* sequence. (a) Color sequence, (b) depth sequence, (c) left view, and (d) right view.

References

Aksay, A., Bici, M. O., Bugdayci, D., Tikanmaki, A., Gotchev, A., and Akar, G. B. 2009. A study on the effect of MPE-FEC for 3-D video broadcasting over DVB-H. In: *Proceedings of the 5th International ICST Mobile Multimedia Communications Conference*, London, United Kingdom.

AVC, M. 2009. Advanced video coding for generic audiovisual services. In: *ITU-T Rec. H.264 and ISO/IEC 14496-10 (MPEG-4 AVC)*, New York, NY.

Barmada, B., Ghandi, M. M., Jones, E. V., and Ghanbari, M. 2005. Prioritized transmission of data partitioned H.264 video with hierarchical QAM. *IEEE Signal Processing Letters*, 12(8), 577–580.

Barmada, B., Ghandi, M. M., Jones, E. V., and Ghanbari, M. 2006. Combined turbo coding and hierarchical QAM for unequal error protection of H.264 coded video. *Signal Processing: Image Communication*, 21(5), 390–395.

Blau, J. 2005. Telephone TV. *IEEE Spectrum*, 42(6), 16–17.

Byers, J. W., Luby, M., Mitzenmacher, M., and Rege, A. 1998. A digital fountain approach to reliable distribution of bulk data. In: *Proceedings of the ACM SIGCOMM '98 Conference on Applications, Technologies, Architectures, and Protocols for Computer Communication*, New York, NY, pp. 56–67.

Chang, Y. C., Lee, S. W., and Komiya, R. 2006. A low-complexity unequal error protection of H.264/AVC video using adaptive hierarchical QAM. *IEEE Transactions on Consumer Electronics*, 52(4), 1153–1158.

Chang, Y. C., Lee, S. W., and Komiya, R. 2009. A low complexity hierarchical QAM symbol bits allocation algorithm for unequal error protection of wireless video transmission. *IEEE Transactions on Consumer Electronics*, 55(3), 1089–1097.

Cherry, S. 2005. Seven myths about voice over IP. *IEEE Spectrum*, 42(3), 52–57.

Chung, W.-H., Paluri, S., Kumar, S., Nagaraj, S., and Matyjas, J. D. 2010. Unequal error protection for H.264 video using RCPC codes and hierarchical QAM. In: *IEEE ICC'10*, Cape Town, South Africa, pp. 1–6.

Fehn, C. 2003. A 3D-TV system based on video plus depth information. In: *Proc. Conference Record of the Thirty-Seventh Asilomar Conference on Signals, Systems and Computers*, Pacific Grove, California, pp. 1529–1533.

Fehn, C. 2004. Depth-image-based rendering (DIBR), compression and transmission for a new approach on 3D-TV. *Proceeding of the SPIE*, 5291(93), 93–104.

Fehn, C., Kau , P., Beeck, M. O. D., Ernst, F., IJsselsteijn, W., Pollefeys, M., Gool, L. V., Ofek, E., and Sexton, I. 2002. An evolutionary and optimised approach on 3D-TV. In: *Proc. International Broadcast Conference*, Amsterdam, The Netherlands, pp. 357–365.

Flack, J., Harrold, J., and Woodgate, G. J. 2007. A prototype 3D mobile phone equipped with a next generation autostereoscopic display. In: *Proceedings of SPIE*, The International Society for Optical Engineering 6490, pp. 502–523.

Flierl, M. and Girod, B. 2007. Multiview video compression. *IEEE Signal Processing Magazine*, 24(6), 66–76.

Grewntsch, S. and Miiller, E. 2004. Sharing of motion vectors in 3d video coding. In: *Proc. International Conference on Image Processing*. ICIP '04', Singapore, Vol. 5, pp. 3271–3274.

Hewage, C. T. E. R., Ahmad, Z., Worrall, S. T., Dogan, S., Fernando, W. A. C., and Kondoz, A. 2009. Unequal error protection for backward compatible 3-D video transmission over WiMAX. In: *Proc. IEEE International Symposium on Circuits and Systems*, ISCAS 2009, Taipei, Taiwan, pp. 125–128.

Hewage, C. T. E. R., Worrall, S., Dogan, S., and Kondoz, A. M. 2008. Frame concealment algorithm for stereoscopic video using motion vector sharing. In: *Proc. IEEE International Conference on Multimedia and Expo*, Hannover, Germany, pp. 1–4.

ISO/IEC JTC1/SC29/WG11. 2008. Overview of 3D video coding. In: *Proc. Doc. N9784*, Archamps, France.

Jiang, H. and Wilford, P. A. 2005. A hierarchical modulation for upgrading digital broadcast systems. *IEEE Transactions on Broadcasting*, 51(2), 223–229.

Kamolrat, B., Fernando, W., Mrak, M., and Kondoz, A. 2008. Joint source and channel coding for 3D video with depth image – based rendering. *IEEE Transactions on Consumer Electronics*, 54(2), 887–894.

Kauff, P., Atzpadin, N., Fehn, C., Muller, M., Schreer, O., Smolic, A., and Tanger, R. 2007. Depth map creation and image based rendering for advanced 3DTV services providing interoperability and scalability. *Signal Processing: Image Communication*, Special Issue on 3DTV, 22(2), 217–234.

Kaup, A. and Fecker, U. 2006. Analysis of multireference block matching for multiview video coding. In: *Proc. 7th Workshop Digital Broadcasting*, Erlangen, Germany, pp. 33–39.

Lee, C.-S., Keller, T., and Hanzo, L. 2000. OFDM-based turbo-coded hierarchical and non-hierarchical terrestrial mobile digital video broadcasting. *IEEE Transactions on Broadcasting*, 46(1), 1–22.

Li, P., Chang, Y., Feng, N., and Yang, F. 2010. A novel hierarchical QAM-based unequal error protection scheme for H.264/AVC video over frequency-selective fading channels. *IEEE Transactions on Consumer Electronics*, 56(4), 2741–2746.

Merkle, P., Morvan, Y., Smolic, A., Farin, D., Muller, K., de With, P. H. N., and Wiegand, T. 2008. The effect of depth compression on multiview rendering quality. In: *Proc. 3DTV Conference: The True Vision – Capture, Transmission and Display of 3D Video*, Istanbul, Turkey, pp. 245–248.

Merkle, P., Muller, K., Smolic, A., and Wiegand, T. 2005. Statistical evaluation of spatiotemporal prediction for multiview video coding. In: *Proc. 2nd Workshop on Immersive Communication and Broadcast Systems*, ICOB 2005', Berlin, Germany, pp. 27–28.

Merkle, P., Muller, K., Smolic, A., and Wiegand, T. 2006. Efficient compression of multiview video exploiting inter-view dependencies based on H.264/MPEG4-AVC. In: *Proc. of IEEE International Conference on Multimedia and Expo, 2006*, Toronto, Canada, pp. 1717–1720.

Merkle, P., Muller, K., and Wiegand, T. 2010. 3D video: Acquisition, coding, and display. *IEEE Transactions on Consumer Electronics*, 56(2), 946–950.

Merkle, P., Smolic, A., and Wiegand, T. 2007. Multi-view video plus depth representation and coding. In: Proc. *IEEE International Conference on Image Processing*, ICIP'07', Vol. 1, Atlanta, GA, pp. 201–204.

Merkle, P., Wang, Y., Muller, K., Smolic, A., and Wiegand, T. 2009. Video plus depth compression for mobile 3d services. In: *3DTV Conference: The True Vision – Capture, Transmission and Display of 3D Video, 2009*, Washington, DC, pp. 1–4.

Microsoft. 2008. Breakdancers sequence. http://research.microsoft.com/en-us/um/people/sbkang/3dvideodownload/

Morvan, Y., Farin, D., and de With, P. H. 2008. System architecture for free-viewpoint video and 3D-TV. *IEEE Transactions on Consumer Electronics*, 54(2), 925–932.

Muller, K. 2009. 3D visual content compression for communications. *IEEE Multimedia Communications Technical Committee, E-LETTER* 4(7), 22–24.

Muller, K., Smolic, A., Dix, K., Kauff, P., and Wiegand, T. 2008b. Reliability-based generation and view synthesis in layered depth video. In: *Proc. IEEE 10th Workshop on Multimedia Signal Processing, 2008*, Cairns, Queensland, Australia, pp. 34–39.

Muller, K., Smolic, A., Dix, K., Merkle, P., Kauff, P., and Wiegand, T. 2008a. View synthesis for advanced 3D video systems. EURASIP *Journal on Image and Video Processing*, 2008(7), 1–11.

O'Leary, S. 1997. Hierarchical transmission and COFDM systems. *IEEE Transactions on Broadcasting*, 43(2), 166–174.

Pourazad, M. T., Nasiopoulos, P., and Ward, R. K. 2006. An H.264-based video encoding scheme for 3D TV. In: *Proc. 14th European Signal Processing Conference (EUSIPCO)*, Florence, Italy, pp. 3271–3274.

Proakis, J. G. 2001. Digital communication. McGraw-Hill 2001, New York, NY.

Shannon, C. E. 1948. A mathematical theory of communication. *The Bell System Technical Journal* 27, 379–423.

Smolic, A., Mueller, K., Stefanoski, N., Ostermann, J., Gotchev, A., Akar, G. B., Triantafyllidis, G., and Koz, A. 2007. Coding algorithms for 3DTV-A survey. *IEEE Transaction on Circuit and Systems for Video Technology*, 17(11), 1606–1621.

Umble, E. A. 2008. Making it real: The future of stereoscopic 3D _lm technology. http://www.siggraph.org/publications/newsletter/volume-40-number-1/makingitreal. Accessed December 20, 2011.

Vetro, A., Wiegand, T., and Sullivan, G. J. 2011. Overview of the stereo and multiview video coding extensions of the H.264/MPEG-4 AVC standard. *Proceedings of the IEEE* 99(4), 626–642.

Vitthaladevuni, P. K. and Alouini, M.-S. 2001. BER computation of 4/M-QAM hierarchical constellations. *IEEE Transactions on Broadcasting*, 47(3), 228–239.

Vitthaladevuni, P. K. and Alouini, M.-S. 2003. A recursive algorithm for the exact BER computation of generalized hierarchical QAM constellations. *IEEE Transactions on Information Theory*, 49(1), 297–307.

Wandell, B. 1995. Foundations of vision. *Sinauer Associates*, Sunderland, MA.

Wang, Z., Bovik, A. C., Sheikh, H. R., and Simoncelli, E. P. 2004. Image quality assessment: From error visibility to structural similarity. *IEEE Transactions on Image Processing*, 13(4), 600–612.

Wei, L.-F. 1993. Coded modulation with unequal error protection. *IEEE Transactions on Communications*, 41(10), 1439–1449.

Yang, L.-L. and Hanzo, L. 2000. A recursive algorithm for the error probability evaluation of M-QAM. *IEEE Communications Letters*, 4(10), 304–306.

Yoon, D., Cho, K., and Lee, J. 2000. Bit error probability of M-ary quadrature amplitude modulation. In: *Proceedings of Vehicular Technology Conference*, Boston, MA, pp. 2422–2427.

Yoon, S.-U. and Ho, Y.-S. 2007. Multiple color and depth video coding using a hierarchical representation. *IEEE Transactions on Circuits and Systems for Video Technology*, 17(11), 1450–1460.

Zitnick, C. L., Kang, S. B., Uyttendaele, M., Winder, S., and Szeliski, R. 2004. High-quality video view interpolation using a layered representation. *ACM Transactions on Graphics*, 23(3), 600–608.

13

2D-to-3D Video Conversion: Techniques and Applications in 3D Video Communications

Chunyu Lin, Jan De Cock, Jürgen Slowack,
Peter Lambert, and Rik Van de Walle

CONTENTS

13.1 Introduction

Three-dimensional (3D) video can be regarded as the next revolution for many image and video applications such as television, movies, video games, video conferences, and remote video classrooms. It provides a dynamic, realistic, and immersive feeling for the viewers, which enhances the sense of presence. However, 3D has not been widely adopted yet. This is because the success of 3D depends not only on technological advances in

3D displays and capturing devices but also on the active deployment of 3D video communication systems and the wide availability of 3D video content. With only a limited number of programs available in 3D format, there is little reason to buy a more expensive 3D device. However, the tremendous production cost and the complicated process for 3D videos result in the lack of 3D content, which hinders the development of 3D technology. To resolve this problem, one effective alternative is to develop new techniques to convert 2D videos into 3D. An efficient 2D-to-3D conversion system can generate more 3D content from existing 2D videos. Most important of all, it can push the popularization of 3D into the general public and consequently stimulate the development of 3D technology. Generally, 2D-to-3D conversion techniques comprise of automatic and human interventional methods (semiautomatic). For the semiautomatic methods, certain key frames of video sequences are assigned depth information by human participation [3], and other frames are converted automatically. Since the semiautomatic methods involve human interaction, its results are better. However, human participation is impractical in many scenarios because it is slow, not real time, and costly. Hence, this chapter will focus on the automatic 2D-to-3D conversion technology mainly.

Except for 3D content creation, 2D-to-3D conversion can also help to reduce the bit rate and enhance the error resilience and error concealment of the 3D video communication system when 2D-plus-depth (2D+Z) or multiview video plus depth (MVD) format is employed. This is very important for the 3D contents transmission in the wireless network because the error-prone feature of wireless network and the packet loss will degrade the visual experience greatly. This chapter will propose such a communication system in the context of 2D-to-3D application.

The objective of this chapter is to provide a comprehensive overview of 2D-to-3D conversion technology and its applications in 3D video communication. Firstly, the principle of 3D display technology is presented in Section 13.2. Then, in Section 13.3, the depth cues that can be used to extract depth are introduced, including the advantage and disadvantage of each cue. Section 13.4 presents the state-of-the-art schemes for the 2D-to-3D conversion. An application of 2D-to-3D conversion in 3D video communication is proposed in Section 13.5. Finally, conclusions and future trends of 2D-to-3D conversion are given in Section 13.6.

13.2 Principle of 3D Display Technology

Because our eyes are spaced apart, the left and right retinas receive slightly different images. This difference in the left and right images is called binocular disparity [26]. The brain integrates these two images into

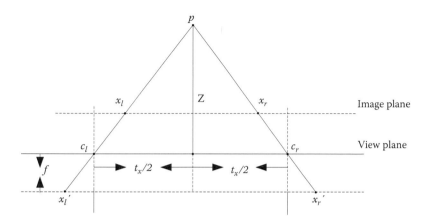

FIGURE 13.1
The relationship between depth map and disparity value.

a single 3D image, allowing us to perceive depth information. In fact, this is the basic idea for 3DTV, where a 3D effect is achieved by generating two slightly different images for each eye.

As Figure 13.1 shows, two viewpoints (cameras/eyes) are located at position c_l and c_r, where f is the focal length of the cameras and t_x is the baseline distance between the two cameras. We assume that the two cameras are on the same horizontal line, just like our two eyes. With this assumption, the left and right cameras generate the images of object p (with distance Z) at the image plane. From the geometry shown in the figure, we have

$$x_l = -\frac{t_x}{2}\frac{f}{Z}$$

$$x_r = \frac{t_x}{2}\frac{f}{Z}$$

(13.1)

The disparity between the two images is $d = x_l - x_r$. After simple calculations, (13.1) can be rewritten as

$$\frac{d}{t_x} = \frac{f}{Z}$$

(13.2)

Formula (13.2) shows that depth is inversely proportional to disparity. Based on this relation, current 3D devices can support the left and right view format or the 2D+Z format. Figure 13.2 shows a general 2D-to-3D conversion system, where depth image–based rendering is the process of synthesizing virtual views of a scene from texture images and its associated per-pixel depth information [7]. The depth recovering part is the most important part for 2D-to-3D conversion and it will be the focus of this chapter.

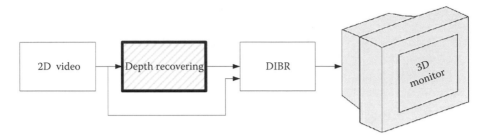

FIGURE 13.2
A general 2D-to-3D conversion system.

13.3 Depth Cues

To generate the 3D version from a 2D image, the basic idea is to discover its corresponding depth value for each pixel. As we all know, we can still get a sense of depth feeling even when we watch 2D images/videos. For example, the farther an object is, the smaller it will appear in the picture. This type of experience is caused by the depth cues in the images/videos. To get the depth information, discovering and understanding the depth cues is a very important step. Hence, the depth cues that can be employed for 2D-to-3D conversion are detailed in this section.

13.3.1 Motion Cues

Imagine taking the train and looking outside, near buildings move faster across the retina than far objects do. The relative motion between the viewing camera and the observed scene provides an important cue to depth perception. Most important of all, the relative motion may be seen as a form of "disparity over time." Generally, the methods using motion cues try to get the motion information from the scene and they can be classified into two groups, which are motion estimation method and structure from motion (SFM).

Motion estimation is extensively used in video coding to get the motion vectors between different macroblocks. To a certain extent, the motion vector can be directly interpreted as disparity. Hence, the simplest way is extracting the motion vector from the video codec (MPEG or H.264/AVC). In Ref. [16], the depth map is obtained from an H.264-encoded sequence, in which different kinds of modes (inter, intra, and skip) are processed differently. Its advantage is the low complexity and its compatibility with broadcasting networks. However, the motion vector from video codec is based on blockwise motion estimation, hence the block artifacts cannot be avoided in the generated depth map. Even though some deblocking filters can be used to reduce block artifacts, the artifacts along object edges still exist, which degrades the

3D experience seriously. Optical flow is another way to estimate the motion vector, which is good on the objects' boundary. However, optical flow is complex and it cannot handle the case when object moves too fast or has large displacement. Energy minimization is employed frequently in computer vision. It could also be used to estimate the motion information by modeling the displacement field as a Markov random field (MRF) and formulating an energy function for the motion information. In the energy minimization function, two constraints are used generally. The first one is the color similarity between two corresponding pixels in two different images. The second is the smoothness for the depth values in a neighborhood. Typically, the energy functions have many local minima and they take enormous computational costs. However, a lot of efficient algorithms, such as that based on max-flow/min-cut, graph cut, and belief propagation [19], can be used to estimate the disparity. For example, with the graph cut, α-expansion moves produce a solution within a known factor of the global minimum of energy [2]. However, energy minimization can only deal with the scenes with sufficient camera moving. Moreover, if there are complex scene changes or multiple independently moving objects, it is difficult to get a good disparity estimation. In addition, the occlusion and textureless region in the scene are difficult to get the correct depth value.

SFM [8] is often employed to recover the camera parameters and 3D scene structure by identifying points in two or more images that are the projections of the same point in space. Using SFM, it is impossible to compare every pixel of one image with every pixel of the other image(s) because of combinatorial complexity and ambiguous matching. In any case, not all points are equally well suited for matching, especially over a long sequence of images. Hence, firstly, SFM extracts feature points, such as Harris corner points that are located at the maxima of the local image's autocorrelation function. Secondly, camera motion and scene structure are estimated based on the detected feature correspondences. Since feature correspondences are used, only a sparse depth map can be obtained from the aforementioned steps. For a dense depth map, Delaunay triangulation or other methods are required [13]. The performance of SFM depends on the feature detection and matching process. In this context, some efficient feature extraction and matching algorithms such as scale-invariant feature transform could be used to improve the performance of SFM. The depth map from SFM is quite good because of the pinhole camera model and epipolar geometry employed. However, SFM assumes that objects do not deform and their movements are linear. It cannot handle scenarios containing degenerated motion (e.g., rotation-only camera) or degenerated structure (e.g., coplanar scene) [13].

In conclusion, the advantage of using motion cues is that it considers more than one frame, hence the depth maps in different frames have a certain consistency, for example, less flickering between consecutive depth maps. However, the motion cues are not effective for static scene or more complex movement. In addition, motion cues require more than one frame to get the depth information, which causes extra delay compared with other depth cues.

13.3.2 Focus/Defocus Cues

Depth-from-defocus (DFD) methods generate a depth map based on the amount of blurring present in the images. In a thin lens system, objects in focus are clearly pictured while objects at other distances are defocused, that is, blurred, as shown in Figure 13.3. The farther the object is, the more blurry it appears. There are two kinds of schemes for the DFD technology. The first one requires a pair of images of the same scene with different focus settings [9]. Providing the camera settings, it recovers the depth by estimating the degree of defocus. If the camera setting is unknown, more images are required to eliminate the ambiguity in blur radius estimation. This approach is reliable and provides a good depth estimation. However, only a small number of 2D video materials will satisfy the aforementioned conditions, that is, images taken from a fixed camera position and object position but using different focal settings. The second scheme extracts the blur information from a single image by measuring the amount of blur. In Ref. [28], a re-blurred scheme is proposed to recover the relative depth from a single defocused image captured by an uncalibrated camera. Firstly, it models the blurred edge in the image as 2D Gaussian blur. Then, the input image is re-blurred using a known Gaussian blur kernel and the gradient ratio between input and the re-blurred image is calculated. Finally, from this ratio the amount of blur at edge locations can be derived and the blur propagation problem is solved as an optimization problem. DFD avoids matching (correspondence) ambiguity problems compared with SFM. Furthermore, its complexity is less than that of SFM. However, it cannot deal with the blur existing originally

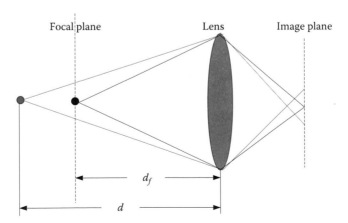

FIGURE 13.3
(See color insert.) Focus and defocus for the thin lens.

in the image as well as textureless regions. Another disadvantage of DFD is that it is only feasible for small depth-of-field images and it generates low precision depth maps.

Depth from focus, as an alternative technique, requires a series of images of the scene with different focus levels by varying and registering the distance between the camera and the scene [23]. This requirement is not easy to meet in practice, and it is not suitable for monocular images or videos.

13.3.3 Geometric Cues

Geometric cues include the familiar size/height of objects in the image, vanishing objects, vanishing lines, and vanishing points. Particularly, vanishing lines and vanishing points are often employed to estimate the depth. Vanishing lines refer to the fact that parallel lines, such as railroad tracks, appear to converge with the distance, eventually reaching a vanishing point at the horizon. The more the lines converge, the farther away they appear to be. Based on vanishing lines slope and origin of the vanishing lines onto the image plane, a set of gradient planes can be assigned, each corresponding to a single depth level. The pixels closer to the vanishing point are assigned a larger depth value and the density of the gradient planes is also higher. Figure 13.4 shows an example of vanishing lines in reality [1].

There are two steps to extract the vanishing point. Firstly, edge detection is employed to locate the predominant lines in the image. Then, the intersection points of these lines are determined by calculating the intersection point of each pair of vanishing lines. If all the intersection points are not located at the same point, a vanishing region whose range covers the most of

FIGURE 13.4
An example of vanishing line and vanishing point.

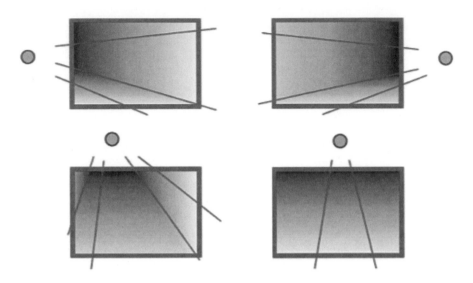

FIGURE 13.5
An example of detected vanishing points. (From Battiato, S. et al., Depth map generation by image classification, in *Proceedings of SPIE 5302*, pp. 95–104, 2004.)

intersection points is defined. Figure 13.5 gives an example for the detected vanishing points. The complexity of this kind of schemes mainly depends on the complexity of detecting vanishing lines, which can be performed in real time. The problem of this method is that it can only be applied to a particular class of images containing geometric objects. In addition, the depth generated by vanishing lines is a qualitative one. It does not take into consideration the prior knowledge of different areas. Accordingly, the points at the same distance away from the vanishing point are assigned the same depth value, regardless of their different semantic areas. In Figure 13.4, for example, the detected vanishing point will be the intersection of the highway. Using the highway as the vanishing line, it will assign the same depth level to the far mountain and sky (with the same gradient level to the vanishing point) without differentiating their semantic knowledge.

13.3.4 Atmospheric Scattering Cues

The propagation of light through the atmosphere is affected in the sense that its direction and power are altered through a diffusion of radiation by small particles in the atmosphere. This leads to the phenomenon called atmospheric scattering, also known as haze. Distant objects appear less distinct and more bluish than objects nearby. In Ref. [6], a physics model is presented to describe the relationship between the radiance of an image and the distance between the object and the viewer. The model used is

$$I = Je^{-\beta d} + A(1 - e^{-\beta d}) \tag{13.3}$$

Here, I is the observed pixel intensity, J is the scene radiance, A is the global atmospheric light, β is the scattering coefficient, and d denotes the depth. In most cases, J is unknown, β has some experimental value, and A can be measured from the sky region in the image. In Ref. [10], a dark prior is discovered to provide a good estimation of J. It is shown in

$$J^{dark}(p) = \min_{c \in \{r,g,b\}} \left(\min_{q \in \Omega(p)} (J^c(q)) \right) \qquad (13.4)$$

where
   $J(p)$ is the pixel intensity
   $\Omega(p)$ represents a block region surrounding pixel p
   c denotes a particular (RGB) color channel

The dark prior is that $J^{dark}(p)$ has a very low intensity value for an outdoor haze-free image [10]. Using this prior and an input image with haze, $J^{dark}(p)$ can be assumed to be zero and the formula about the depth is [10]

$$e^{-\beta d} = 1 - \min_{c \in \{r,g,b\}} \left(\min_{q \in \Omega(p)} \left(\frac{I^c(q)}{A^c} \right) \right) \qquad (13.5)$$

The atmospheric light A can be estimated from the most haze-opaque pixel [6], for example, the pixel with the highest intensity.

   Using atmospheric cue can provide a good estimation of the depth map for an outdoor image, and the complexity of this algorithm depends on the estimation of J and A, which can meet the real-time requirement. However, the estimation of A is not easy for an indoor image, which limits the application of atmospheric cues. In addition, generally, there is no big difference between the observed scene I and the scene radiance J, unless haze, fog, and smoke are strong in the capturing environment.

13.3.5 Other Cues

In the previous sections, we commented on the most important cues, but there are a lot of other cues that can be exploited to estimate depth information. These cues will be briefly summarized in this section. Gradual variation of surface shading in an image provides information about the shape of the objects in the image. These cues are referred to as shading cues. Shape-from-shading (SFS) reconstructs 3D shapes from intensity images using the relationship between surface geometry and image brightness [27]. However, SFS is a well-known ill-posed problem just like SFM, in the sense that the solution may not exist or the solution is not unique and requires additional constraints.

Shape-from-texture tries to recover the 3D surface in a scene by analyzing the distortion of its texture projected in an image [4]. The texture cue offers a good 3D impression because of two key ingredients: the distortion of individual texels and the change rate of texel distortion across the texture region [23]. The latter is also known as the texture gradient. To recover the 3D surface from the texture variations in the image, the texture must be assumed to have some form of spatial homogeneity on the surface. Then, the texture variations are only produced by the projective geometry. In general, the output of shape-from-texture algorithms is a dense map of normal vectors. This is feasible for recovering the 3D shape under the assumption of a smooth textured surface. As a rule, the shape of a surface at any point is completely specified by the surface's orientation and curvatures. However, it is an under-constrained problem. Most algorithms are designed to tackle specific group of textures, which limits their application.

Occlusion cues imply a depth relation between objects. An object which overlaps the view of another object is considered to be closer. In Ref. [15], specific points, such as T-junctions, are detected to infer the depth relationships between objects in a scene. The results in Ref. [15] indicate that occlusion cues are a good feature for relative depth ordering. However, in some cases, it is insufficient to classify depth planes and the lack of sufficient T-junctions may result in incorrect depth order.

13.4 State-of-the-Art Scheme for 2D-to-3D Conversion

From the last section, we know that all previously discussed cues are only suitable for a certain group of images. Generally, one particular cue alone does not produce a good depth map for all kinds of images or video sequences. Normally, effective algorithms will fuse different cues together to recover the depth map. In this section, some classical 2D-to-3D conversion schemes are presented.

13.4.1 Image Classification Techniques

In Ref. [1], images are first under-segmented into color regions using the mean-shift technique [5]. Then, the semantic regions are detected by comparing a group of columns of the regions with a set of typical sequences of a landscape. The typical semantic regions are characterized as *sky, farther mountain, far mountain, near mountain, land,* and *others*. A qualitative depth map can be generated by assigning different depths to these semantic regions. After that, the vanishing lines and vanishing points are detected by further classifying the semantic regions into *landscape, outdoor with geometric elements,* and *indoor*. For landscape without geometric elements, the lowest point is located at the boundary between the land and the other regions. Using such a boundary point, the vanishing point is obtained. For the indoor and outdoor with geometric elements, the edge or line detecting scheme is employed to get the vanishing lines firstly. Then, the intersections of the vanishing lines are located as vanishing points. After vanishing points detection, the gradient plane is generated and the depth is assigned based on the gradient. Finally, the qualitative and geometric depth maps are combined to provide a more accurate depth map. The whole scheme can be regarded as an approach using the semantic information and geometric cues together. With limited computation, it can generate a relative depth map from a single image. However, the accuracy of this approach depends on the detection of the vanishing lines and points as well as the identification of semantic regions. When the scene is more complex, the detection and identification will be difficult, which affects the performance of depth extraction.

13.4.2 Bundle Optimization Techniques

In Refs. [24,25], a bundle optimization model exploiting color constancy and geometric coherence constraints is proposed for estimating the depth map. It solves problems such as image noise, textureless pixels, and occlusions in an implicit way through an energy minimization model.

The method is composed of four steps. In step 1, the camera parameters for each frame are obtained by epipolar geometry with SFM. In step 2, the disparity is initialized for each frame independently by minimizing the initial energy function that considers the color similarity for two corresponding pixels and spatial smoothness constraint in the flat regions. Taking the possible occlusions into account, the temporal selection method is used to only select the frames in which the pixels are visible for matching. In order to deal with textureless regions, mean-shift color segmentation [5] is used to generate segments for each frame and this segmentation information is used during initialization. After that, each disparity segment is modeled with a 3D plane. In step 3, it refines depth information by minimizing the energy function (13.6) frame by frame.

$$E(\hat{D};\hat{I}) = \sum_{t=1}^{n}(E_d(D_t;\hat{I},\hat{D}\backslash D_t) + E_s(D_t)) \qquad (13.6)$$

where D_t is the disparity map for time t. The variables with hat are vectors that contain the corresponding variable at a different time. For example, $\hat{D} = \{D_t | t = 1,...,n\}$ and $\hat{I} = \{I_t | t = 1,...,n\}$. The item E_d represents the color constraint and geometric constraint, and E_s is used for spatial smoothness. This process is called bundle optimization. It is a typical energy minimization function that considers both the geometric constraints and spatial smoothness. Finally, in step 4, the space–time fusion is employed to further reduce the noise in the depth. The fusion considers the spatial continuity, temporal coherence, and sparse feature correspondences (computed from SFM) together to provide a more accurate and stable depth map.

The good performance of this scheme is explained by considering the different constraints together. Since it takes the geometric coherence constraint associated with multiple frames and refines the results iteratively, the obtained depth maps in the sequence are stable. However, it also generates large delay because more frames are considered together. Therefore, its complexity is not suitable for real-time application unless some simplifications are adopted. In addition, if there is not sufficient camera movement, the obtained depth is probably less accurate. Even when the segmentation information is employed, the depth value for textureless region could still be incorrect due to the ambiguity for the depth inference.

13.4.3 Machine Learning Techniques

In Refs. [17,18], MRF is used to model monocular cues and the relations between various parts of the image. For each small homogeneous patch in an image, MRF is used to infer a set of plane parameters capturing both the 3D location and 3D orientation of the patch. The MRF, trained via supervised learning, models depth cues and the relationships between different parts of the image. Since it does not make any explicit assumptions about the image structure, it can capture more details. Firstly, the input image is oversegmented into different superpixels (patches). Then, the image features are calculated for each superpixel. The features include the texture-based summary and statistic features as well as the patch shape and location-based features. These schemes use 17 filters at 3 different spatial scales to get the features. The local feature alone is not enough. Hence, it also attempts to capture global information by including features from neighboring patches as well as from neighbors at different scales. The model considers properties such as connected structure, coplanar structure, and colinearity. All these features are used to derive the plane parameters for each patch. The fractional depth errors between the estimated depth and the ground-truth depth are used to train the model. Since the semantic context such as sky or grass field is recovered by the machine learning to predict the depth map accurately, it does put an enormous burden on the learning process. The parameter learning process is very complex; however, it can be performed offline. Apart from this step, obtaining the features with 524 dimensions is also very computationally intensive. This scheme creates 3D models which are both quantitatively and visually pleasing. With 588 images downloaded from the Internet, it provides 64.9% qualitatively correct 3D structure. Since this scheme is designed just for a single image, no motion information in the sequence is considered. If some scene structure is missed during the training phase, the results will be unreliable. In addition, if this scheme is applied to all the frames in a sequence, there could be some flickering in the depth map even though the depth of each frame may be good enough. A solution is using temporal filters to reduce the flickering.

13.4.4 Schemes Using Surrogate Depth Maps

Except for schemes using the aforementioned cues, other simple algorithms can also be used to get the disparity/depth map for application. In Refs. [21,22], one color component is adopted as the surrogate depth map, including some adjusting and scaling of the pixel value for the specific color component. This surrogate depth map provides the depth information needed for 2D-to-3D conversion. Specifically, the Cr chroma component from the YCbCr is used as the surrogate in Ref. [22] because this component has good features to be the depth map. For example, the skin generally consists of Cr component and will be put in the foreground. In the meantime, blue sky and green grass are placed in the background which applies to most of the scene. This kind of technique can be implemented in real time and does not have a large dependence on the existence of the cues. However, its performance is affected by the color information in the image. The bright red regions and spots will be indifferently rendered in the front of the scene and are very visible, which provides the wrong depth information and annoying visual experience.

13.5 Application of 2D-to-3D Conversion in 3D Video Communications

The application of 2D-to-3D in 3DTV is one of the most important application cases. Since there is not enough 3D content yet, 2D-to-3D conversion is especially important. In fact, almost all the 3DTV companies make 2D-to-3D conversion function available in their products, such as Philips, Sony, Samsung, and LG. Besides 3DTV, 2D-to-3D conversion is also employed in 3D smartphone and 3D video games. 2D-to-3D conversion will play its important role to provide enough 3D content before the 3D era really comes. As a lot of existing 2D content cannot be recaptured into 3D version, this technology will work even though there is enough 3D content.

As more and more 3D content will become available, efficient 3D compression and transmission is required. In this context, an error-resilient 3D wireless communication system assisted with 2D-to-3D is proposed here. For the stereoscopic 3D video, the left view and right view should be provided and encoded. Even though the interview prediction can be used, the bit rate is still quite high compared with 2D video coding. As an alternative, the 2D+Z format stores the texture frame and its corresponding depth map, which requires an additional bandwidth <10% compared to 2D video format. In addition, this format decouples the content creation and visualization process. Most important of all, it offers flexibility and compatibility with existing production equipment and compression tools. With so many advantages, this format has been standardized in MPEG as an extension for 3D filed under ISO/IEC 23002-3:2007 [11]. Due to the bandwidth of wireless network, the 2D+Z format is the better candidate compared with the stereo view format and we will focus on this format in this chapter.

With the 2D-to-3D conversion technology, the depth map coding efficiency and error concealment ability can be improved, which is very important for the transmission of 3D content in wireless network. In the 2D+Z format, the general input is a texture sequence and a depth sequence. This pair of sequences could be encoded separately with some existing video codecs, such as MPEG2 or H.264/AVC. Since our scheme will only affect the performance of depth information, the texture sequence will not be discussed here. For the depth map encoding, there will be motion prediction between the current depth map and its previous depth maps to exploit the temporal information. The coded stream will be transmitted through the network. If some packets containing one depth map are lost, the error will happen and it will propagate to other following depth maps due to the motion estimation. The previous study [14] has shown that coding artifacts on depth data can dramatically influence the quality of the synthesized view. Hence, the error robustness of the whole system is necessary. Generally, some error-resilient tools such as flexible macroblock ordering and data partitioning can be used to mitigate this problem [12]. However, these tools will definitely degrade the compression performance. For the proposed communication system, the compression efficiency can be improved and the error robustness is also enhanced.

In Figure 13.6, the new 3D video communication system with the 2D-to-3D conversion technology is shown. In the encoding end, firstly, the texture

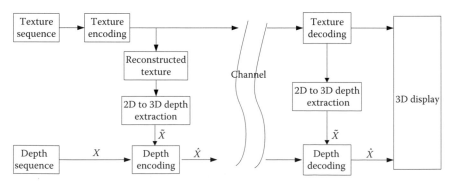

FIGURE 13.6
3D video communication system with 2D-to-3D conversion.

video frames are encoded and the reconstructed texture frames are used to extract the estimated depth map. Then, the estimated depth map is combined with the original depth map to provide more efficient depth map coding. With the estimated depth map, the current depth map (X_n) can select the previous depth map $(\hat{X}_{n-1})$ or its estimated depth map $(\tilde{X}_n)$ as its predicted version according to rate distortion function. From the information theory, it is reasonable to believe that more efficient compression can be achieved, that is, $H(\hat{X}|\tilde{X}) \leq H(\hat{X})$. Here, X represents the original depth map, $\hat{X}$ denotes the reconstructed or decoded depth map and $\tilde{X}$ is the estimated depth map from texture images. Here, we use the reconstructed texture to estimate the depth because it is available at both the encoder and decoder end.

Most important of all, the estimated depth map could contribute to the error resilience and error concealment for the whole system. Generally, the error robustness is in conflict with compression efficiency. However, with the help of estimated depth map, both the error robustness and compression efficiency can be achieved for the depth map coding. In fact, the estimated depth map is available at both the encoder and decoder side, as shown in Figure 13.6. Hence, when one packet containing depth map $\hat{X}_{n-1}$ is lost, the depth map $\hat{X}_n$ that uses $\hat{X}_{n-1}$ as prediction will be affected. However, if some macroblocks in depth map $\hat{X}_n$ use the estimated depth map $\tilde{X}_n$ as prediction, then these macroblocks will not be affected because their predicted version, the estimated depth map, can still be acquired at the decoder end. In addition, the lost content in depth map $\hat{X}_{n-1}$ can be error concealed by $\tilde{X}_{n-1}$, which helps to mitigate the error propagation to a certain extent. The only assumption here is that the encoded texture sequence can be obtained at the decoder. However, if the encoded texture sequence is lost, the depth map is useless because it will not be displayed. Hence, the dependence on the encoded texture sequence is reasonable.

Besides the 2D+Z format, 2D-to-3D conversion can also be applied to the MVD format to provide a more robust and efficient coding scheme. Moreover, there is some research on novel predictions in depth map coding, such as edge-aware prediction [20]. Since depth maps typically consist of smooth regions separated by edges, the schemes that can describe the edge information efficiently will reduce the bit rate and artifact caused by quantization. With the estimated depth map, some coarse edge information can be obtained. Then together with the original depth map, the coarse information could be used to reduce the bit rate and represent the edge information efficiently.

It should be noted that the aforementioned applications rely on the correctness of the estimated depth map. The more accurate the estimated depth map is, the more compression and robustness it could provide for the aforementioned systems. In order to obtain more accurate estimated depth maps, some information of the original depth map could be used, such as the maximum and minimum depth value. This information could be transmitted as parameters if they cannot be available at the decoder end.

13.6 Conclusion and Future Trends

This chapter provides an overview of the main techniques for 2D-to-3D conversion and its application in 3D video communications. Currently, most of the 2D-to-3D conversion techniques focus on some particular cues in the images/videos, which are not easy to be generalized. Even though some algorithms use combined cues, they are often computation intensive and still limited to some kinds of images/videos. Hence, 2D-to-3D conversion is still a difficult task. To make it more effective, other information should be considered in the future research. The available information includes the camera parameters, the capturing environment, the known context of the scene, the video subtitle, and sounds information corresponding to the images/videos. In fact, the current 2D-to-3D conversion does not rely on this information. However, it could help to improve the performance and reduce the computation a lot. For example, if the context of the scene is known as landscape, there is no need to classify the scene as *outdoor* or *indoor*. Especially for the latest generated 2D material, most of the aforementioned information could be obtained. Hence, the future trend should focus on how to employ this information.

When there is less 3D content, 2D-to-3D conversion is very important for 3D content creation. With the development of 3D technology, there will be more and more 3D content. With such a large amount of data, efficient 3D video coding and transmission are required. In this case, 2D-to-3D conversion can be applied to reduce the bit rate and enhance the error resilience and error concealment of the whole 3D video communications system when 2D+Z or MVD format is employed. Such a system is proposed in this chapter. How to extract the depth map with the help of original depth map and how to exploit the estimated depth map in 3D video communications will be another hot topic in the future.

References

1. Battiato S., S. Curti, M. La Cascia, M. Tortora, and E. Scordato. Depth map generation by image classification. In *Proceedings of SPIE Electronic Imaging-Three-Dimensional Image Capture and Applications VI*, San Jose, CA, pp. 95–104, 2004.
2. Boykov Y., O. Veksler, and R. Zabih. Fast approximate energy minimization via graph cuts. *IEEE Transactions on Pattern Analysis and Machine Intelligence*, 23(11):1222–1239, November 2001.
3. Cao X., Z. Li, and Q. Dai. Semi-automatic 2D-to-3D conversion using disparity propagation. *IEEE Transactions on Broadcasting*, 57(2):491–499, June 2011.

4. Clerc M. and S. Mallat. The texture gradient equation for recovering shape from texture. *IEEE Transactions on Pattern Analysis and Machine Intelligence*, 24(4):536–549, April 2002.
5. Comaniciu D. and P. Meer. Mean shift: A robust approach toward feature space analysis. *IEEE Transactions on Pattern Analysis and Machine Intelligence*, 24(5): 603–619, May 2002.
6. Cozman F. and E. Krotkov. Depth from scattering. In *Proceedings of IEEE Conference on Computer Vision and Pattern Recognition*, San Juan, Puerto Rico, pp. 801–806, June 1997.
7. Fehn C. Depth-image-based rendering (DIBR), compression, and transmission for a new approach on 3d-tv. In *Proceedings of SPIE Stereoscopic Displays and Virtual Reality Systems XI*, San Jose, CA, pp. 93–104, January 2004.
8. Forsyth D.A. and J. Ponce. *Computer Vision: A Modern Approach*. Prentice Hall, Upper Saddle River, NJ, 2002.
9. Hasinoff S.W. and K.N. Kutulakos. Confocal stereo. *International Journal of Computer Vision*, 81(1):82–104, January 2009.
10. He K., J. Sun, and X. Tang. Single image haze removal using dark channel prior. *IEEE Transactions on Pattern Analysis and Machine Intelligence*, 33(12):2341–2353, December 2011.
11. ISO/IEC JTC1/SC29/WG11. Text of ISO/IEC FDIS 23002-3 Representation of Auxiliary Video and Supplemental Information. Doc. N8768, Marrakech, Morocco, January 2007.
12. Kumar S., L. Xu, M.K. Mandal, and S. Panchanathan. Error resiliency schemes in h.264/avc standard. *Journal of Visual Communication and Image Representation*, 17(2):425–450, 2006.
13. Li P., D. Farin, R.K. Gunnewiek, and P.H.N. de With. On creating depth maps from monoscopic video using structure from motion. In *Proceedings of IEEE Workshop on Content Generation and Coding for 3D-television*, Eindhoven, Netherlands, pp. 85–92, 2006.
14. Merkle P., A. Smolic, K. Muller, and T. Wiegand. Multi-view video plus depth representation and coding. In *IEEE International Conference on Image Processing (ICIP)*, San Antonio, TX, Vol. 1, pp. I201–I204, October 2007.
15. Palou G. and P. Salembier. Occlusion-based depth ordering on monocular images with binary partition tree. In *IEEE International Conference on Acoustics, Speech and Signal Processing (ICASSP)*, Prague Congress Centre, Prague, Czech Republic, pp. 1093–1096, May 2011.
16. Pourazad M.T., P. Nasiopoulos, and R.K. Ward. Generating the depth map from the motion information of H.264-encoded 2D video sequence. *Journal of Image and Video Processing*, 2010:4:1–4:13, 2010.
17. Saxena A., M. Sun, and A.Y. Ng. Make3D: Learning 3D scene structure from a single still image. *IEEE Transactions on Pattern Analysis and Machine Intelligence*, 31(5):824–840, May 2009.
18. Saxena A., S.H. Chung, and A.Y. Ng. 3-D depth reconstruction from a single still image. *International Journal of Computer Vision*, 76(1):53–69, 2008.
19. Scharstein D. and R. Szeliski. A taxonomy and evaluation of dense two-frame stereo correspondence algorithms. *International Journal of Computer Vision*, 47: 7–42, April 2002.

20. Shen G., W.-S. Kim, A. Ortega, J. Lee, and H. Wey. Edge-aware intra prediction for depth-map coding. In *17th IEEE International Conference on Image Processing (ICIP)*, The Hong Kong Convention and Exhibition Center, Hong Kong, pp. 3393–3396, September 2010.
21. Tam W.J. and C. Vazquez. Generation of a depth map from a monoscopic color image for rendering stereoscopic still and video images, United States Patent Application 12/060,978 Filed April 2, 2008.
22. Tam W.J. and C. Vazquez. CRC-CSDM: 2D to 3D conversion using colour-based surrogate depth maps. In *International Conference on 3D Systems and Applications (3DSA 2010)*, Tokyo, Japan, pp. 1194–1205, 2010.
23. Wei Q. Converting 2D to 3D: A survey. TU Delft, the Netherlands, 2005.
24. Zhang G., J. Jia, T.-T. Wong, and H. Bao. Recovering consistent video depth maps via bundle optimization. In *IEEE Conference on Computer Vision and Pattern Recognition (CVPR)*, Anchorage, AK, pp. 1–8, June 2008.
25. Zhang G., J. Jia, T.-T. Wong, and H. Bao. Consistent depth maps recovery from a video sequence. *IEEE Transactions on Pattern Analysis and Machine Intelligence*, 31:974–988, June 2009.
26. Zhang L., C. Vazquez, and S. Knorr. 3D-TV content creation: Automatic 2D-to-3D video conversion. *IEEE Transactions on Broadcasting*, 57(2):372–383, June 2011.
27. Zhang R., P.-S. Tsai, J.E. Cryer, and M. Shah. Shape-from-shading: A survey. *IEEE Transactions on Pattern Analysis and Machine Intelligence*, 21(8):690–706, August 1999.
28. Zhuo S. and T. Sim. On the recovery of depth from a single defocused image. In *Proceedings of the 13th International Conference on Computer Analysis of Images and Patterns, CAIP'09*, Münster, Germany, pp. 889–897, 2009.

14

Combined CODEC and Network Parameters for an Enhanced Quality of Experience in Video Streaming

Araz Jahaniaval and Dalia Fayek

CONTENTS

14.1 Introduction

The shift toward IP-based infrastructure and the availability of more bandwidth on both the wire-line and the wireless and the satellite communication channels provide the capability to support video transmission over the Internet. As a result, today's networks transport multimedia data with large bandwidth requirements. However, the challenges to transmit high quality video over the Internet with low latency still exist. For example, if many viewers tune in with their internet protocol television (IPTV) systems or their third generation wireless (3G) handhelds to follow a major event, the quality of the video may significantly deteriorate due to the overloaded networks, which will unavoidably result in the loss and/or the late arrival of video packets with latency higher than the maximum decoder play out deadline. One solution is that the video packets can be given a higher priority by using premium quality of service (QoS) in a differentiated services environment (Blake et al., 1998). However, ensuring this special level of QoS treatment is rather costly in terms of network management overhead inherent that is in TCP-based protocols. The current approach is to use a mix of prioritized packet transmission schemes and error features at the decoder (Shao and Chen, 2011).

The move toward IP convergence has certainly many advantages; however, there are challenges with transmitting video data over the Internet without specific QoS guarantees. As the objective of video transmission is to maintain a satisfactory quality of video and an increased quality of experience at the decoder (Zapater and Bressan, 2007; Piamrat et al., 2009; Wiegand et al., 2009), the compressed video is streamed over the Internet using the real-time transport protocol (RTP) over the connectionless user datagram protocol/internet protocol (UDP/IP) protocol, thus reducing the acknowledgments or retransmissions overheads (Belda et al., 2006). This type of video transmission is referred to as the conversational application (such as video telephony and video conferencing) where the acceptable delay is stringent. As a result of this short delay, in the case of network congestion, if a video packet exceeds the predefined time limit or play out deadline, the routers will discard the packet. Moreover, if the router buffer overflows, the router may discard the extra packets. The loss of packets may have adverse effect on the quality of the video at the decoder, especially if key video packets are discarded. In addition, as there is no flow control, the overall video quality can be improved by dynamically scheduling and prioritizing these key video packets in response to the varying traffic load of the network and its end-to-end (E2E) delay and loss probability (Kuo et al., 2011).

Hence, we are proposing the reordering of the packets in the queue of the ingress edge routers, so that those packets, whose presence at the decoder is absolutely crucial to maintaining adequate video quality, are transmitted first. The E2E distortion model we first presented in Jahaniaval and Fayek (2007) is based on error-resilience features in the H.264 codec: flexible macroblock ordering (FMO) and DP. Our model describes distortions as a result of inter- and intra-coding as well as the loss probability from the network. This mathematical distortion model is scalable for S video streams. This model is then used to minimize the overall E2E distortion of all live streams, hence producing a prioritized order of transmission (Jahaniaval, 2010; Jahaniaval et al., 2010). In this chapter, we will present our distortion model and the linear optimization we developed that minimizes the total distortion experienced by the set of active streams at any given time. Then, we will demonstrate that our implemented formulation results in higher video quality for the live streams.

The rest of this chapter is organized as follows. In Section 14.2, we highlight some of the research conducted in video streaming to date and give an overview of the resilience features in H.264/AVC. Section 14.3 describes our proposed E2E distortion model (Jahaniaval and Fayek, 2007) that takes into account the lossy nature of the transport network and embeds the corresponding distortion parameters in enhanced video traces. Section 14.4 describes our optimization model (Jahaniaval, 2010; Jahaniaval et al., 2010) that builds on the distortion model in Section 14.3. In Section 14.5, we validate our mathematical formulation by presenting the results obtained when we used our optimization algorithm to minimize the overall distortion of four video streams. Finally, we present our concluding remarks and future direction in Section 14.6.

14.2 Background and Related Research

Video compression involves the reduction of the data size contained in the raw video frames so that the video content can be transmitted and stored more efficiently. Although the current network bandwidth and the storage capacity have increased tremendously and are offered at a cheaper cost, video compression is still a definite necessity. For example, the amount of data required to store (or transmit) 1 s of a raw common intermediate format (CIF) is 4.35 MB of network or storage capacity, thus justifying the need for an encoder/decoder system. The main function of the enCOder/DECoder (CODEC) is to encode the video sequence and represent the video signal with lower bit representation in comparison to the raw (YUV) video and later decompress, and play the encoded video.

The Motion Picture Expert Group (MPEG) is a member body of the International Organization for Standards (ISO) that has developed standards for video and audio compression and processing. Starting with MPEG-1 standard, this body of the ISO has developed the high broadcasting quality for video and audio signal for transmission and storage by introducing the MPEG-2, which is primarily used in DVD video, digital video broadcasting along with the MPEG layer 3 audio coding (MP3). The MPEG-2 standard replaced the MPEG-1 standard used in video CDs (VCD) and VHS system.

The MPEG-2 standard introduced different profiles and levels of coding as well as improvements in coding efficiency (higher bit rate of 19.4 Mbps in comparison with the 1.5 Mbps of MPEG-1) and support for interlace and progressive video.

The standard has evolved to MPEG-4 and the H.264, which improves the video quality at a lower bit rate in comparison to the previous MPEG-2 standard.

H.264 (MPEG-4 Part 10, which is introduced in the next section) was designed to address higher coding efficiency as well as improved support for reliable transmission for applications such as videoconferencing, video telephony, and IPTV (Richardson, 2003). This is still the standard that is evolving to adapt to wireless networks video streaming applications.

14.2.1 H.264 Video Encoder/Decoder

The components of an H.264 video encoder/decoder are illustrated in Figure 14.1. As we can see, a group of pictures (GOP) is composed of an I-frame (spatially or intra-coded frame) followed by a number of predicted P frames (temporally or inter-coded frames). There also could be B frames or bidirectionally encoded frames. The input to the encoder is a raw video bit stream, usually in the YUV format. These streams undergo block-based compression, quantization, and encoding to produce a compressed bit stream

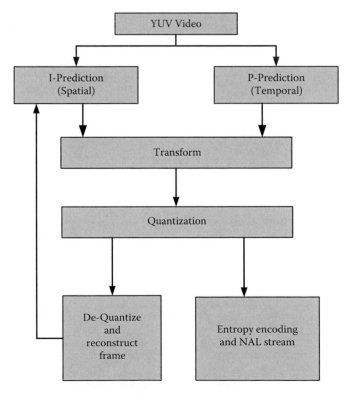

FIGURE 14.1
H264 CODEC components.

that is then packaged into network abstraction layer (NAL) packets that are ready for transmission over digital data networks.

The I-prediction block takes advantage of the similar pixels within the frame (image), which demonstrate high spatial correlation or similarities. The compression is achieved by first de-correlating the image so that there are minimal interdependencies between the pixels. This step is referred to as the "intra-coding" in H.264 and involves the compression of the images by using transform coding, quantization, and entropy coding. The goal of the transform coding is to de-correlate the data so that most of the energy of the image is concentrated to smaller values. An example of this type of transform is the wavelet and discrete cosine transform (DCT), which is used in the H.264 CODEC (Richardson, 2003).

Typically, the video information is concentrated in a few low spatial frequency components, which allows for a more compact representation of the video frame. Once the DCT is applied to the image and the residual information, the resultant information (the transform coefficients) is quantized.

The quantization step size is of importance and a higher quantization step size would result in highly compressed video information (lower bit representation) at a cost of loss in quality. Unlike the DCT, this step is not lossless, and during reconstruction, the data at this stage cannot be fully recovered. The quantization step for a constant bit rate (CBR) video varies and adjusts to maintain the overall bit rate at a prescribed level. In variable bit rate (VBR) encoding, the quantization step remains constant throughout the encoding operation.

Once the data have been quantized, entropy encoding is used to represent the video sequence into a compressed bit stream. The bit code will contain markers and headers describing the synchronization period and image/ sequence headers, respectively. The resultant of this step is the bit stream that can be transmitted or stored (Richardson, 2003).

14.2.2 H.264 Concealment and Error Correction Techniques

As video data are transported over the Internet and wireless networks, losses in video data and signal deterioration are expected. Error detection and concealment schemes attempt to minimize the distortion effect in video quality and prevent image artifacts during the decoding of the transmitted video. The error concealment in H.264 is classified into (1) intra-frame/ intra-slice interpolation and (2) inter-frame/inter-slice concealment or interpolation.

In intra-slice interpolation, also referred to as spatial error concealment, the values of the missing pixels are estimated from the surrounding pixels of the same slice. The weighted pixel interpolation method is used to estimate each pixel from the pixel boundaries of the four adjacent healthy or concealed macroblocks (MBs) (Xu and Zhou, 2004; Xiang et al., 2009).

However, in the inter-slice concealment, also referred to as temporal error concealment, the missing data are calculated by using the available motion vectors (MVs). The damaged MB is replaced from the reference frame that had been referenced by the MV during the encoding process. In the case that the MVs are lost and not available, the damaged MB can be replaced by the MB in the reference frame that is located at the same spatial location (hence, the damaged MB is replaced by the collocated MB in the reference frame). Another alternative in the case of the unavailability of the MVs is the prediction of the lost data by taking the median of the surrounding MVs and replacing the MB from the reference frame using the predicted MV. Other methods of predictions are the boundary matching algorithm (BMA) and the absolute sum algorithm for MV prediction, which are further discussed in Agrafiotis et al. (2006), Kumar et al. (2006), and Xu and Zhou (2004).

14.2.3 Flexible Macroblock Ordering

The FMO is an error-resilience tool in H.264 that allows the assignment of MBs to different slice groups and divides the raw video frame into two or more slices to increase the probability of accurate concealment at the decoder (Stockhammer et al., 2003). As a result, encoding is only performed within the slice and the slice data are encoded independently from data in other slices within the same frame (Dhondt et al., 2006). The objective of FMO is to scatter possible errors to the entire frame to avoid error accumulation in a certain region. The MBs are assigned in different patterns to slice groups. In this work, we have chosen the dispersed (checkerboard) arrangement shown in Figure 14.2.

In this arrangement, each black and white colored MB is assigned to slice groups 1 and 2, respectively. The slices are then divided into DPs that are

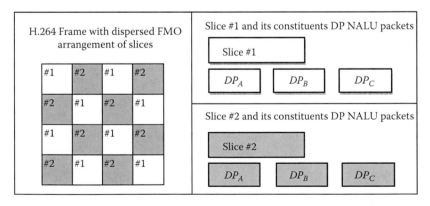

FIGURE 14.2

Dispersed FMO of QCIF (176×144) frame. (From Jahaniaval, A., Video quality enhancement through end-to-end distortion optimization and enriched video traces, M.Sc. thesis, University of Guelph, Guelph, Ontario, Canada, 2010.)

packaged into NAL Unit (NALU) packets. Each slice group is transmitted independently. If one slice is lost in transmission, each of its MBs can then be reconstructed (concealed) from its four neighboring MBs. In the case of temporal concealment when the MV is lost, the decoder will utilize the temporal replacement (TL) or BMA for concealment (Agrafiotis et al., 2006; Chen et al., 2006a,b). Simulation results in Wenger (2003) have demonstrated that with 10% channel loss probability, the visual distortion is unnoticeable and can only be spotted with a trained eye when using FMO.

14.2.4 Data Partitioning

DP classifies the encoded video data to three DPs with unequal importance, thereby providing unequal error protection during transmission. In the usual case of encoding with H.264 (without the use of DP), each MB in a slice is encoded in a single bit stream. On the other hand, DP divides the encoded data into three partitions per slice (instead of one) based on the importance of the encoded data (Figure 14.2). The video header, MVs, and quantization parameters (QPs) are encoded and classified as data partition A (DP_A). The intra-coefficients and inter-coefficients are classified as DP_B and DP_C, respectively. The DP_A has the highest priority since DP_B and DP_C will become useless if the DP_A is lost. In the case when both the DP_B and the DP_C are lost, the MVs and the header can conceal the lost MB. For other loss scenarios, Table 14.1 outlines the concealment strategies adopted in H.264.

14.2.5 Current Related Research

The emergence of the fourth generation wireless networks and digital TV poses great challenges in video transport. Two of these prominent challenges

TABLE 14.1

Decoder Concealment Based on the Arrival of Data Partitions

Availability	Concealment
DP_A, DP_B, and DP_C	Full recovery of video data
DP_A	Concealment using the header and MV from surrounding MBs
DP_A and DP_B	intra-concealment or concealment using MV from DP_A and from intra-residual from DP_B
DP_A and DP_C	inter-concealment or concealment using MV from DP_A and from inter-residual from DP_C

Source: Data adapted from Kumar, S. et al., Error resiliency schemes in H.264/AVC standard, *J. Visual Commun. Image Represent.*, 17(2):425–450, 2006; Jahaniaval, A. and Fayek, D., Combined data partitioning and FMO in distortion modeling for video trace generation with lossy network parameters, in *IEEE International Symposium on Signal Processing and Information Technology, ISSPIT'07*, pp. 972–976, 2007.

are (1) providing and maintaining visually acceptable video quality given variations in network conditions and (2) the accurate measurement of this quality objectively by the networking researchers who use the data in the video trace files, which closely mimic the video encoding statistics to measure the quality after lossy transmission (Ke et al., 2007). Wireless networks complicate this task by the time-varying and location-dependent channel characteristics thereby increasing the probability of packet loss and hence affecting the received video quality (Seeling and Reisslein, 2005a,b). The most commonly used objective metric in video traces is the PSNR. There have been extensive research efforts to improve both the subjective and the objective quality at the receiver by utilizing techniques such as scalable video coding (SVC) (Seeling and Reisslein, 2005a,b; Ke et al., 2007; Seeling et al., 2007; Wang et al., 2007). In SVC, the video is encoded as a base and one or more enhancement layer(s). Decoding only the base layer represents the minimum visually acceptable quality. Adding the enhancement layer to the base layer improves the quality further. One of the most widely used SVC in wireless transmission is the fine grain scalability (FGS) coding. In FGS coding, the base layer is encoded at a coarser quality in comparison with the original video. The enhancement layer is encoded as bit-planes, which can be discarded at the Bit-Plane granularity level. Also, further improvement in quality is possible with the advent of the H.264 video codec that achieves efficient encoding at lower bit rates (Stockhammer et al., 2003; Ksentini et al., 2006). This chapter focuses on encoding mechanisms applied to the base layer only.

There has been extensive research in the area of rate-distortion (R-D) optimization and rate control including but not limited to the research in Dai et al. (2006), Kim and Kim (2002), Kim et al. (2003); Maani et al. (2008), Mansour et al. (2011), Xiong et al. (2005), and Zhang et al. (2006).

In Zhang et al. (2006), the authors have proposed a R-D model to predict the overall decoder distortion of the video frames. The concept of their model is to optimally select the encoding mode so as to reduce the distortion at the decoder. The optimal mode selection takes into account the decoder distortion including the propagation distortion when multiple prediction sources are selected. The encoder then changes the encoding mode in terms of MB to ensure lower distortion at the decoder. Their work did not include multiple streams nor error-resilience features. In addition, it uses flow control and packet retransmission mechanisms that are bandwidth expensive for multiple concurrent streams.

In Maani et al. (2008), the channel model is combined with the R-D model based on slice level encoding. The channel information is utilized by the scheduler to prioritize the packets for transmission. Their proposed method is based on the packetization of each encoded slice, without the consideration of FMO or DP error-resilience features. The GOP distance is set to 16 frames that reduce the error propagation effect (Maani et al., 2008), and this is the GOP size used in our work.

14.3 E2E Distortion Model

Video traces characterize video encoding in a simple text format by including information such as the video frame number, the frame size, the time, and the YUV-PSNR values of individual video frames. The traces are widely used in networking research since video size, copyright issues, and video encoding equipment are the typical problems associated when experimenting with real video data (Seeling and Reisslein, 2005a,b). However, experimentation with actual video bit streams is necessary to obtain more accurate objective and subjective quality rating after lossy transmission (Seeling et al., 2004). In a video trace, if part or all of the video frame is lost, the entire frame is considered to be dropped and therefore the PSNR of that frame is considered as null (Seeling et al., 2007). This method does not take into account the partial frame data losses nor the concealment functions at the decoder and hence portrays high variation in video quality assessment. In addition, current traces do not provide enough parameters for experimental and fine-tuning of network protocols during the design and simulation phases.

In order to enable a more precise assessment, we proposed the addition of extra information based on the distortion model that we developed (Jahaniaval and Fayek, 2007). In this distortion model, we combined FMO with DP so that each slice group is partitioned with DP_A, DP_B, and DP_C as shown in Figure 14.2. The additional parameters reflect the expected distortion as a function of the current network performance, namely, current E2E packet loss probability modeled in Section 14.3.1. These parameters are then included in the trace file to equip network designers with more information when designing and assessing priority-based protocols that aim to minimize the overall distortion of received video. In this work, we classify the distortion for intra- and inter-mode coding (in Sections 14.3.2 and 14.3.3, respectively) and we measure the E2E distortion in terms of the mean square error (MSE) of the original MB and the reconstructed MB at the encoder. Section 14.3.4 presents the overall distortion introduced in a video stream due to the encoding and network conditions. The distortion parameters identified there are then used in our proposed enriched video trace format presented in Section 14.3.5.

14.3.1 E2E Loss Probability

Let $Pb_{A,i}$ denote the E2E mean loss probability for packets belonging to DP_A of slice i; we evaluate it using the steady-state distribution of a Gilbert extended model (Liang and Liang, 2007). We characterize the state transition diagram depicted in Figure 14.3 at the receiver's end. The primary state, state-(0), represents a successful packet reception. State-(1) denotes no reception (or one packet loss) at the following time unit. State-(2) is reached if still no DP packet is received after two time units since the last successful reception,

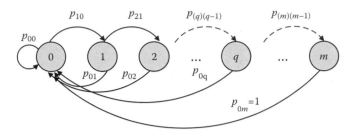

FIGURE 14.3
Extended Gilbert state diagram for data partition transmission. (From Jahaniaval, A. and Fayek, D., Combined data partitioning and FMO in distortion modeling for video trace generation with lossy network parameters, in *IEEE International Symposium on Signal Processing and Information Technology, ISSPIT'07*, pp. 972–976, 2007. With permission.)

and so on up to state-(m). We define the time unit to be the amount of time of the reception of at most one packet. Let state-(q) denote a typical state where $1 \leq q \leq m$ at which q consecutive time slots have elapsed since the last successful reception.

Let:

X:	Random variable describing the number of lost packets
$\bar{s}(n)$:	Probability distribution vector at time step n
s_q:	Steady state probability of state q
$P_{(i)(j)}$:	Transition probability from state j at $(n-1)$ to state i at (n)

The transition from state $\bar{s}(n-1)$ to $\bar{s}(n)$ at discrete time-step n occurs according to Equation 14.1:

$$\bar{s}(n) = A \, \bar{s}(n-1) \tag{14.1}$$

where A is the state transition matrix,

$$A = \begin{bmatrix} P_{00} & P_{01} & P_{02} & \cdots & P_{(0)(m-1)} & 1 \\ P_{10} & 0 & 0 & \cdots & \cdots & 0 \\ 00 & P_{21} & 0 & \cdots & \cdots & 0 \\ 0 & 0 & P_{32} & \cdots & \cdots & 0 \\ \cdot & \cdot & \cdot & \cdot & \cdots & \cdots \\ 0 & 0 & 0 & \cdots & P_{(m)(m-1)} & 0 \end{bmatrix} \tag{14.2}$$

For the state transition matrix A in Equation 14.2, we adopted the notation used in Gebali (2008), where the probability transition matrix is the transpose of the one defined in most literature. Since A is a column-stochastic matrix, the steady state solution of Equation 14.1 will be provided by the convergent state vector $\bar{s}(n)$ as $n \to \infty$ (Gebali, 2008):

$$\bar{s} = A\,\bar{s} \tag{14.3}$$

Since we are interested in the E2E mean loss probability $Pb_{A,i}$ for packets belonging to DP_A of slice i, we evaluate it using the steady-state distribution as follows:

$$Pb_{A,i} = \sum_{q=1}^{m} q \cdot s_q = 1 - s_0 \tag{14.4}$$

In a similar way, we can compute the respective mean loss probability $Pb_{B,i}$ and $Pb_{C,i}$ of DP_B and DP_C, respectively. However, we would like to explore the modeling of the latter quantities conditionally based on the successful reception of the DP_A packets during a time period T that is less than the play out deadline and above which the DP_A is considered completely lost ($Pb_{A,i} = 1$). This will pose an upper bound on m: $m = \lfloor playout/T \rfloor$. Otherwise, for an intra-coded slice (I-slice), the DP_A packets are received with $Pb_{A,i} = 0$.

14.3.2 Intra-Mode E2E Distortion Cases

The proposed E2E distortion model is comprised of the H.264 encoder characteristics as well as the channel loss probability and delay information in order to select the most optimum order of packet deployment from the queue of the ingress router. This model combines the FMO and the DP (Zhang et al., 2006; Jahaniaval and Fayek, 2007) so that each slice group is divided to DP and further packetized for transmission. The model mathematically describes the distortion of the slices at the decoder based on the availability and the reception of the packets carrying the slice information. The model also combines the loss probability (presented in Section 14.3.1) as additional parameters so as to quantify the distortion as a function of the current network performance as we will show in the following discussion.

As an example, if an I-Slice (which is usually packetized and transmitted as DP_A and DP_B packets) is available at the decoder, then the reconstruction of the slice information will be fully recoverable with the exception of the information lost during quantization. However, if due to an increased loss probability or delay beyond the decoder's play out deadline, the packet containing DP_B is lost, the video header information in the DP_A packet will be sufficient to conceal the slice. On the other hand, if an inter-coded slice, which consists of DP_A and DP_C packets, lacks the DP_C information at the decoder, then the motion vectors and quantization information carried in the DP_A packet will be sufficient to locate the lost information in the reference frame and replace the missing MB.

In this model, the distortion is classified based on inter- and intra-mode encoding and the E2E distortion is measured in terms of the MSE of the original MB from the raw (YUV) video and the reconstructed MB at the decoder. Hence, in this case, the encoding quantization distortion is also considered in the model.

Let $X_{i,j}$ be the MB belonging to slice i and frame j where $i=1, 2$ indicating the slice number and $j=1, 2, ..., J$ indicating the frame number in the GOP. Let also $X_{i,j}^{Q}$ and $X_{i,j}^{Conc}$ represent the reconstructed MB at the encoder and the decoder, respectively. The reconstructed MBs at the decoder ($X_{i,j}^{Conc}$) contain both the quantization distortion and the distortion as a result of the MB concealment due to the possible loss of video data during transmission. As the FMO is integrated in our model, $(Pb_{A,i})(1 - Pb_{A,k})$ denotes the probability of a lost or corrupted MB, which is reconstructed according to the information from its four neighboring MBs in slice k, where $k \neq i$.

The intra-mode distortion model describes the possible distortion scenarios related to the intra-coded MBs. As the availability of both DP_A and DP_B contributes to the full reconstruction of the intra-MBs, the presence of these DPs at the receiver is essential to decode a visually acceptable video quality (Maani et al., 2008). In the first two cases, the DP_A must be received timely at the decoder, so we set $Pb_{A,i}=0$ in Equations 14.5 and 14.6.

Case 1: If DP_A and DP_B are successfully received within the play out deadline at the decoder and are not corrupted, then the distortion is only due to quantization:

$$D(i,j) = E\left[\left(X_{i,j} - X_{i,j}^{Q} \right)^2 \right] \cdot \left(1 - Pb_{A,i} \right) \cdot \left(1 - Pb_{B,i} \right) \tag{14.5}$$

The successful reception of DP_A and DP_B is of high importance, not only for ensuring the acceptable visual quality for the decoded MB but rather to be used as an accurate reference for the reconstruction of other inter-coded MBs.

Case 2: If DP_A is received correctly but DP_B is corrupted or lost, the spatial concealment that takes place results in the following distortion:

$$D(i,j) = E\left[\left(X_{i,j} - X_{i,j}^{Conc} \right)^2 \right] \cdot \left(1 - Pb_{A,i} \right) \cdot \left(Pb_{B,i} \right) \tag{14.6}$$

The total distortion for the intra-coded MBs in slice i and frame j is obtained by adding up the two previous equations and resulting in the following:

$$D(i,j)_{Intra} = E\left[\left(X_{i,j} - X_{i,j}^{Q} \right)^2 \right] \cdot \left(1 - Pb_{B,i} \right) + E\left[\left(X_{i,j} - X_{i,j}^{Conc} \right)^2 \right] \cdot \left(Pb_{B,i} \right) \tag{14.7}$$

14.3.3 Inter-Mode E2E Distortion Cases

The inter-mode distortion cases explain the possible scenarios due to the loss and availability of inter-coded information. The inter-coded information exists in DP_A and DP_C, which contain the motion vectors and inter-coded residuals, respectively. Following the same notation in Section 14.3.2, we define the following inter-coding distortion cases:

Case 1: If DP_A and DP_C are received at the decoder and have not been corrupted nor delayed:

$$D(i,j) = E\left\{\left[X_{i,j} - \left(res_{i,j}^Q + X_{i,j-1}^Q\right)\right]^2\right\} \cdot \left(1 - Pb_{A,i}\right) \cdot \left(1 - Pb_{C,i}\right) \qquad (14.8)$$

In Equation 14.8, the residual contained in $DP_C\left(res_{i,j}^Q\right)$ is added to the location of the reference MB in the previous frame $X_{i,j-1}^Q$ and the resultant is used to calculate the MSE in comparison to the MB in the raw video $X_{i,j}$.

Case 2: DP_A is received successfully at the decoder, however, DP_C is lost:

$$D(i,j) = E\left[\left(X_{i,j} - X_{i,j-1}^Q\right)^2\right] \cdot \left(1 - Pb_{A,i}\right) \cdot \left(Pb_{C,i}\right) \qquad (14.9)$$

In this scenario, the decoder will use the MV in DP_A to identify the MB in the reference frame (previous frame in which the motion prediction was made). We need to note here that our model assumes that the motion prediction and compensation is accomplished using a reference point with a distance of one frame, hence the notion of $X_{i,j-1}^Q$ and not $X_{i,j-2}^Q$. Equation 14.9 represents the quantization distortion of the MB located in the previous frame as the decoder uses the available motion vector in DP_A to simply just replace the current lost MB with the collocated one from the previous frame.

Case 3: Both DP_A and DP_C are lost or corrupted:

$$D(i,j) = E\left[\left(X_{i,j} - X_{i,j-1}^{Conc}\right)^2\right] \cdot \left(Pb_{A,i}\right) \cdot \left(Pb_{C,i}\right) \cdot \left(1 - Pb_{A,k}\right) \quad \text{where } k \neq i \qquad (14.10)$$

This specific scenario is permitted as a result of the FMO arrangement; a lost MB can be concealed from its neighboring MBs belonging to a different slice. As discussed earlier, the BMA algorithm can be used to estimate the lost MB. Alternatively, it is more common practice that the lost slice will be estimated and concealed using the MV information from the surrounding MBs belonging to the second FMO slice.

In Equation 14.10, the MB is concealed from the colocated MB in the previous frame or concealed from the surrounding MBs provided that the MB belonging to the second slice is not lost with probability $(1 - Pb_{A,k})$.

Again, for the realization of these scenarios, $Pb_{A,i}=0$ in Equations 14.8 and 14.9, whereas $Pb_{A,i}=1$ in Equation 14.10. As a result, the total inter-distortion is the sum of the quantization distortion and concealment distortion in slice i belonging to frame j given in Equations 14.8 through 14.10:

$$D(i,j)_{Inter} = E\left\{\left[X_{i,j} - \left(res_{i,j}^Q + X_{i,j-1}^Q\right)\right]^2\right\}\cdot(1 - Pb_{C,i})$$

$$+E\left[\left(X_{i,j} - X_{i,j-1}^Q\right)^2\right]\cdot(Pb_{C,i}) \qquad \text{where } k \neq i \qquad (14.11)$$

$$+E\left[\left(X_{i,j} - X_{i,j-1}^{Conc}\right)^2\right]\cdot(Pb_{C,i})\cdot(1 - Pb_{A,k})$$

14.3.4 Parametric Distortion Model

The total distortion $D(i,j)_{total}$ of slice i in frame j is therefore given by the sum of Equations 14.7 and 14.11:

$$D(i,j)_{total} = D(i,j)_{Intra} + D(i,j)_{Inter}$$

By examining Equations 14.7 and 14.11, we can formulate the total distortion as a function of the E2E loss probabilities (Equation 14.4) as follows:

$$D(i,j)_{total} = K_1(i,j)\cdot(1 - Pb_{B,i}) + K_2(i,j)\cdot(Pb_{B,i})$$

$$+K_3(i,j)\cdot(1 - Pb_{C,i}) + K_4(i,j)\cdot(Pb_{C,i}) \qquad (14.12)$$

$$+K_5(i,j)\cdot(Pb_{C,i})\cdot(1 - Pb_{A,k})$$

where $k \neq i$ and the distortion coefficients (the K parameters) are as follows:

$$K_1(i,j) = E\left[\left(X_{i,j} - X_{i,j}^Q\right)^2\right]$$

$$K_2(i,j) = E\left[\left(X_{i,j} - X_{i,j}^{Conc}\right)^2\right]$$

$$K_3(i,j) = E\left\{\left[X_{i,j} - \left(res_{i,j}^Q + X_{i,j-1}^Q\right)\right]^2\right\} \qquad (14.13)$$

$$K_4(i,j) = E\left[\left(X_{i,j} - X_{i,j-1}^Q\right)^2\right]$$

$$K_5(i,j) = E\left[\left(X_{i,j} - X_{i,j-1}^Q\right)^2\right]$$

TABLE 14.2

Standard Verbose FGS Base Layer Trace File

Frame Information				PSNR		
Frame No.	Time	Type	Size	Y	U	V
0	0.01	I	10,000	40	35	35
1	0.33	P	4,096	38	34	33
2	0.66	P	2,046	32	30	30
3	0.99	P	1,024	28	31	25
⋮	⋮	⋮	⋮	⋮	⋮	⋮

Source: Data from Seeling, P. et al., *Video Traces for Network Performance Evaluation,* Springer, Heidelberg, Germany, 2007; Jahaniaval, A., Video quality enhancement through end-to-end distortion optimization and enriched video traces, M.Sc. thesis, University of Guelph, Guelph, Ontario, Canada, pp. 511–516, 2010.

14.3.5 Enriched Trace File Format

The video traces are an abstraction of the actual encoded video which consists of video information such as video frames, data size, and PSNR values. The standard verbose trace format for the FGS encoder would not portray an accurate quality measurement when used in conjunction with the optimizer described in Section 14.4. In fact when using a standard verbose file, if a frame is partially or completely lost, the entire frame is considered as lost. As a result, the quality of the decoder concealment of a lost MB is not taken into consideration. The commonly used verbose trace format for FGS base layer is illustrated in Table 14.2 (Seeling et al., 2007; Seeling and Reisslein, 2010).

As previously described, the data in DP_A has higher precedence in comparison to the data in DP_B and DP_C, and hence the network will assign a higher priority to DP_A packets. Each slice within a frame will then contain three priority packet types. The proposed trace file format explores a higher level of granularity by including details at the DP level per slice. Table 14.3 shows the enriched trace file format* we used in our work.

We embed the K parameters (Equation 14.13 and Table 14.3) in each packet header. These parameters are used by the optimizer to estimate the overall distortion experienced by the active flows as explained in the following section.

14.4 Optimization Model

In this section, we describe our distortion optimization formulation that utilizes the level of granularity described in the previous section. In network

---

* So far we focused only on enriching of the FGS Base-Layer trace file.

TABLE 14.3

Proposed Verbose Trace Format for FGS Base Layer

Packet Seq.	Frame No.	Slice No.	DP	Size (Bytes)	Distortion Coefficients
0	0	0	A	20	$K_1\,(0,0) \ldots K_5\,(0,0)$
1	0	0	A	16	$K_1\,(0,0) \ldots K_5\,(0,0)$
2	0	1	A	2902	$K_1\,(0,1) \ldots K_5\,(0,1)$
3	0	1	A	2844	$K_1\,(0,1) \ldots K_5\,(0,1)$
4	1	0	A	62	$K_1\,(1,0) \ldots K_5\,(1,0)$
5	1	0	B	275	$K_1\,(1,0) \ldots K_5\,(1,0)$
6	1	0	C	26	$K_1\,(1,0) \ldots K_5\,(1,0)$
7	1	1	A	63	$K_1\,(1,1) \ldots K_5\,(1,1)$
⋮	⋮	⋮	⋮	⋮	⋮

Source: Data from Jahaniaval, A. and Fayek, D., Combined data partitioning and FMO in distortion modeling for video trace generation with lossy network parameters, in *IEEE International Symposium on Signal Processing and Information Technology, ISSPIT'07*, pp. 972–976, 2007; Jahaniaval, A., Video quality enhancement through end-to-end distortion optimization and enriched video traces, M.Sc. thesis, University of Guelph, Guelph, Ontario, Canada, pp. 511–516, 2010.

simulations, the a priori knowledge of the distortion model in Equation 14.12 and its distortion coefficients (Equation 14.13) enables the design of network protocols that make informative resource allocation so as to maximize the quality of the decoded video at the receivers.

According to Equation 14.12, the overall distortion of all the streams is calculated based on the number of distortion parameters and the probability of loss for the packets belonging to each stream. The order of transmission plays an important role since the chance of distortion will increase as more data are sent. The optimization outcome is a reordering of the streams so that the overall distortion of all active streams is minimized. In other words, the optimizer selects the stream whose packets have a highest chance of loss or delay in comparison to all the other existing streams to be transmitted first, followed by the stream that will experience the second highest and the third highest, and so on. Without optimization, the stream with the highest distortion is anticipated to experience a lower received quality that degrades further over time if no corrective measure is taken.

Our optimization model is described mathematically as follows. Let d_s denote the distortion for stream s, then for all $s \in S$,

$$d_s = \sum_{j}^{J}\sum_{i}^{N} \left(\begin{array}{l} K_1(i,j,s)\cdot(1-Pb_{B,i,s}) + K_2(i,j,s)\cdot(Pb_{B,i,s}) + \\ K_3(i,j,s)\cdot(1-Pb_{C,i,s}) + K_4(i,j,s)\cdot(Pb_{C,i,s}) + \\ K_5(i,j,s)\cdot(Pb_{C,i,s})\cdot(1-Pb_{A,k,s}) \end{array} \right) \quad \forall s \in S \quad (14.14)$$

where
$k \neq i$

N is the number of slices per frame ($N=2$ in our implementation)

J is the number of frames per traffic stream

$K(i,j,s)$ is the distortion associated with the ith slice in the jth frame in the sth stream as given in Equations 14.12 and 14.13

We define an ordering factor (ODF) ζ_s for stream s. The increased chance of distortion is represented by multiplying ζ_s by the stream distortion value and we select the ODF value ζ_s to be increasing exponentially as a function of the order by which the stream s is sent. In a sense, ζ_s represents the added distortion factor caused by ordering the streams.

Now, let us define the total distortion D_{Total} incurred by all streams taking into account their transmission ordering as follows:

$$D_{Total} = \begin{bmatrix} d_1 \cdots d_s \end{bmatrix} \begin{bmatrix} b_{11} & \cdots & b_{1S} \\ \vdots & \ddots & \vdots \\ b_{S1} & \cdots & b_{SS} \end{bmatrix} \begin{bmatrix} \zeta_1 \\ \vdots \\ \zeta_S \end{bmatrix} \tag{14.15}$$

Expanding the matrix notation in Equation 14.15, we have

$$D_{Total} = \begin{aligned} & \zeta_1(d_1 b_{11} + d_2 b_{21} + \cdots + d_S b_{S1}) \\ & + \zeta_2(d_1 b_{12} + d_2 b_{22} + \cdots + d_S b_{S2}) \\ & \vdots \\ & + \zeta_S(d_1 b_{1S} + d_2 b_{2S} + \cdots + d_S b_{SS}) \end{aligned} \tag{14.15a}$$

The binary coefficient matrix B is utilized to solve the minimization of the total distortion in Equation 14.15 where d_s, $s \in S$, is defined by Equation 14.14 and the binary values b_{ss}, $s \in S$, are constrained such that the sum of each row and each column in matrix B equals 1. Now we define our optimization problem as

$$Minimize \quad D_{Total}$$

$$Subject \ to:$$

1. $b_{rc} - \{0,1\} \quad \forall r, c \in S$

2. $\sum_r^S b_{rc} = 1 \quad \forall c \in S$

3. $\sum_c^S b_{rc} = 1 \quad \forall r \in S$

4. $\zeta_s = (s+1)^s$

$$\tag{14.16}$$

Constraint 4 shows that the ODF for stream *s* increases exponentially with its transmission order.

For each stream *s*, the optimization algorithm extracts the information carried by all the packets belonging to the stream and using the *K* parameters for each DP belong to stream *s*, the distortion incurred by the stream is evaluated using the current network conditions in terms of the E2E loss probability. Thus, the values of the loss probabilities for each stream are continuously measured during simulation and are fed back to the optimizer in the ingress router.

During the simulation, the loss probability for each stream is calculated at its corresponding receiver using the weighted moving average (WMA) method (Ross, 2009). Hence, for the *n*th measurement with window size equal to *m* (where $m = [playout/T]$)

$$WMA_n = \frac{n \cdot LP_m + (n-1) \cdot LP_{m-1} + (n-2) \cdot LP_{m-n+2} + \cdots + LP_{m-n+1}}{n + (n-1) + (n-2) + \cdots + 2 + 1} \tag{14.17}$$

Recalling that the ODF ζ_s multiplies the distortion value of each stream and the optimization binary variables are restricted to be equal to either 0 or 1, now by solving the system, the optimization results in the current live streams reordering that achieves the minimum overall distortion D_{Total} (Equation 14.15).

14.5 Implementation, Analysis, and Results

In this section, we describe the experimental and simulation environment we developed to validate the mathematical modeling presented in the previous sections. In Section 14.5.1, we explain how the video traces were generated before feeding the encoded bit streams into a network topology explained in Section 14.5.2. In Section 14.5.3, we present the analytical results of our work.

14.5.1 Video Traffic and Trace Generation

To produce the enhanced video trace files (Table 14.3), the H.264/AVC JM 12.0 (Sühring, 2008) encoder was modified to encode four video sequences: Akiyo, Carphone, Foreman, and Silent (Seeling and Reisslein, 2010). Table 14.4 shows the properties of these four video sequences. The selected videos are commonly used in the video research domain as they portray a high degree of temporal and spatial variation, high and low motion sequences (Ekmekci et al., 2006), and different texture and motion characteristics (Ruolin, 2008).

To calculate the *K* parameters for each slice at the encoder, the MSEs in Equation 14.13 are evaluated by computing the difference between the collocated pixels in the raw YUV FMO slice and its corresponding encoded–decoded slice. To evaluate K_5, the DP_A packets belonging to slice 1 in frame

TABLE 14.4

Video Sequences Used in Simulation

Video Sequence	Akiyo	Foreman	Carphone	Silent
Frame dimension	QCIF[a]	QCIF	QCIF	QCIF
Number of frames	298	396	381	298
Sampling	4:2:0	4:2:0	4:2:0	4:2:0
Avg. packet size (kB)	5.13	12.9	10.9	8.74
Max packet size (kB)	45	42	51	64
Avg. rate (kbps)	307.55	774.54	654.2	524.8
Max rate (kbps)	2693	2530	3083	3821

Source: Data adapted from Jahaniaval, A., Video quality enhancement through end-to-end distortion optimization and enriched video traces, M.Sc. thesis, University of Guelph, Guelph, Ontario, Canada, 2010.

[a] 176×144.

j are removed, then frame j is decoded using the DP_A packets of its other slice, slice 2. The same process is repeated when calculating K_5 for slice 2, but now the DP_A packets of slice 1 are used.

To reduce the error propagation, a 16-frame GOP IPPP structure was used. Bidirectional predicted frames (or the B-frames) were not used (Seeling et al., 2007) as it will increase the chance of error propagation. Each video sequence was encoded at a rate of 30 frames per second and a constant QP and hence the output of the encoder is a stream of VBR video data as opposed to a CBR. The QP has been set to 20 with a range between 0 (lossless) and 51 (heavily distorted, as a result of reduced bit rate). The average PSNR value was set at approximately 65 dB. This PSNR value produces a high quality video so that any loss in video information can be detected by visual comparison due to the high contrast between the high quality (received FMO slices) and low quality (concealment) areas, when the video is decoded, which we experimentally observed during the trace generation phase. Statistically, the averages were calculated for each video sequence and are summarized in Table 14.4.

14.5.2 Network Simulation

In this work, the OMNeT++ network simulator (OMNeT++, 2010) was used to validate the optimization model. OMNeT++ is a discrete-event simulator, which provides an object-oriented with an open-architecture network simulation environment. The optimization algorithm was implemented using CPLEX™ (CPLEX, 2010), which is an optimization library originally developed by ILOG and now is part of the product suite of IBM™. The C++ version of CPLEX was used within the network simulator.

Figure 14.4 displays the network topology we designed and used to carry the video traffic. Five traffic sources stream video and non-video data to

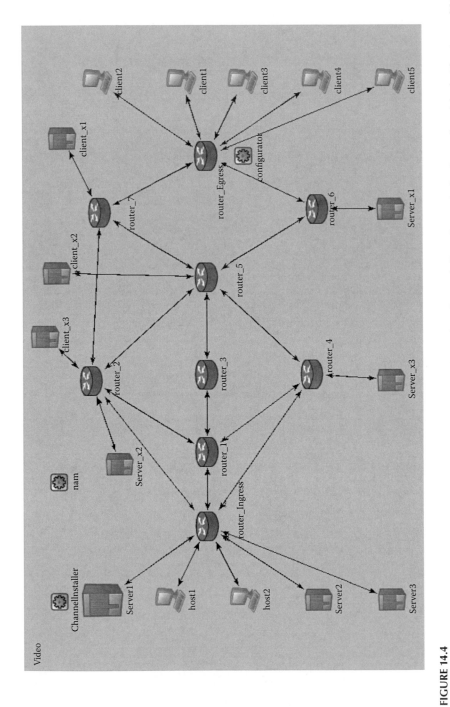

FIGURE 14.4
Network topology. (From Jahaniaval, A., Video quality enhancement through end-to-end distortion optimization and enriched video traces, M.Sc. thesis, University of Guelph, Guelph, Ontario, Canada, 2010.)

five destinations through the lossy Internet cloud represented by routers 1 through 7 in Figure 14.4. The ingress and egress Edge routers interface the sources and destinations to the Internet cloud, respectively. The traffic used in our simulations from source to destination is as follows:

Source	Receiver	Video Sequence
Server 1	Client 1	Stream 1: Akiyo
Server 2	Client 2	Stream 2: Carphone
Server 3	Client 3	Stream 3: Foreman
Host 1	Client 4	Stream 4: Silent
Host 2	Client 5	Stream 5: Non-video network traffic

The implementation of the optimization model occurs in the ingress router. Based on the information provided to the optimizer, namely the K distortion values (Equation 14.13) and the calculation of the loss probability, the optimizer will schedule the deployment of the packets so as to minimize the predicted distortion at the receivers for all the streams. Figure 14.5 shows the edge router architecture that we used in our simulations, in which four video traffic sources are incident on an ingress router that has one queue for each DP type.

The rearrangement of the packets is first applied on the DP_A queues, then on queues with DP_B and DP_C packets, respectively. This rearrangement will give priority to the packets of the streams that are experiencing higher deterioration at the current evaluation period in order to ensure that the decoded quality is not jeopardized as a result of the overall latency and loss probability. As will be demonstrated in the following section, the optimizer is triggered under different conditions and at different GOP intervals.

After reordering, the packets sequence belonging to the same stream is preserved. It has been shown (Maani et al., 2008) that, as in the event of a

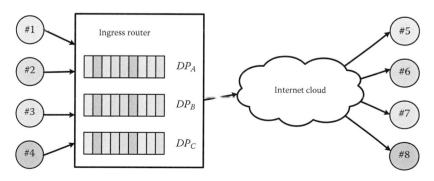

FIGURE 14.5
(See color insert.) DP queues in the ingress router. (From Jahaniaval, A., Video quality enhancement through end-to-end distortion optimization and enriched video traces, M.Sc. thesis, University of Guelph, Guelph, Ontario, Canada, pp. 511–516, 2010.)

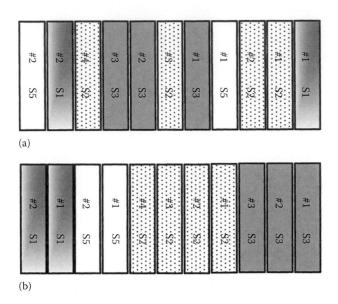

(a)

(b)

FIGURE 14.6
Order of video packets in DP_A. (a) Before re-ordering and (b) after reordering. (Adapted from Jahaniaval, A., Video quality enhancement through end-to-end distortion optimization and enriched video traces, M.Sc. thesis, University of Guelph, Guelph, Ontario, Canada, 2010.)

packet loss, the concealment will be more effective if the packets are ordered according to their original sequence. Figure 14.6 demonstrates the reordering in the DP_A queue.

The proposed model does not take into consideration the effect of error propagation as it assumes that the GOP is no more than 16 frames, hence the error is only confined to the 16 frames until the next I-slice is received at the beginning of each GOP. The scheduler at the ingress router compares the current loss probability and the current delay to the decoder's delay and loss probability thresholds for each stream. In the event that the current delay and loss probability exceed their respective thresholds, the optimizer is activated. In our implementation, we did not account for any lag between the measurements of the average loss probability and the delay per stream, which is calculated at each end receiver, and the measurements availability at the ingress router's optimization algorithm before each invocation.

With regard to the transport of the video packets, the RTP/UDP/IP protocols are used to transport the packets in a connectionless fashion where no acknowledgment is required due to the stringent delay constraints and to reduce the control overhead. The K distortion parameters are packetized along with the DPs and are available for the optimizer's processing and analysis. The overall functionality is outlined in Figure 14.7.

The maximum threshold proposed in Pinson et al. (2007) is 2% for E2E loss probability and 1 ms of maximum E2E delay. Since we used a combination

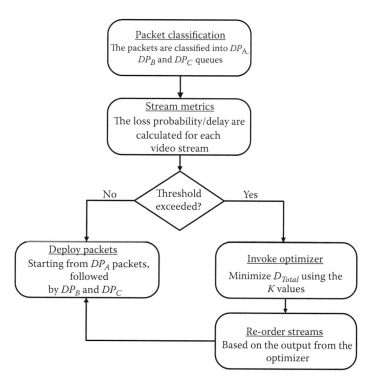

FIGURE 14.7
Operation sequence in the ingress router. (Modified and adapted from Jahaniaval, A., Video quality enhancement through end-to-end distortion optimization and enriched video traces, M.Sc. thesis, University of Guelph, Guelph, Ontario, Canada, 2010.)

of FMO and DP error-resilience features, we utilized a higher value of 2.5% for the loss probability threshold. This value has been obtained as we experimentally observed that dropping packets over 2.5% loss probability reveals noticeable visual artifacts. Similarly, an observable deterioration happens with a 1.18 ms delay with the encoding setup we have used.

We experimented with the optimizer invocation under different conditions. Four different invocation patterns are summarized in Table 14.5.

In the baseline case, the packets arrival will follow a FIFO fashion in the queues of the ingress router. The processing delay added by the optimizer (due to its quadratic complexity) has no adverse effect on the E2E delay of the four video streams. But we suspect that it will introduce an unforgivable delay for a large number of streams. Our ongoing work is focusing on this area. In addition to the rearrangement of the packets, the receivers record the arrival of all the packets in a raw trace format similar to the original trace file, but with the addition of the arrival time-stamp information. This trace file is further processed to calculate the PSNR values of the received video slices for each stream.

TABLE 14.5

Optimizer Invocation Patterns

Baseline	The Optimizer is not Invoked
Optimizer #1	The optimizer is invoked when the delay and/or the loss probability thresholds are exceeded (1.18 ms and 2.5%, respectively) for any of the stream. After the rearrangement of the packets in the queues, the DP_A packets are deployed until no packets remain in the queue, followed by deployments of all the packets in the DP_B queue, and finally the deployment of all the packets in the DP_C queue
Optimizer #2	Similar to Optimizer #1, except that the reordering outcome is applied only to DP_A queue whose packets are deployed first until no packets remain in it, followed by the packets in the DP_B queue, and finally the packets in the DP_C queue
Optimizer #3	Similar to Optimizer #2, the optimization occurs when the thresholds are exceeded, however, with an allowable activation frequency of every five GOP.
Optimizer #4	Similar to Optimizer #3, but the activation frequency is 10 GOP instead of 5 GOP.

Source: Data adapted from Jahaniaval, A., Video quality enhancement through end-to-end distortion optimization and enriched video traces, M.Sc. thesis, University of Guelph, Guelph, Ontario, Canada, 2010.

14.5.3 Results

In this section, we present the results obtained when we used the enriched traces based on our E2E distortion model presented in Section 14.3. During the real-simulation time, the E2E loss probability and delay were monitored for each video stream. We then invoked the optimizer based on our model in Section 14.4.

In Figure 14.8, we present the results of our comparative simulation runs for Stream 1. When no optimization is performed (Baseline in Figure 14.8a), we can observe the undesirable high fluctuations in PSNR values that are reset to a high value at the beginning of each GOP (with the I-frame), degrading as the transmission of the GOP progresses. The other graphs show the respective results for Optimizers #1 through #4. With each I-frame transmission, the network metrics, namely, the current E2E loss probability and delay, are input to the optimizer, which then produces the best reordering of the live streams that minimizes the total distortion (Equation 14.15). As expected, the optimizer introduced a significant amount of control overhead in the network, giving preference to Optimizer #2 whose frequency of invocation is based on threshold violation; this can be observed in Figure 14.8b. Optimizers #3 and #4 still perform better than the baseline (used as a reference run) but at the expense of lesser improvement in the PSNR values than those achieved by #2. Even though, Optimizer #1 has the same frequency of invocation as #2, its queue management causes the DP_A packets arriving during optimization to be delayed until all pre-optimization packets have been deployed from the

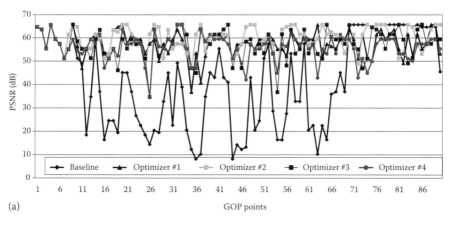

(a)

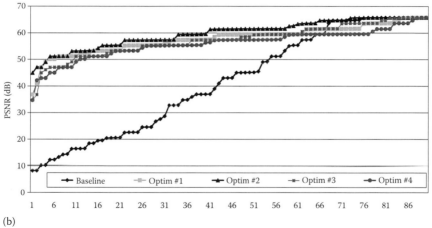

(b)

FIGURE 14.8
(See color insert.) Received PSNR for Stream 1. (a) Received PSNR for Stream 1 (Akiyo). (b) Reordered with ascending PSNR value. (From Jahaniaval, A., Video quality enhancement through end-to-end distortion optimization and enriched video traces, M.Sc. thesis, University of Guelph, Guelph, Ontario, Canada, pp. 511–516, 2010.)

three queues in the ingress router. This handling of the queue has a minor negative impact on the PSNR values as shown in Figure 14.8b.

In comparison with the baseline simulation, in the latter cases (with optimization), we can observe a smoothing effect on the PSNR values that eliminates much of its fluctuations with an overall higher value throughout the total simulation time. To provide a better visualization, we sorted the results in ascending order of the PSNR values, as opposed to following the simulation time line.

The discussion just presented for Stream 1 is equally applicable to the other three video streams. Figure 14.9 shows the sorted PSNR results for video streams 2, 3, and 4. Overall, the variation in the PSNR values is greatly

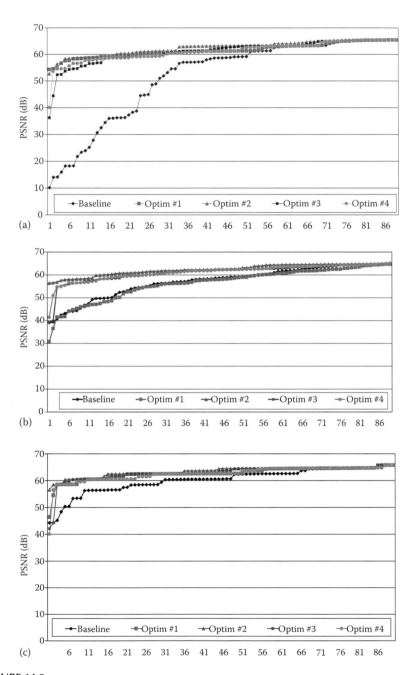

FIGURE 14.9
(See color insert.) Received PSNR values in ascending order for (a) Stream 2 (Carphone), (b) Stream 3 (Foreman), (c) Stream 4 (Silent). (From Jahaniaval, A., Video quality enhancement through end-to-end distortion optimization and enriched video traces, M.Sc. thesis, University of Guelph, Guelph, Ontario, Canada, 2010.)

TABLE 14.6

PSNR Comparison: Optimizers versus Baseline

PSNR (dB)	Baseline	#1	#2	#3	#4
Average	51.61	61.52	62.43	61.03	61.11
Max.	65.5	65.49	65.5	65.49	65.49
Min.	10.16	54.35	52.75	36.26	40.08
Median	58.66	61.25	63.21	62.46	61.25

Source: Data adapted from Jahaniaval, A., Video quality enhancement through end-to-end distortion optimization and enriched video traces, M.Sc. thesis, University of Guelph, Guelph, Ontario, Canada, 2010.

TABLE 14.7

GOP Comparison Count for Optimizers versus Baseline

	#1	#2	#3	#4
Higher	49	64	52	54
Lower	31	16	28	26
Same	9	9	9	9

Source: Data adapted from Jahaniaval, A., Video quality enhancement through end-to-end distortion optimization and enriched video traces, M.Sc. thesis, University of Guelph, Guelph, Ontario, Canada, 2010.

reduced and maintained in an acceptable range when the optimization is activated in the ingress router. Table 14.6 summarizes the ranges of PSNR values averaged on the video streams, indicating that the main conclusion is that the optimization improves the PSNR values at the receivers.

In Table 14.7, we report the number of GOP which averaged higher, lower, and the same PSNR values as in the baseline run for all four video streams.

Expanding on the averages reported in Table 14.6, Figure 14.10 gives a detailed view of the Gaussian (Normal) distribution and cumulative distribution of the PSNR values for each of the optimizers and for the baseline. From Figure 14.10, we can observe that

- The baseline (no optimization) has the worst performance among all simulations.
- In all optimization cases, we obtain significantly higher video quality than in the no optimization case.
- The best ranking in performance goes to Optimizer #2, followed by #1, then #4, and finally #3. This is observed by narrower (lower standard deviation) and higher average PSNR values in the Gaussian distributions of these curves.
- The cumulative distribution curves also support this analysis.

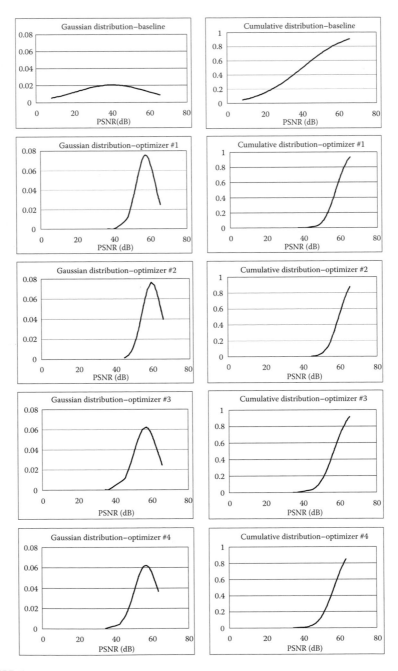

FIGURE 14.10

Probability distribution of PSNR values. (Modified and adapted from Jahaniaval, A., Video quality enhancement through end-to-end distortion optimization and enriched video traces, M.Sc. thesis, University of Guelph, Guelph, Ontario, Canada, 2010.)

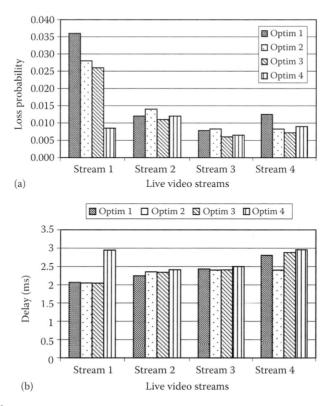

FIGURE 14.11
Optimization effect on (a) loss probability and (b) delay. (From Jahaniaval, A., Video quality enhancement through end-to-end distortion optimization and enriched video traces, M.Sc. thesis, University of Guelph, Guelph, Ontario, Canada, pp. 511–516, 2010.)

In general, we found that we obtain better results when the optimization is invoked on a threshold violation basis than when it is invoked on a constant GOP interval basis.

Moreover, up to this point, our approach indicates that there is naturally a trade-off between the frequency of the optimizer invocation and the overhead placed on the network's control plane.

Figure 14.11 displays the effect of each of the optimizers on the video streams in terms of (a) loss probability and (b) delay.

14.6 Conclusion

As the demand for multimedia services has increased over the recent years, the challenge still remains in delivering high quality multimedia over the

Internet in a competitive manner to the conventional technologies such as cable and satellite TV. The work presented in this chapter delivers some bridging research aspects between the video compression domain and the video delivery over lossy networks.

In this work, we presented a new mathematical model for the E2E distortion of video streams that takes into account QoS specific parameters, namely, the E2E loss probability (Jahaniaval and Fayek, 2007). Two error-resilience features of H.264 (DP and FMO) have been used to accurately model the slice expected distortion of the base layer as a function of E2E packet loss probability. Then we proposed an augmentation for the standard verbose file by creating representative video quality parameters that are associated with the loss probability to enable a more accurate assessment of distortion at run time.

Based on this model, we devised a linear optimization model (Jahaniaval, 2010; Jahaniaval et al., 2010) that makes use of the enhanced trace file format and incorporates the measurements of the loss probability at real simulation time to quantify the distortion of each video stream and take corrective actions to overcome the degradation in video quality. The objective of the optimization algorithm was to minimize the overall distortion of the live video streams at the moment when the optimizer is invoked. The algorithm is implemented in the ingress routers.

In order to validate our modeling, we ran several simulation tests with and without optimization. Several modes of optimization invocation were used to reach a trade-off where the optimization overhead does not negatively impact video quality but rather improves it for the live streams. Our experiments indicated that running the optimization on a constant interval basis is outperformed by the threshold violation mode of invocation in terms of overall PSNR improvement and network overhead. The optimization ordering outcome was applied to the DP packets.

Our findings so far have indicated that the optimization significantly improves the video quality at the receivers with an acceptable small cost of increased loss probability and delay due to the overhead introduced in the network's control plane. The statistical analysis conducted on the data samples from the simulation runs also demonstrates improvement in the received quality in comparison to the baseline run.

So far we successfully concluded that it is beneficial to use our optimization model to improve the video quality at the receivers when only the PSNR for each slice is used. However, we would also like to explore other objective video quality metrics in addition to PSNR, such as the structure similarity index metric (Wang et al., 2004), the hybrid image quality metric (Engelke et al., 2006), and others. Moreover, we will investigate our approach when applied to both the base and enhancement layers in FGS coding.

References

Agrafiotis D., D. R. Bull, and C. Nishan. Enhanced error concealment with mode selection. *IEEE Transactions on Circuits and Systems for Video Technology*, 16(8):960–973, 2006.

Belda A., J. C. Guerri, and A. Pajares. Adaptive error resilience tools for improving the quality of MPEG-4 video streams over wireless channels. In *IEEE 32nd EUROMICRO Conference on Software Engineering and Advanced Applications, SEAA'06*, Cavtat, Dubrovnik, pp. 424–429, 2006.

Blake S., D. Black, M. Carlson, E. Davies, Z. Wang, and W. Weiss. An architecture for differentiated services. Technical report, RFC 2475, 1998.

Chen Y., O. Au, C. Ho, and J. Zhou. Spatio-temporal boundary matching algorithm for temporal error concealment. In *IEEE International Symposium on Circuits and Systems, ISCAS'06*, Island of Kos, Greece, vol. 29, p. 4, 2006.

Chen Y., K. Xie, F. Zhang, P. Pandit, and J. Boyce. Frame loss error concealment for SVC. *Journal of Zhejiang University - Science A*, 7(5):677–683, 2006.

CPLEX, licensed to D. Fayek, SoE, University of Guelph, IBM ILOG CPLEX optimizer. http://www-01.ibm.com/software/integration/optimization/cplex-optimizer/, accessed from January 2008–August 2010.

Dai M., D. Loguinov, and H. M. Radha. Rate-distortion analysis and quality control in scalable Internet streaming. *IEEE Transactions on Multimedia*, 8(6):1135–1146, 2006.

Dhondt Y., P. Lambert, and R. Van de Walle. A flexible macroblock scheme for unequal error protection. In *IEEE International Conference on Image Processing*, Atlanta, GA, vol. 8, pp. 829–832, 2006.

Ekmekci S., T. Sikora, and P. Frossard. Unbalanced quantized multi-state video coding. *EURASIP Journal on Applied Signal Processing*, 2006:1–10, 2006.

Engelke U., T.-M. Kusuma, and H.-J. Zepernick. Perceptual quality assessment of wireless video applications. In *International ITG-Conference on Source and Channel Coding*, Munich, Germany, pp. 1–6, 2006.

Gebali F. *Analysis of Computer and Communication Networks*. Springer, Heidelberg, Germany, 2008.

Jahaniaval A. Video quality enhancement through end-to-end distortion optimization and enriched video traces, M.Sc. thesis, University of Guelph, Guelph, Ontario, Canada, 2010.

Jahaniaval A. and D. Fayek. Combined data partitioning and FMO in distortion modeling for video trace generation with lossy network parameters. In *IEEE International Symposium on Signal Processing and Information Technology, ISSPIT'07*, Cairo, Egypt, pp. 972–976, 2007.

Jahaniaval A., D. Fayek, and R. J. Brown. Distortion optimization in enriched video traces for end to-end video quality enhancement. In *IEEE/IFIP 6th International Conference on Networks and Services Management, CNSM'10*, Niagara Falls, Canada, pp. 511–516, 2010.

Ke C.-H., C.-K. Shieh, W.-S. Hwang, and A. Ziviani. Improving video transmission on the internet. *IEEE Potential*, 26(1):16–19, 2007.

Kim I.-M. and H.-M. Kim. A new resource allocation scheme based on a PSNR criterion for wireless video transmission to stationary receivers over Gaussian channels. *IEEE Transactions on Wireless Communications*, 1(3):393–401, 2002.

Kim I.-M. and H.-M. Kim. An optimum power management scheme for wireless video service in CDMA systems. *IEEE Transactions on Wireless Communication*, 2(1):81–91, 2003.

Ksentini A., M. Naimi, and A. Gueroui. Toward an improvement of H.264 video transmission over IEEE 802.11e through a cross-layer architecture. *IEEE Communications Magazine*, 44(1):107–114, 2006.

Kumar S., L. Xu, M. K. Mandal, and S. Panchanathan. Error resiliency schemes in H.264/AVC standard. *Journal of Visual Communication and Image Representation*, 17(2):425–450, 2006.

Kuo W.-H., W. Liao, and T. Liu. Adaptive resource allocation for layer-encoded IPTV multicasting in IEEE 802.16 WiMAX wireless networks. *IEEE Transactions on Multimedia*, 13(1):116–124, 2011

Liang G. and B. Liang. Balancing interruption frequency and buffering penalties in VBR video streaming. In *26th IEEE International Conference on Computer Communications INFOCOM*, Anchorage, AK, pp. 1406–1414, 2007.

Maani E., P. V. Pahalawatta, R. Berry, T. N. Pappas, and A. K. Katsaggelos. Resource allocation for downlink multiuser video transmission over wireless lossy networks. *IEEE Transactions on Image Processing*, 17(9):1663–1671, 2008.

Mansour H., P. Nasiopoulos, and V. Krishnamurthy. Rate and distortion modeling of CGS coded scalable video content. *IEEE Transactions on Multimedia*, 13(2):165–180, April 2011.

OMNeT++. OMNeT++ network simulation framework, http://www.omnetpp.org/, accessed August 2010.

Piamrat K., C. Viho, J.-M. Bonnin, and A. Ksentini. Quality of Experience measurements for video streaming over wireless networks. In *Third International Conference on Information Technology: New Generations*, pp. 1184–1189, 2009.

Pinson M. H., S. Wolf, and R. B. Stafford. Video performance requirements for tactical video applications. In *IEEE Conference on Technologies for Homeland Security*, Woburn, MA, pp. 85–90, 2007.

Richardson I. E. G. *H.264 and MPEG-4 Video Compression*. John Wiley & Sons, New York, 2003.

Ross S. *Probability and Statistics for Engineers and Scientists*, 4th edn. Elsevier Academic Press, Amsterdam, the Netherlands, pp. 567–570, 2009.

Ruolin R. A novel intra refreshment algorithm for ROI. In *International Conference on Multimedia and Information Technology*. IEEE Computer Society, Washington, DC, pp. 62–65, 2008.

Seeling P., F. Fitzek, and M. Reisslein. *Video Traces for Network Performance Evaluation*. Springer, Heidelberg, Germany, 2007.

Seeling P. and M. Reisslein. Evaluating multimedia networking mechanisms using video traces. *IEEE Potential*, 24(4):21–25, 2005a.

Seeling P. and M. Reisslein. Video coding with multiple descriptors and spatial scalability for devices diversity in wireless multi-hop networks. In *IEEE Consumer Communications and Networking Conference, CCNC'05*, Las Vegas, NV, pp. 278–283, 2005b.

Seeling P. and M. Reisslein. Video trace library. Arizona State University, http://trace. eas.asu.edu/yuv/index.html, accessed in 2010.

Seeling P., M. Reisslein, and B. Kulapala. Network performance evaluation using frame size and quality traces of single-layer and two-layer video: A tutorial. *IEEE Communications Surveys and Tutorials*, 6(3):58–78, 2004.

Shao S.-C. and J.-H. Chen. A novel error concealment approach based on general regression neural network. In *International Conference on Consumer Electronics, Communications and Networks, CEC-Net 2011*, Xianning, China, pp. 4679–4682, 2011.

Stockhammer T., M. M. Hannuksela, and T. Wiegand. H.264/AVC in wireless environments. *IEEE Transactions on Circuits and Systems for Video Technology*, 13(7):657–673, 2003.

Sühring K. H.264/AVC software JM 12.0, http://iphome.hhi.de/suehring/tml/, accessed in 2008.

Wang Y.-K., M. M. Hannuksela, S. Pateux, A. Eleftheriadis, and S. Wenger. System and transport interface of SVC. *IEEE Transactions on Circuits and Systems for Video Technology*, 17(9):1149–1163, 2007.

Wang Z., L. Lu, and A. C. Bovik. Video quality assessment based on structural distortion measurement. *Elsevier Signal Processing: Image Communication*, 19(2):121–132, 2004.

Wenger S. H.264/AVC over IP. *IEEE Transactions on Circuits and Systems for Video Technology*, 13(7):645–656, 2003.

Wiegand T., L. Noblet, and F. Rovati. Scalable video coding for IPTV services. *IEEE Transactions on Broadcasting*, 55(2):527–538, 2009.

Xiang X., Y. Zhang, D. Zhao, S. Ma, and W. Gao. A high efficient error concealment scheme based on auto-regressive model for video coding. In *27th Conference on Picture Coding Symposium*, Chicago, IL, pp. 305–308, 2009.

Xiong H., J. Sun, S. Yu, J. Zhou, and C. Chen. Rate control for real-time video network transmission on end-to-end rate-distortion and application-oriented QoS. *IEEE Transactions on Broadcasting*, 51(1):122–132, 2005.

Xu Y. and Y. Zhou. H.264 video communication based refined error concealment schemes. *IEEE Transactions on Consumer Electronics*, 50(4):1135–1141, 2004.

Zapater M.-N. and G. Bressan. A proposed approach for quality of experience assurance for IPTV. In *First International Conference on the Digital Society*, Gaudeloup, French Caribbean, pp. 25–30, 2007.

Zhang Y., W. Gao, and D. Zhao. Joint data partition and rate-distortion optimized mode selection for H.264 error-resilient coding. In *IEEE International Workshop on Multimedia Signal Processing, MMSP'06*, Victoria, British Columbia, Canada, pp. 248–251, 2006.

15

Video QoS Analysis over Wi-Fi Networks

Rashid Mehmood and Raad Alturki

CONTENTS

15.1 Introduction

According to the recently published report (June 1, 2011) "Cisco Visual Networking Index: Forecast and Methodology, 2010–2015," global Internet video amounts to over 40% of the consumer Internet traffic, predicted to reach 50% mark by the end of 2012. The use of advanced video communications will increase in the enterprise sector causing business IP traffic to grow by 2.7 times between 2010 and 2015 (the forecast period). Business video conferencing will increase six times over the forecast period. The collective video traffic, such as TV, video on demand, and P2P, will reach approximately 90% of global consumer traffic by 2015. On the wireless and mobile side, Internet traffic originating with non-PC devices will grow five times over the forecast period; and by 2015, traffic from wireless devices will exceed traffic from wired devices. Globally, mobile data traffic will increase 26 times over the forecast period. Furthermore, machine-to-machine traffic will grow at 258%.

These trends substantiate the view that a range of video-based multimedia services* will be delivered over multiple heterogeneous network platforms, comprising fixed, wireless, and ad hoc networks,† all employing IP-based technologies. In the recent past, it was not feasible to deploy many of these emerging multimedia services due to the lack of business models and the huge investments that were required. The rising demands for mobility and anytime, anywhere communications (e.g., military battlefield operations, emergency, and disaster applications) and the increasing amount of mobile

---

* See, for example the various multimedia services offered by TrafficLand, Inc (http://www. trafficland.com/). They provide services based on aggregation and delivery of live traffic video over the Internet and on TV.
† An ad hoc network is usually defined as a collection of nodes dynamically forming a network without the use of any existing network infrastructure or centralized administration.

traffic and devices have transpired and accelerated the developments of ad hoc networks allowing shorter network deployment times at lower costs. Therefore, future communication systems will increasingly integrate and rely on ad hoc networks in order to support anytime, anywhere seamless mobility.

While multimedia forms high data rate traffic with stringent quality of service (QoS) requirements, wireless ad hoc networks characterize frequent topology changes, unreliable wireless channel, network congestion, and resource contention. We are motivated by the many challenges and opportunities in supporting multimedia over heterogeneous networks with scalable QoS and have developed a range of design and analysis techniques for wireless infrastructure and ad hoc networks in order to support multimedia applications with scalable QoS.

15.1.1 Objectives

The aim of this chapter is to present an overview of our work on multimedia wireless networks. Firstly, an extensive review of the literature on multimedia ad hoc network design and QoS analysis is presented. Based on the review, we structure the literature on multimedia ad hoc networks design and QoS into 10 dimensions, including routing algorithms, QoS, medium access control (MAC) design, scalability, security, green networking, and so on. The literature on each dimension is discussed in some detail. Secondly, a detailed analysis of multimedia applications over wireless networks, both infrastructure and ad hoc, is presented. For ad hoc networks multimedia QoS analysis, four routing schemes are used: ad hoc on demand distance vector (AODV); optimized link state routing (OLSR); hierarchical clustering, provisioning, and routing (HCPR); and geographic routing protocol (GRP). Several networking scenarios have been carefully configured with variations in network sizes, applications, codecs, and routing protocols to extensively analyze the ad hoc networks performance. All the network analyses presented in this chapter are based on simulations using the OPNET simulator. The HCPR scheme is implemented as an extension (module) to the OPNET simulation software.

We note that the vast majority of networks are populated with multimedia traffic, as opposed to video alone, and therefore we have included voice and data (HTTP traffic) in our network analyses. Nevertheless, we will present network analysis with focus on video application alone. Our approach is to simulate and evaluate realistic multimedia application scenarios for networks with a sufficiently large number and type of applications, both elastic and nonelastic.

The rest of the chapter is organized as follows. Section 15.2 briefly provides the necessary background material related to this chapter. Section 15.3 provides the literature review on multimedia ad hoc networks design and QoS structured into 10 dimensions. Section 15.4 presents the multimedia QoS analysis over wireless networks. Finally, Section 15.5 concludes this chapter with some directions for future work.

15.2 Wireless Networking Technology, QoS, and Routing

This section provides the background material related to this book chapter. We briefly introduce IEEE 802.11 local area wireless network (WLAN), the wireless networking technology that we have used in our network analysis work. Subsequently, we introduce QoS, its measuring metrics, and routing schemes used in this chapter.

The IEEE 802.11 WLAN, also known as Wi-Fi, is widely used these days because of its flexibility and the low cost of equipment and deployment. The IEEE 802.11 is a family of standards that specifies the MAC and Physical layers of a WLAN (O'Hara and Petrick, 2005). The main purpose of 802.11 standard, according to the published standard (IEEE 802.11 Working Group, 2007), is to provide wireless connectivity for mobile and stationary devices within a certain range. WLAN can be configured in two modes: infrastructure mode or ad hoc mode. In this chapter, the WLAN performance in both modes is analyzed in detail.

15.2.1 QoS Metrics

QoS is the performance level of a service offered by the network to the application (Murthy and Manoj, 2004). Applications may vary in their requirements from the network; some of them need to have fast packet delivery but could be flexible in security or confidentiality of the packets. The QoS is the service that prioritizes users' requirements and then aims to guarantee satisfying their requirements without effecting other users' requirements. Some common QoS parameters for applications include delay, jitter (variation in delay), throughput, and (tolerance in) packet losses. Networks create delay, jitter, and packet losses due to, for example, limited bandwidth availability, buffering and switching delays, lack of buffer space, and transmission errors. An application can use its QoS parameters to negotiate QoS with the underlying network. The network designer or operator can consider these QoS parameters for network design purposes to negotiate service level agreements or to develop and implement policies and procedures required to guarantee service level agreements. This has led to many developments and standardization activities to provide end-to-end QoS over the Internet, including the IPv4 to IPv6 evolution. Best effort networks are not enough for some applications where, for example, delay cannot be tolerated, and therefore there have been many methods and techniques used to overcome such problems. Each QoS-enabling method has its own pros and cons and there is no technique that can satisfy all QoS requirements. Well-known QoS techniques include buffering, shaping, resource reservation, admission control, multipath routing, packet scheduling, and provisioning. QoS architectures developed to support end-to-end QoS include integrated services (Braden et al., 1994), differentiated services (Blake et al., 1998), and multi-protocol label switching (Rosen et al., 2001).

A number of QoS metrics exist to measure network performance. These are described in the following. We mainly focus on those that we have used in the analysis of results presented in this chapter. Traffic sent is the traffic that has been successfully sent from the application layer to the next lower layer (transport layer). Traffic intended to send is the traffic that has been planned to send from application layer to lower layer. However, not all traffic is intended to send is sent because some layers (e.g., physical) will not be able to handle the traffic for any reason. Throughput is the percentage of the average traffic received per second to the traffic that was intended to send per second in the network. In results presented in this section; throughput is used for every individual application: video, voice, and HTTP, besides having throughout of total traffic in network. Delivery ratio (DR) is the percentage of the average traffic received per second to the traffic sent per second in the network. In this section, there will be DR for every individual application: video, voice, and HTTP, besides having the total DR of all traffic in network. End-to-end delay is the average time spent in seconds for the packet to reach destination. The video and voice end-to-end delay have slightly different definitions by OPNET. The video end-to-end delay is defined as the average time spent in seconds to transfer video packets from source's application layer to destination's application layer in the network. It is measured from the time of creating the packet until the time of receiving it. The voice end-to-end delay is the average time in second that voice packet takes to reach destination from time of encoding the analogue signals at source to the time the packet is decoded at destination. This is called mouth-to-mouth delay, which include encoding delay, compression delay, network delay, decompression delay, and decoding delay (OPNET Technologies Inc, 2008). Delay variation is the average variance in packet end-to-end delay. In OPNET, user can collect delay variation for video and voice packets besides having jitter for voice. Voice jitter is the variation of delay in second in received voice packets (Cisco Systems, 2006). If two packets left the source voice application at t_1 and t_2 times consequently and arrived at destination's voice application at t_3 and t_4 consequently, the jitter is the result of this following equation: $(t\_4 - t\_3) - (t\_2 - t\_1)$ (OPNET Technologies Inc, 2008). Voice mean opinion score (MOS) "is a subjective measurement representing the quality of digital multimedia including video, voice, or audio" (Mehmood et al., 2011) and it represents the quality in a numerical format ranging from 1 as reference to bad quality to 5 as reference to excellent quality. In the results presented in this chapter, the average of voice MOS in the network per second was taken and used. HTTP page response time is the average time in second required to retrieve the entire page with all objects (e.g., image) in the page.

15.2.2 Routing Protocols

In data networking, the process of identifying and selecting network routes to direct the network traffic is known as routing. A large number (over 30, see

Liu and Kaiser [2003], for instance) of routing protocols exist which mainly differ in the way they select the network path; however, most of them fall into two major protocol categories: Distance vector protocol and link state protocol. Moreover, routing protocols can also be classified on the basis of how ready they are to send the packets into reactive and proactive protocols. There are also some other classification methods in the literature, such as whether the routing protocol is flat or hierarchical, source routing or hop-by-hop, location based or not, uniform or nonuniform (Liu and Kaiser, 2003), and single or multiple channels.

In this chapter, we have used four routing schemes to analyze and compare video performance over wireless networks. These are the AODV (Perkins et al., 2003); the OLSR (Clausen and Jacquet, 2003); the GRP (Takagi and Kleinrock, 1984); and the HCPR scheme (Alturki and Mehmood, 2012; Mehmood and Alturki, 2011). The AODV protocol was first published as an Internet draft by Charles Perkins in 1997, then in 2003 published as an RFC (Perkins et al., 2003). It is a reactive routing protocol that is based on a distance vector routing approach. Nodes are kept quiet until a connection is required. The node that requires the connection will broadcast a route request message asking for a route to destination. A reply to this message is expected from any node that has the route to destination or from the destination itself. Subsequently, it will choose the route to destination with the least number of hops. The OLSR protocol (Clausen and Jacquet, 2003) is a link state proactive routing protocol that maintains its routing and forwarding table with disregard to packet arrivals. It sends a periodic hello message to exchange information about network. In order to avoid extensive routing maintenance messages from consuming the limited wireless bandwidth, it has some chosen nodes called Multi point relays (MPRs) to reduce packets. Those MPR nodes are selected nodes that do two main jobs: (1) generate topology control messages and (2) act as data packet forwarders for other nodes. The advantage of using MPRs is clearly observed when a network is big and dense in comparison to pure link state routing or other ad hoc networks routing protocols. Also, when a network is small and nodes are sparse, OLSR becomes a pure link state routing protocol. Packets are routed in a hop-by-hop basis— that is, nodes in the route will make the forwarding decision according the local routing table. GRP is a class of protocols that use geographical position for routing packets, firstly proposed in Takagi and Kleinrock (1984). We have used the implementation of the GRP protocols as available in OPNET.

The HCPR was proposed in Alturki (2011), Alturki and Mehmood (2012), and Mehmood and Alturki (2011). HCPR is a cross-layer scheme that works on application, transport, and network (routing) layers. The HCPR was designed with particular emphasis on scalability and in order to address QoS challenges to deliver multimedia over ad hoc networks. The HCPR scheme is based on intelligent protocols that optimize multimedia QoS provisioning by enabling and exploiting interactions between application, transport, and network layers. Most of the cross-layer schemes for multimedia ad hoc networks have focused on interfacing or merging, the physical and MAC layers; MAC, physical, and network layers; MAC, application, and physical layers, etc. We believe that no other work has identified the potential of, and have considered, the delicate balance of layered interaction between application, transport, and network layers, as is the case in HCPR. The interfacing of application, transport, and network layers in HCPR allows HCPR to scale well, and provide QoS, to large networks under heavy multimedia traffic. The HCPR scheme comprises two major phases: the network formation phase and the network operational phase. The network formation phase prepares the network in order for the network to move to the operational phase. The network formation phase enables the nodes to find each other, their positions, to form node clusters and an overlay network connecting the cluster heads (i.e., the head or leader of a cluster). The network formation phase is divided into the network discovery phase and the cluster overlay formation phase. The network discovery phase allows the nodes to know the number of nodes, their respective positions, and the geographic boundaries of the network. The cluster overlay formation phase is used to build multiple network clusters, elect cluster heads, and build a QoS routing overlay network. Subsequent to the formation of the HCPR-enabled network, the network moves to its operational phase: a node in the network can now make requests to connect to a destination node with some application. A node wishing to receive a service, however, does not talk to the destination node directly; rather it requests the cluster head of its cluster to make a request to the destination on its behalf. This request is initiated from the application layer, which will call the corresponding function of HCPR (i.e., at the transport layer) that makes the reservation request to the cluster head. Essentially, in an HCPR-enabled network although the requests are made locally, these are propagated and provisioned by a hierarchical structure made up of cluster head nodes. Consequently, the network provides high reliability and scalability. In a mobile ad hoc networking environment, the network will continue to toggle between the network formation and the network operational phases because it will have to reconfigure itself according to the changing network conditions. Further details of the HCPR scheme, its protocols, and its implementation as a separate module of the OPNET simulator can be found in Alturki (2011) and Alturki and Mehmood (2012).

15.3 Multimedia Ad Hoc Network Design and QoS—10 Dimensions

We now provide an extensive review of the literature on the main topic of this chapter, that is, multimedia and ad hoc network design and QoS. A huge amount of literature is available on the design and performance analysis of multimedia ad hoc networks. We structure the literature on ad hoc networks design into 10 dimensions. The 10 dimensions are routing algorithms, QoS, MAC design, scalability, cross-layer design, security, green networking, topology control, wireless sensor networks (WSNs), and methodologies for multimedia performance analysis over ad hoc networks. We shall define and discuss each dimension in a separate section and, wherever available, we shall include a review of the literature on each dimension in relation to video or multimedia QoS support and analysis.

15.3.1 Cross-Layer Design in Multimedia Ad Hoc Networking

The traditional layered protocol network architecture is based on a series or stack of layers, each layer provides services to the higher layers, allowing an abstraction, and hiding implementation complexities, for the higher layer. This layered approach reduces complexity through modularization of the network and allows manageable design, interoperability, extensibility, scalability, and standardization. There are, however, well-known disadvantages of the layered design. The so-called cross-layer design techniques have emerged in the recent years as a potential solution to address the multimedia QoS provision challenges plaguing the development of heterogeneous networks. The cross-layer ideology is to optimize the design and performance by increased and effective interaction between layers and optimization at a global level (multiple layers) rather than at the local level (single layer).

Ramanathan and Redi (2002) have described the cross-layer approach as a promising method that could solve problems related to QoS. By allowing layers to use other layers' information and accessing them freely, layers could adapt to the changes using other layers' information. Therefore, the QoS requirements can be satisfied accordingly. Lee and Song (2010) have proposed a cross-layer algorithm for video streaming over ad hoc network. The algorithm chooses the efficient PHY mode and the retransmission limit of WLAN for each node in a distributed way. The algorithm makes the choice depending on the information available at other layers, application, MAC, and PHY. Pompili and Akyildiz (2010) have studied the multimedia underwater delivery in the acoustic sensor networks. They mentioned that the previous underwater communication works have followed a traditional stack layered approach, which was firstly designed for wired networks. Authors claim that the cross-layer approach will improve the multimedia and delay sensitive applications in underwater communication environments. They studied the interactions of underwater communication functionalities like modulation, forward error correction, MAC, and routing. In addition, they proposed a new distributed cross-layer solution that allows devices to share the underwater acoustic medium fairly and efficiently which is characterized by having high delay and bandwidth limitations. Gharavi and Ban (2003) have proposed a cross-layer feedback control scheme for video transmission over ad hoc networks. The proposed scheme was designed to overcome the current problems of variations of delay and bandwidth. The current schemes like real-time transport protocol have limitations in mobile environments which causes the network to be congested. The proposed scheme is zero-delay and traffic free channel assessment scheme that allow the sending node to adapt to the current network conditions. The scheme takes the critical information (such as number of hops) from the underlying routing layer. Those extracted information are also used in other proposed schemes like packet recovery and redundant packet transmission.

Melodia and Akyildiz (2010) have proposed new cross-layer communication architecture for multimedia delivery over WSN based on time hopping impulse radio ultra wide band. The architecture aims to deliver heterogeneous application traffic over WSN reliably and with flexibility by controlling the interactions between the layers. The resulted simulations show that the architecture achieved the aimed target without affecting the modularity of the stacked design. Setton et al. (2005) have investigated the potential that cross-layer approach can offer to support real video by allowing layers to exchange information and showed how end-to-end performance was optimized by adapting to such information. The results show that the video-stream performance has improved substantially. Mundarath et al. (2009) have proposed a new cross-layer approach for multiple antenna ad hoc networks that consider QoS and reduce power consumption called QoS-aware smart antenna protocol. The proposed approach adapts the degree of freedom present in the multiantenna in each node to achieve less energy consumption while keeping application's QoS requirements assured. The results show that using cross-layer approach in the design saves considerable energy consumptions in comparison to other QoS schemes that use strict layering approach.

15.3.2 Routing Algorithms

Routing in ad hoc networks has received significant attention from the research community, most probably because ad hoc networks' main concept is based on routing, where nodes are responsible for routing the packets for other nodes. Despite extensive work in the field of routing in ad hoc networks, routing is still a fundamental problem in ad hoc networks' design. The first problem is the scalability of routing. Most current routing protocols have only been tested for relatively small size networks, yet none of the current routing protocols have showed reasonable performance with large networks and heavy network traffic. Another problem in designing routing algorithms for ad hoc networks is the large number of design parameters that are supposed to be optimized such as mobility of nodes, range of applications and their intensity, low power consumption, etc. We now review some of the most notable literature on multimedia ad hoc networking where the focus of the researchers has been to either design/improve routing protocols or study routing performance against multimedia traffic.

Considerable amount of research in ad hoc networks has been done to compare and analyze various routing protocols. Layuan et al. (2007) have carried out a multimedia performance analysis of four routing protocols using QoS metrics, routing load, and connectivity. QoS metrics include delay, jitter, throughput, and loss ratio. They have used different network sizes of 10, 20, 40, 50, finishing with 100 nodes, and placed randomly in 1×1 km area. They have used NS2 as a simulation tool and used user datagram protocol (UDP) constant bit rate (CBR) traffic with 512 byte as the packet size and pairs of 6, 12, 24, 30, and 60 UDP streams. Ng and Liew (2007) have studied and indentified the maximum throughput in 802.11 multi-hop networks. They mention that having the network overloaded leads to two problems: packet loss and rerouting instability. They used NS2 to show their results, which is based on having eight stationary nodes placed in string or chain topology. The transmission range was 250 m and UDP stream traffic was sent from the first node on the edge to the last node on the other edge with 1460 byte packet size. The main metric of evaluation was throughput in percentage. Their results show that 1.18 Mbps is the optimal sending rate. Kumar et al. (2009) have compared two reactive routing protocols, DSR and AODV, by varying the traffic load, mobility, and type of traffic using NS2 simulator. They used packet delivery, packet loss, and end-to-end delay as metrics to measure the performance. They had generated four traffic patterns with randomly varying the sources and destinations using two type of traffics CBR and TCP. They used 50 nodes placed in a 1.5×0.3 km area. The packet size was 512 byte. A random waypoint mobility model was used during those emulations.

Pucha et al. (2007) have studied the performance of ad hoc networks under different traffic patterns. The authors argued that previous performance analysis studies cannot give the same results with different traffic patterns. For this reason, the authors studied the impact of traffic patterns on the performance of 112 nodes. They used their new connection models to allow multiple connections per source. They studied three routing protocols: DSDV, DSR, and AODV using NS2 simulator, where 112 nodes were populated in 2250×450 m area with 250 m transmission range and 2 Mbps bandwidth. Around 80% or 90% of nodes were CBR traffic sources of 60 packets/s and the packet size chosen was 64 byte. They used packet DR and delay as valuation metrics. Qadri and Liotta (2010) surveyed the performance analysis studies of ad hoc network routing. The study gives a good background of ad hoc routing classification and categories and descriptions of some common routing protocols. This is followed by an overview of ad hoc network requirements and metrics used for performance analysis. We note here that all papers and work discussed in this chapter (Qadri and Liotta, 2010), which was published fairly recently (in 2009) were general in their scenario settings and analysis and there were no detailed specifications in the scenarios for video, voice, and data together such as the one presented by us. Chatzistavros and Stamatelos (2009) have examined the behavior of ad hoc network routing protocols through the simulation using CBR traffic with different packet sizes and different topologies. They have used QualNet to compare DBF, DSR, and ZRP routing protocols and to study the mobility effect on performance using 802.11 standard. They started by simulating a chain of nodes and one flow of CBR traffic between the two ends and then made parallel chains of nodes with a CBR flow with every chain.

15.3.3 Multicasting and Quality of Service

Satisfying QoS in wireless infrastructure networks still has some challenges to meet (Murthy and Manoj, 2004) due to, in particular, the unpredictable nature of the wireless medium. Because ad hoc networks use wireless technology, it inherits this open problem. What makes achieving QoS requirements in ad hoc networks harder is its lack of infrastructure and its multi-hop functionality. It is proposed in literature that to improve the QoS in ad hoc networks, the function of QoS should be taken from the upper layers to be placed in the MAC layer (Ramanathan and Redi, 2002). Also, some others researchers propose that QoS cannot be achieved in the traditional structured layering approach but, instead, by collaborations between the layers by allowing layers to interact with each other and use other layers' information.

de Morais Cordeiro et al. (2003) presented a comparison of well-known multicasting algorithms in ad hoc networks. Morgan and Kunz (2005) proposed a QoS gateway that acts as an interface between the two types of networks. Their motivation was that ad hoc networks sometimes need to connect to some public service on the Internet which probably requires guaranteeing the QoS requirements. Their proposed work is based on having an existing QoS solution in both network types, such as SWAN/ESWAN in Ad hoc networks and DiffServ in structured networks. The main job of the gateway is to map the QoS in the ad hoc network to the QoS in the infrastructure network. The proposed gateway has shown smoother bandwidth when the gateway is used between the two different networks in comparison to not having any QoS gateway. A description of the available approaches for group communication in MANET, like multicasting and broadcasting, is covered in Mohapatra et al. (2004). It has been mentioned that QoS in group communication is an open problem. The standard QoS protocols are required to guarantee some measurable performance, including delay, bandwidth, packet loss, and delay variance. MANETs add two more attributes—power consumption and service coverage. Those attributes and some characteristics of MANET make the QoS in MANET a complex process to be achieved. Not only is QoS desirable for MANET to guarantee general aspects of its communication but also group communication needs QoS support that can be customized to suit and satisfy the group communication protocols. One of the proposed QoS protocols is QoS-aware core migration for the multicasting algorithms. It uses a group of shared multicast tree in which the leaves achieve the required multicast quality.

15.3.4 Medium Access Control Design

MAC layer has two main functions: addressing and access control. WLAN 802.11 provides MAC functionality for wireless local area networks. The exiting MAC protocols, like 802.11, do not very well suit ad hoc networks because the existing protocols have been firstly developed for specific types of networks; and all of them are infrastructure networks. Designing a MAC layer for ad hoc networks requires the consideration of many issues particularly related to its lack of infrastructure. Some of these issues have been discussed separately in this literature review as separate problems, such as security, QoS, and power consumption. Other issues are related to MAC design only and these include collision control, channel optimization, and channel fairness. The integration of the MAC sub-layer with the other network layers in MANET has been discussed in the literature. This area of literature falls under cross-layer protocols and has been discussed in Section 15.3.1.

15.3.5 Scalability

Scalability in ad hoc networks is an extremely important and active research area. The network scalability can be defined in terms of various network parameters such as the number of nodes in the network, the number of multimedia flows, and their intensities, etc. Scalability in ad hoc networks design is difficult due to its infrastructure-less attributes (e.g., nodes acting as routers, multi-hop), and hence we can say, in general, that the network performance has an inverse relationship with the number of nodes in the network. Researchers in the ad hoc networks area are trying to answer the question: to what extent can ad hoc networks grow and how can that be extended while maintaining an acceptable level of performance?

15.3.6 Security in Ad Hoc Networks

Security has emerged as one of the most important topics in every aspect of computing and communications, from operating systems to networking and programming languages. The level of security required differs from one application to another. For example, the level of security required for a gaming PC is much smaller than the requirement for securing internet banking web server. Similarly, in ad hoc networks, applications differ in their security requirements, for instance, securing an ad hoc network system for military use is not comparable to securing an ad hoc network for cooperative gaming. The main security goals as described in Lidong and Haas (1999) are: availability, confidentiality, integrity, authentication, and non-repudiation. There are many security methods in practice to achieve security goals. These methods include, among others, authentication protocols, digital signatures, and encryptions. Although the traditional security technologies do play a positive role in achieving many of the security goals; however, they are not sufficient to achieve the desired highest levels of security in ad hoc networks. Two other methods used in ad hoc networks to provide security are route redundancy and distribution of trust over the nodes. Route redundancy promotes the availability of the network connection. If one route has fallen down, packets are able to go to the destination using another route. This feature makes the network connection more available. The distribution of trust over the network nodes promotes availability and the authentication. Because of the low physical security and availability of the connection, nodes in MANET are not trustworthy. Consequently, trust can be distributed over the network nodes and the aggregation of the nodes' trust will be trustworthy.

15.3.7 Energy and Computational Efficiency: Green Networking

Until recently, there has been a lack of literature on energy and computational efficiency in ad hoc networks design. However, due to the increasing emphasis on environmental issues and "green" approaches to everything, analysis and efficiency of energy usage is becoming a critical topic. Video and multimedia uses huge bandwidth and, to date, there is virtually no effective work available on minimizing energy usage in multimedia ad hoc networking. However, research in this area is expected to grow rapidly in the near future.

15.3.8 Topology Control

This challenge is a direct result of having mobile nodes in the network although topology can also be affected in a static network due to nodes becoming faulty or unavailable. The challenge is to design a network that maintains its performance as if there is no change in the network topology. This challenge could be achieved by predicting the movement of nodes and their neighbors. As a result, many researchers are investigating modeling of nodes' mobility while others are focusing on the methods and techniques used to control the mobility of the nodes to give an acceptable performance. Camp et al. (2002) have given comprehensive coverage of mobility models in ad hoc networks. Broch et al. (1998) in their work in comparing the performance of four well-known routing protocols have shown how these protocols are affected directly by the movement pattern of the nodes. When nodes are highly mobile, protocols showed the worst performance, and when nodes have less movement, the performance increases dramatically. Hong et al. (1999) have surveyed the mobility models in cellular and ad hoc networks. They have proposed a group mobility model called reference point group mobility. Regarding a cellular network, the authors have mentioned Random Walk and Random Gauss–Markov models which have been used in some people's work to simulate node movements. In ad hoc networks, Random Walk, Random Waypoint, Chiang's Markovian, Pursue and Column models have all been used in simulating the individual nodes.

15.3.9 Wireless Sensor Network

WSNs are a type of ad hoc networks, which inherit the ad hoc challenges in addition to some challenges of its own. WSN is an ad hoc network that has the ability to sense some environmental variables. WSN nodes are used to collect data about environmental variables and then inform a sink node or other nodes about their sensed data. WSNs are growing in their applications and scope, their most recent applications relevant to the topic of this chapter (i.e., multimedia over ad hoc networks) is the use of mobile devices as sensor networks in emergency situations such as in natural or manmade disasters, data search and analysis, etc. (Alazawi et al., 2011; Satyanarayanan, 2010).

15.3.10 Multimedia Ad Hoc Networks: Analysis Methodologies

This section presents the various works from the literature that present an analysis of multimedia performance over ad hoc networks by studying the existing networks and protocols or proposing new ones. We begin with the various approaches that attempt to improve multimedia performance over ad hoc networks by new proposals.

Literatures that have proposed new methods to improve multimedia over ad hoc networks and evaluate it include Hongqi et al. (2008), Li et al. (2000), Li and Cuthbert (2005), Utsu et al. (2010), and Xue and Ganz (2003). In Hongqi et al. (2008), the authors have proposed a new method to reduce packet loss, whereas the authors in Utsu et al. (2010) propose a new approach to make video transmission smoother. Li and Cuthbert (2005) claim that they have made improvements to the ad hoc performance for multimedia. Xue and Ganz (2003) propose a new end-to-end QoS protocol, while the authors of Li et al. (2000) propose a scalable location service. Each of these works will now be discussed in more detail. Hongqi et al. (2008) have studied the performance of voice over wireless ad hoc networks and proposed a new scheme to improve the performance by reducing packet loss. They have used MOS as the main measure for the performance, besides packet loss and jitter. Utsu et al. (2010) have studied the problem of video over ad hoc networks and suggest new methods to make video transmission smoother. In order to evaluate their results, they used NS2 as a simulation tool with some experimental work. The number of nodes simulated in NS2 scenarios was 20 nodes placed in 1×0.6 km and a random waypoint mobility model was used. The bandwidth used was 2 Mbps with a 250 m transmission range. They used DSR as routing protocol and a video traffic application of 128 kbps and frame size of 1024 byte, and bit sequence traffic to simulate that. Throughput was the main evaluating metric. Li and Cuthbert (2005) have proposed a new QoS multipath routing method that improves the performance of multimedia. They have shown some performance improvements by simulating 50 nodes in 1×1 km area and 250 m transmission range using OPNET 8.1 modeler. Traffic was populated with 512 byte CBR data packets. Five to 20 sources were chosen randomly to send 80 kbps expedited forwarding packets. Another 20 nodes sent 8 kbps background packets in best effort manner. The source-destination pairs were chosen randomly with variety in the number of connections for EF packets. They used packet DR and packet end-to-end delay as metrics for performance evaluation. The simulation was run for 800 s.

Li et al. (2000) propose a distributed location service for mobile ad hoc network routing that scales well when the network gets bigger. They use a NS2 simulator with up to 600 nodes and 2 Mbps bandwidth with a 200 m transmission range. The traffic is CBR with number of connections half the number of nodes. For each connection, four 128-KB data packets per second are sent to a destination for 20 s. Connections are generated at random times in the simulation time. Packet DR metric is used to view the traffic performance. Xue and Ganz (2003) have introduced a new resource reservation–based routing and signaling protocol called ad hoc QoS on-demand routing, which provides end-to-end QoS support and has been implemented in OPNET modeler. They have used 802.11 DCF-enabled nodes with 2 Mbps and 250 m transmission range, and they have used small and large network sizes to see how the protocol behaves. Small networks are used to discover the QoS recovery time which consists of five nodes—four statics and one mobile with one flow (400 kbps) from source to destination. The large network is used to study the performance of the proposed protocol and consists of 50 nodes placed in a 1×0.5 km area and moving randomly. To simulate multimedia streaming, they have used CBR traffic with 512 packet size and sent 10 packets a second. They have populated 10 and 15 traffic flows with random sources and destinations. In order to study performance, they used the following metrics: traffic admission ratio, end-to-end DR, average end-to-end delay, ratio of late packet, and normalized control overhead. They defined end-to-end DR as the "ratio between the number of data packets received at the destinations and the number of data packets sent from the sources. This metric indicates the reliability of the admitted flows." The average end-to-end delay defined as "the latency incurred by the packets between their generation time and their arrival time at the destination. This metric indicates the performance of the admitted flows."

We now turn to the second category of works that focus on the analysis of multimedia performance over wireless and ad hoc networks. Sondi et al. (2010) have done a performance evaluation of OLSR-QoS extension by using OPNET simulator. They populated 50 nodes in a 1×1 km area and nodes were equipped with an 802.11 g network card. They have simulated scenarios with statics and mobile nodes, with the aim of evaluating voice communication with some background. Therefore, they have made one voice connection between two nodes using G.711 codec (64 kbps) with the two nodes having FTP traffic as background traffic. For evaluation, they have used IP number of hops, jitter, packet end-to-end delay, MOS, and traffic sent and received as evaluation metrics. It has been observed that voice packet end-to-end delay were around 0.06 s in most scenarios. Gottron et al. (2009) show the feasibility of voice communication in larger scale ad hoc networks if the right settings are used. They have used Jist/SWANS simulation tools to conduct their studies and scenarios with 50,100,200, and 500 nodes. They have evaluated the results according to (1) Voice quality where MOS is used as the measure, (2) Transmission delay, and (3) Packet loss. They have varied the network load between 5 and 20 voice streams of G.711 voice codec. The transmission range used was 250 m and nodes were surrounded by an average of eight nodes. There is no clear description of how the source and destination were chosen.

Jeong et al. (2009) have studied the observation that only five simultaneous VoIP calls can be made in 802.11 using NS2 and show the reason behind that. Then, they proposed a new algorithm at the MAC sub-layer to improve performance. They have used jitter, loss rate, delay, and call capacity as metrics for measurement. They used a network consisting of wireless and wired nodes. Two laptops communicated a single VoIP application over the wireless network and the rest of the communications were within the wired network. They report that theoretically 85 calls can be made in 802.11b but, in practice, only five can be supported with QoS.

Hofmann et al. (2007) have carried out a performance analysis study of voice over a static wireless ad hoc network. They used throughput, delay, and jitter as metrics for the evaluation using simulation (NS2) and experimental work. Eight nodes have been placed to form a ring topology and a source is sent to the destination node which is the last node in the ring and should be accessible through the other six nodes. They used 1 Mbps and G.711 codec. Santos (2009) has studied the performance of VoIP over ad hoc networks using OLSR routing protocol. It varied node intensities, number of data streams, and mobility in the simulations to obtain 18 different scenarios using OPNET simulator. It focuses on studying the impact of the number of nodes, number of streams, and node mobility on the performance of ad hoc networks using end-to-end delay and packet loss. The author has simulated 10–50 nodes placed randomly in a 1×1 km area to send streams of VoIP with fixed packet lengths of 200 byte using the G.711 codec. The author has defined the end-to-end delay but not packet loss. However, the author did not show clearly how end-to-end delay was obtained and whether voice application or IP statistics were used. Their definition of end-to-end delay is defined as delay measured from the instant a packet leaves the sender's Network Interface Card (NIC) to the instant it is received at the destination's NIC. Still, it is not clear how that has been obtained from the OPNET simulator, and from our experience, there is nothing defined as such in voice application.

An extensive literature review that we have carried out have confirmed our view that the existing methodologies to evaluate multimedia networking studies lack sufficient details and fail to provide assessment of realistic multimedia networking environments. To address this, we have developed a novel analysis methodology for multimedia networks (the methodology is used in our research, see Alturki and Mehmood, 2008; Mehmood and

Alturki, 2011; Mehmood et al., 2011; and this chapter for the QoS analysis presented in Section 15.4). The approach is to simulate and evaluate realistic multimedia applications scenarios for networks with a sufficiently large number and type of applications, both elastic and nonelastic. The network QoS is explored for a set of different applications (as may typically be found in realistic situations), so that the mutual effects of network applications can be taken into account. Specifically, the networks have been populated with VoIP, video, and HTTP applications. Networks typically have to share multiple types of applications including voice, video, and data. We have seen in the literature review that most of the studies that exist in the literature have focused on studying capacity and performance of multimedia applications in isolation to other applications, that is, the networks are populated by the traffic that only belongs to one application type. These studies are therefore unable to capture the dynamics of realistic networking environments, because they do not take into account the mutual effects that various multimedia applications will have on each other, while sharing a single network resource. While there are some studies that have reported on multimedia performance while sharing the network resources with other applications, those studies are limited in their approach to setting up the applications and networking scenarios (e.g., the number and types of applications, and the analyses are limited).

15.4 Multimedia over Ad Hoc Networks: QoS Analysis

We now analyze video performance over wireless ad hoc networks using the four routing scheme as mentioned in Section 15.2; these are AODV, OLSR, GRP, and HCPR. We have simulated well over 500 different networking scenarios by varying the routing protocols, traffic intensities, mix of multimedia applications configurations, and numbers of nodes. Results show that different protocols act differently under different network conditions, and some protocols perform better at low traffic intensities while others give better performance for networks with higher numbers of nodes. In this section, we give a selection of scenarios and results from our simulation experiments, with one section devoted to video performance analysis. Our aim is to explore how the various protocols are affected by increasing the number of network nodes, traffic intensities, and the mix of multimedia traffic applications.

This section is organized as follows. Section 15.4.1 describes the performance analysis methodology including the simulation scenarios, traffic profiles and network topologies. Section 15.4.2 presents an analysis on the collective network behavior when the network is populated with multimedia applications. The results are presented separately for networks of different sizes, from a 5-node network up to a network containing 50 nodes. Section

15.4.3 gives a detailed depiction and discussion of video performance over ad hoc networks for different traffic profiles. In Section 15.4.4, we summarize and conclude with an overview of the QoS analysis for ad hoc networks presented in Sections 15.4.2 and 15.4.3. Subsequently, in Section 15.4.5, we analyze video performance over infrastructure networks. As expected, we will see that infrastructure networks are able to deliver much higher intensity of video.

15.4.1 Methodology for Analysis

We describe here our performance analysis methodology that includes: the four application profiles that we have created using a mix of multimedia (voice, video, and HTTP) applications (Section 15.4.1.1); the variation in the network size and geographical structure (Section 15.4.1.2); and the network topology of multimedia applications (Section 15.4.1.3).

15.4.1.1 Applications Traffic Profiles

In order to study the effect of how routing protocols are behaving with different traffic intensities, we have configured voice, video, and HTTP applications into four traffic intensities. These are listed in Table 15.1. Rows 1–4 give details for each traffic intensity beginning with Low (L) traffic at 100 kbps up to the High (H) intensity at 20 Mbps. Columns 1–4 list bit rate for the voice, HTTP, and video including the codec name for voice. Column 5 lists the total bit rate for each traffic intensity. The bit rates for each application also give the number of connections (this will be further explained in Section 15.4.1.3). For example, in the third row, which represents the details of Medium–High (MH) traffic, the first and second columns show the voice codec used and the total voice bit rate. GSM-FR codec is the codec used which has a bit rate of 12.3 kbps. The total voice bit rate is 52.3 which is a bit rate for two connections and each connection has two ways of voice.

Columns 3 and 4 show the total HTTP and video traffics which are 43.2 kbps and 9900 kbps=9.9 Mbps, respectively. The last column is the total

TABLE 15.1

Traffic Profiles and Their Respective Component Intensities and Attributes

Traffic Profile	Voice		HTTP (kbps)	Video (kbps)	Total (kbps)
	Codec	kbps			
Low (L)	Low quality (G.723.1 5.3 K)	21.312 (5.3 K × 2 × 2)	10.2	68.488	100
Medium–low (ML)	IP telephony (G.729 A)	32 (8 K × 2 × 2)	20.4	947.6	1,000
Medium–high (MH)	GSM quality (GSM-FR)	52.8 (13.2 K × 2 × 2)	43.2	9904	10,000
High (H)	PCM quality (G.711)	256 (64 K × 2 × 2)	99.2	19,644.8	20,000

traffic populated in the network, including voice, HTTP, and video, which is 10,000 kbps = 10 Mbps. The last row shows the H intensity chosen at 20,000 kbps = 20 Mbps because this is usually the maximum achievable bandwidth for 802.11 g at 54 Mbps. As can be seen from the table, traffic populated were a mix of HTTP, voice, and video in all the four traffic intensities. Having a mix of data and multimedia applications in the same network was chosen to show how applications are performing with each other to simulate more realistic networking scenarios.

15.4.1.2 Network Geography and Size

Varying the network size against fixed traffic intensities was intricate and required careful configuration, as it is difficult to compare the performance of two scenarios unless they are exactly the same, except one parameter, which is the one intended to study. In the case of this chapter, all parameters should be fixed and network sizes will be varied. The problem here is that increasing the number of nodes requires other parameters to change, which is the scenario of application sessions, that is which node is connected to which. For example, if a network has five nodes (A, B, C, D, and E) and there is a one video streaming from two nodes A and B. If number of nodes increases to 10, there will be 5 nodes added to the network. The challenge here is how to make use of the increase to 10 nodes, to assess that increase, with keeping the same application scenario. Since the goal of this study is to see the effect of number of hops increase when increasing number of nodes. In order to study the effects of increasing the number of hops on the network performance, nodes are fixed and distributed as follow. The study will start by having five nodes organized in one row and having distance of 200 m between each other. As number of nodes increases to 10, other row of 5 nodes will be added below and parallel to the first row. This is done consequently for 20, 30, 40, and 50 node scenarios. Figures 15.1 and 15.2 illustrate the geographical structure of a network with 5 and 20 nodes, respectively. The figures are explained further in the next section where we discuss applications topology.

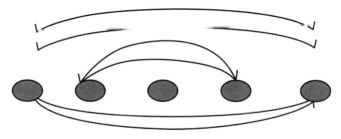

FIGURE 15.1
The network geographical structure and applications topology (5 nodes).

15.4.1.3 Topologies

Having applications deployed similarly across network sizes is very important to study the effects of network size and hops increase. In all scenarios used in this study, there were two HTTP connections, two video streaming connections, and two voice connections. In case of five nodes, the pairs of connections of HTTP and video were initiated between nodes that are located in the edges of the network. The pair connections of Voice were between nodes that are located just to the next of the edge nodes. Figure 15.1 shows applications' connections details for f5- node scenario. The red lines represent video, purples for HTTP, and green lines for voice. As the number of nodes increases to 10, the video connections will be between the two corners of the network. One video connection is between two facing corners which will be like a diagonal line. The Voice connection will be between nodes that are next to the two corners toward the middle of every row. The HTTP connections will be between the middle nodes in the first and second rows and between the two edges of the first row. The settings for 20-, 30-, 40-, and 50-node scenarios will follow the same patterns. Figure 15.2 depicts the applications topology for a network with 20 nodes.

15.4.2 Collective Multimedia Network Behavior against Application Profiles

Firstly, we study collective (overall) QoS behavior of ad hoc networks populated by multimedia applications (i.e., overall behavior of network hosting a range of video, voice, and HTTP applications). The performance is studied using four different routing protocols (HCPR, OLSR, AODV, and GRP), a set of four traffic intensities (100 kbps, 1 Mbps, 10 Mbps, and 20 Mbps) and six network sizes (network with 5, 10, 20, 30, 40, and 50 nodes). As mentioned earlier, in a later section, we will specifically look at the performance of vide traffic over ad hoc networks.

15.4.2.1 A 5-Node Network

Figures 15.3 and 15.4 show the overall DR and throughput for four routing protocols across four traffic intensities for a network containing five nodes. It can be observed that the two figures depict similar trends. For low traffic intensities (100 kbps and 1 Mbps), the network is able to sustain 100% throughput and DR. For higher traffic intensities, the network is unable to sustain the traffic load and starts dropping packets, resulting in lower DR and throughput, dropping to around 40% for 10 Mbps overall traffic and reaching to near 0 for 20 Mpbs. All four protocols show similar performance. AODV performs slightly better, we believe, due to its on-demand features: when other routing protocols consume part of their bandwidth by sending periodical routing messages, AODV uses the bandwidth to send the

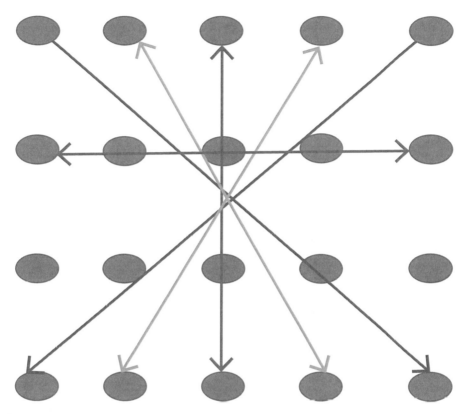

FIGURE 15.2
The network geographical structure and applications topology (20 nodes).

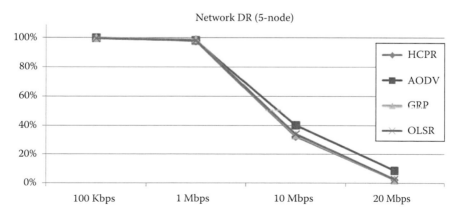

FIGURE 15.3
Delivery ratio (DR) for 5-node network across four traffic intensities.

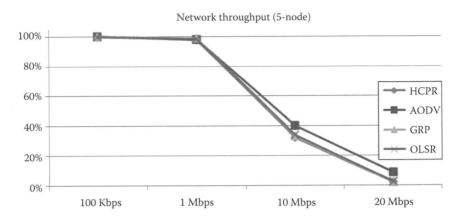

FIGURE 15.4
Throughput for 5-node network across the four traffic intensities.

applications traffic. In the rest of this section, due to limited space, we will only present DR results as these show trends similar to throughput with some differences in the actual values.

15.4.2.2 A 10-Node Network

Figure 15.5 shows the total DR, as before, for four routing protocols across four traffic intensities, however, this time for a network containing 10 nodes. The figure depicts that DR for 100 kbps intensity varied from 81.8% for OLSR and GRP to 100% for HCPR protocol where AODV was in the middle of the two values with around 92.6% DR. As intensity was increased, HCPR was affected slightly with the increase by having 97.3% DR while GRP and OLSR

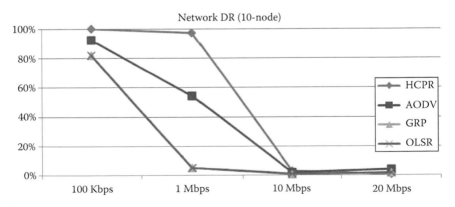

FIGURE 15.5
Delivery ratio (DR) for 10-node network across the four traffic intensities.

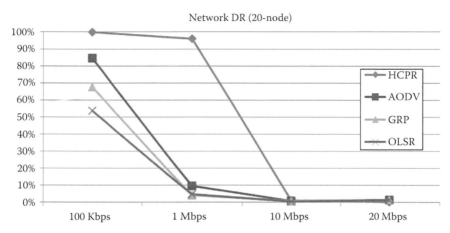

FIGURE 15.6
Delivery ratio (DR) for 20-node network across the four traffic intensities.

affected badly to reach below 5.2% DR. AODV was not affected as OLSR and GRP but showed a big drop to reach around 54.4%. By increasing the traffic to 10 Mbps, all routing protocols have reached to below 3.2% DR with HCPR being the highest among the others.

15.4.2.3 A 20-Node Network

Figure 15.6 shows the total DR and throughput for a network with 20 nodes for 4 routing protocols across 4 traffic intensities. The figure depicts that different routing protocols have different DR when traffic was the lowest with values of 100%, 84.4%, 66.6%, and 53.8% for HCPR, AODV, GRP, and OLSR consequently. As traffic increased, all routing performances have fallen sharply except HCPR which has shown only 4% drop in performance. When traffic intensity reached 10 Mbps, all routing protocols' performances went down to reach blow 1.2% with HCPR still being the highest among the other routing protocols. As traffic intensity reached the 20 Mbps, HCPR has gone to the lowest value of 0.3% while the other routing protocols has shown slight improvement of around 1% in comparison to the DR at 10 Mbps intensity, this is we believe due to the larger overhead that HCPR incurs due to clustering and reservations.

15.4.2.4 A 30-Node Network

Figure 15.7 shows the DR and throughput of 30 nodes scenarios for 4 routing protocols across 4 traffic intensities while network contains 30 nodes. The DR results show that all routing protocol performances were between 59.3% and 83.4% for 100 kbps traffic intensity, with HCPR in best performance position

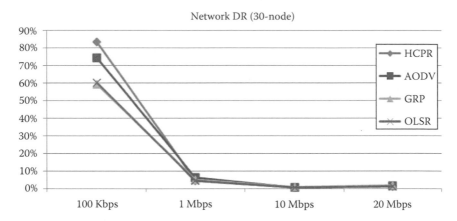

FIGURE 15.7
Delivery ratio (DR) for 30-node network across the four traffic intensities.

and OLSR and GRP at lowest position. As traffic increased to 1 Mbps, all routing protocols have shown a big drop in the performance to reach below 6.2%, and that trend continued to go down to reach below 1.4% when traffic was 10 Mbps. As traffic reached 20 Mbps, there has been slight improvement of around 1% for all routing protocols. The relatively poor DR performance for the 30-node network in comparison to 20-node network scenario is explained as follows. Firstly, this could be due to the MAC layer collisions caused by the increased number of transmissions attempts; this will in turn result in potential MAC buffer overflow caused by the packets arriving from higher layers. Secondly, it is because the increase in the number of network nodes has resulted into the increase in the geographical area of the network and hence a larger number of clusters. Additional nodes cause higher interference at the physical layer and add to the packet-dropping rate. Furthermore, a higher number of clusters can cause higher overhead and hence higher packet drops. These aspects of the HCPR scheme—that is, the relationship between the number of nodes, geographical area size, the traffic configuration, and the number/size of clusters—need further investigation. This will form our future work and further discussion on this topic is given in Section 15.5.

15.4.2.5 A 40-Node Network

Figure 15.8 shows the DR of a network with 40 nodes for 4 routing protocols across 4 traffic intensities. As can be observed from the figure that the results follow trends similar to the 30-node scenarios except the slight improvement of GRP over other routing protocols when traffic was 1 Mbps. The four routing protocols had DR between 63.2% and 83.4% when traffic was 100 kbps and between 1.2% and 1.5% when traffic was 20 Mbps.

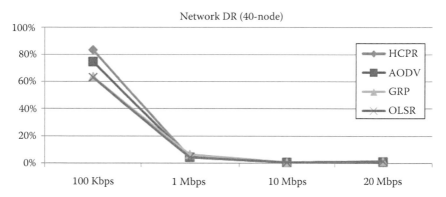

FIGURE 15.8
Delivery ratio (DR) for 40-node network across the four traffic intensities.

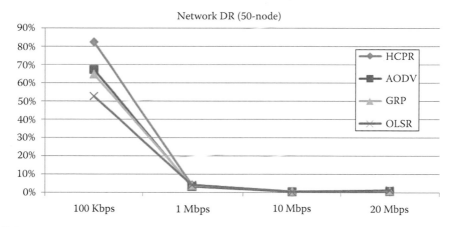

FIGURE 15.9
Delivery ratio (DR) for 50-node network across the four traffic intensities.

15.4.2.6 A 50-Node Network

Figure 15.9 shows the DR of a network with 50 nodes for 4 routing protocols across 4 traffics intensities. As can be observed from the figure, the results follow almost the same trend as for 30-node and 40-node scenarios. The four routing protocols had DR between 52.8% and 82.4% when traffic was 100 kbps, which was slightly lower than scenarios of 40 nodes.

15.4.3 Video Performance

This section studies the performance of video applications over ad hoc networks. As before, the video application shared the network with voice and HTTP traffics. The study of video performance is for varying network sizes and a set of four traffic intensities; L, Medium–Low (ML), MH, and H. We plot results for DR, delay, and delay variation.

15.4.3.1 Traffic Profile: Low

Figure 15.10 depicts the performance of video DR when the network is populated with L overall traffic of 100 kbps and video traffic was around 68.5 kbps of the total traffic. The overall view of DR figure is a decrease with the increase of number of nodes except HCPR which showed excellent results at 10 and 20 nodes. That made HCPR to be the best achiever among routing protocols followed by AODV, which was in clear pattern and decreased gradually. OLSR dropped to around 34% when number of nodes was 20, rose at 30- and 40-node network sizes, and again dropped again at 50-node network size. GRP went down gradually until 30-node network size, and then started increasing slightly at 40- and 50-node scenarios.

Figure 15.11 shows the video packet end-to-end delay across network sizes when the networks are populated with L overall traffic of 100 kbps. The figure started by having L delay of around zero for all routing protocols and then started increasing at 30 nodes. AODV increased gradually from 30-node network size while HCPR increased sharply from 40-node network size. Although, HCPR was the worst at 50 nodes, it has the least lost ratio than others and it was the same for AODV as well. That means sometimes having a poor performance for one metric does not necessarily mean all metric performances are poor but it could indicate good performance in terms of other metrics, in this case DR was achieving good. Figure 15.12 shows the video delay variation across network sizes when the networks are populated with L overall traffic of 100 kbps. The variation was almost zero for all routing protocols across network sizes except for AODV and HCPR when number of nodes was 50 where the variation rose to just below 0.5 and 5 s consequently.

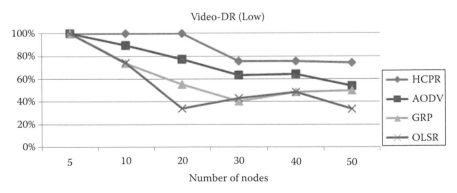

FIGURE 15.10
The video delivery ratio (DR) for L traffic across network sizes.

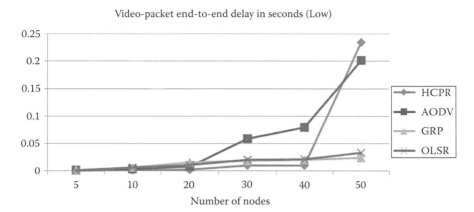

FIGURE 15.11
Video packet end-to-end delay for L traffic across network sizes.

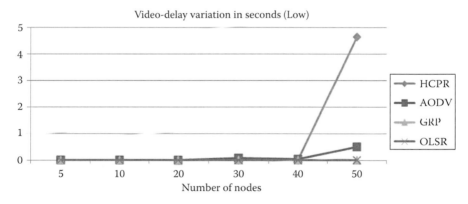

FIGURE 15.12
Video packet delay variation for L traffic across network sizes.

15.4.3.2 Traffic Profile: Medium–Low

Figure 15.13 depicts the performance of video DR when the networks are populated with ML overall traffic of 1 Mbps, and video traffic was around 947 kbps of the total traffic. The overall view of DR figure is a decreasing trend with the increase of number of nodes. All routing protocols started with DR above 97.7% and then started decreasing with differences of the degree of the decrease. HCPR decrease was slightly and gradually to have 95.9% DR when number of nodes was 20 and then declined sharply to reach 4.5% at 30 nodes and kept under 5% for 50-node network size. AODV came after HCPR at 10 and 20 nodes. OLSR and GRP acted similarly by dropping to below 6% and remained like that except GRP where it has shown a slight improvement at 40 nodes. By comparing this figure with the overall traffic

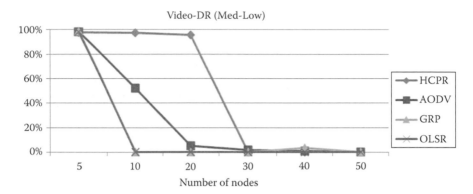

FIGURE 15.13
Video delivery ratio (DR) for ML traffic across network sizes.

DR figure, we could conclude that video DR acted similar to the overall DR for L traffic profile.

Figure 15.14 shows the video packet end-to-end delay across network sizes when the networks are populated with ML overall traffic of 1 Mbps. AODV was the only routing protocol for which packet end-to-end delay across all network sizes was recorded. It started like the other protocols by having low delay of around 0.03 s and kept the trend of low delay until it reached 30-node network size by having delay of around 0.09 s, and then rose sharply to reach 17 and 29 s at 40 and 50 nodes consequently. The delay recorded for HCPR was for 5-, 10-, and 20-node network sizes with delay between 0.02 and 0.07 s. GRP and OLSR recorded low delay at 5-node network size of around 0.02 s and at 40 nodes of around 60 and 76 s consequently. The reason for OPNET not to have recorded delays for some cases, we believe, is that the DR in those cases was zero (equivalently, it meant that the delay was infinite because the DR was zero).

Figure 15.15 shows the video delay variation across network sizes when the networks are populated with ML overall traffic of 1 Mbps. The delay

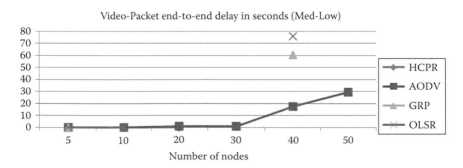

FIGURE 15.14
Video packet end-to-end delay for ML traffic across network sizes.

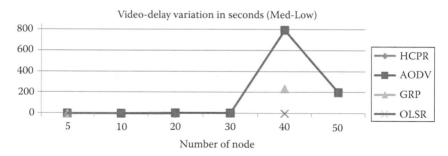

FIGURE 15.15
Video packet delay variation for ML traffic across network sizes.

variation results followed a trend similar to the end-to-end delay results; AODV was the only routing protocol that had values for this metric recorded across network sizes. It started by a delay value of near zero variation and kept it until 40-nodes network size where it increased to very high values and then went down at 50-node network size. As we said earlier, a low value of delay variation or delay is to be seen in the context of DR or throughput.

15.4.3.3 Traffic Profile: Medium–High

Figure 15.16 depicts the performance of video DR when the networks are populated with MH overall traffic of 10 Mbps, and video intensity is 9.9 Mbps. As expected, the video DR decreases with increase in the number of nodes in the network. In general, the DR reaches to a near zero value for the network with 10 nodes. As the figure shows, AODV is slightly better than other routing protocols when the number of nodes in the network was 5: HCPR performed slightly better for 10- and 20-node networks; however, the difference in performance is negligible. Figure 15.17 shows video packet end-to-end delay results across the various network sizes when the networks are populated with MH overall traffic of 10 Mbps. HCPR had increasing trend

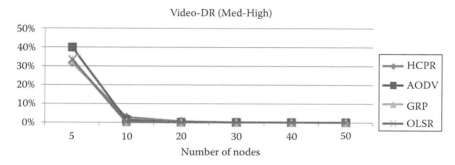

FIGURE 15.16
Video delivery ratio (DR) for MH traffic across network sizes.

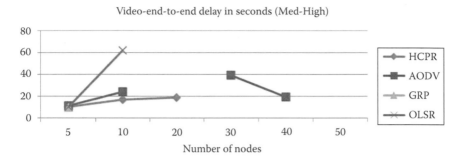

FIGURE 15.17
Video packet end-to-end delay for MH traffic across network sizes.

moving from 5- to 20-node network size and did not record values for higher network sizes while others had discrete values. AODV had delays recorded between 5- and 10-node and between 40- and 50-node network sizes. GRP recorded one value at 5-node scenario while OLSR recorded values at 5- and 10-node network sizes. The delay of all routing protocols at 5-node network size was below 10 s which is the minimum values achieved at this intensity rate. Not having some values recorded at some points means that there is no video traffic being delivered to the destination.

Figure 15.18 plots video delay variation results across network sizes when the networks are populated with MH overall traffic of 10 Mbps. The figure followed the same direction as the packet end-to-end delay for MH traffic. AODV showed a high variation at 5 nodes; a very high variation at 10- and 30-node network sizes; 0 variation at 40-node network size. HCPR variations started by having around 5 s at 5-node network size, above 20 s for 20- and 30-node network sizes, and then stopped recording for higher network sizes. OLSR and GRP had around 5 s variations at 5 nodes and then only OLSR had a variation of 18 s at 10 nodes.

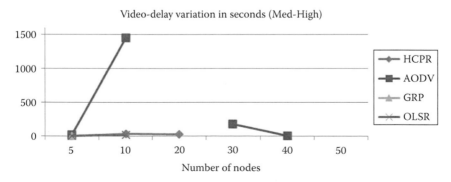

FIGURE 15.18
Video packet delay variation for MH traffic across network sizes.

15.4.3.4 Traffic Profile: High

Figure 15.19 plots video DR results when the networks are populated with H overall traffic of 20 Mbps and video was 19.6 Mbps. The figure shows trends similar to the previous section, though the DR value is lower, obviously due to the higher traffic intensity. Figure 15.20 shows the video packet end-to-end delay. Some values were recorded for the protocols at 5- and 10-node network sizes, with very high delay between 7 and 11 s for 5-node network size. This high delay will not give a good interactive video streaming service but may be used for downloading noninteractive video streams. Figure 15.21 shows the video delay variation. The figure followed the same trend as the packet end-end-delay for H traffic, which was normal and expected. The values for recorded variations were from 5 to 120 s for 5- and 10-node network sizes, which are not acceptable for real-time video streaming. No values were reported by OPNET for network sizes greater than 10 nodes.

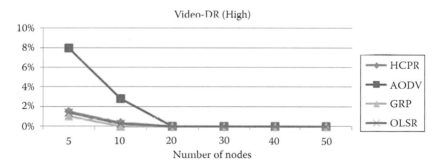

FIGURE 15.19
Video delivery ratio (DR) for High (H) traffic across network sizes.

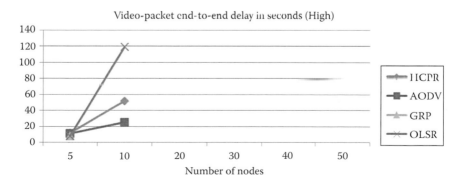

FIGURE 15.20
Video packet end-to-end delay for H traffic across network sizes.

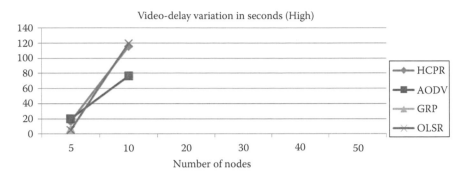

FIGURE 15.21
Video packet delay variation for H traffic across network sizes.

15.4.3.5 Video Performance for All Traffic Profiles: An Overview

We summarize here the results of video applications. A conclusion of each of the four traffic profiles is drawn using four metrics: DR, throughput, delay, and delay variation as follows. Generally, the performance of video DR was similar to overall DR performance because video traffic intensity was over 60% of the total traffic in case of L traffic and over 94% of ML, MH, and H intensities. As mentioned earlier, the video application shared the network with voice and HTTP traffic; therefore, there is more to the video behavior than the traffic intensity alone (e.g., the effects of VoIP applications on the network behavior). At L traffic intensity profile, all routing protocols achieved around 100% DR at 5-node network. HCPR was the only routing protocol that kept its high performance of 100% DR at 10- and 20-node networks. Packet end-to-end delay was below 0.25 s for all network sizes and below 0.02 when network sizes were between 5 and 20 nodes. At ML traffic intensity profile, all nodes had over 98% DR at 5-node network and then dropped to below 55% for larger networks except HCPR, which kept its high performance at 10- and 20-node networks to reach around 97.3% and 95.8%, respectively. In all cases where DR was above 95%, we had a maximum of 0.07 s delay and 0 delay variation. At MH and H traffic intensities, none of the routing protocols achieved acceptable DR at any network sizes. In conclusion, HCPR had the best performance in L and ML traffic intensities.

15.4.4 Ad Hoc Networks: Performance Overview

In this section (Section 15.4), by now, we have presented an extensive analysis of the HCPR performance for multimedia delivery over ad hoc networks. A number of scenarios were configured by varying different routing protocols, different intensities, and different numbers of nodes. It showed how protocols are affected by increasing the number of nodes and traffic intensities and how HCPR was performing in comparison to other routing protocols

(AODV, GRP, and OLSR) under different scenario settings. Network sizes were varied by the number of nodes from 5-node to 50-node networks. In addition, traffic intensities were set in four main intensity profiles, L, ML, MH, and H. Each intensity profile was a mix of three multimedia applications video, voice, and HTTP. We have studied the overall performance by fixing the network size and looked into as to how networks were behaving under different intensity loads and mix of applications. That was followed by a performance analysis of multimedia applications by fixing traffic intensity and varying network sizes. While studying the effects of varying traffic intensities across network sizes, HCPR was the only routing protocol that showed a good overall performance at L and ML traffic intensities for 10- and 20-node networks. When studying the effect of varying network sizes for each traffic intensity profile, the collective network performance and video performance for HCPR outperformed other routing protocols at 10- and 20-node networks for L and ML traffic profiles. The high bitrate of video and HCPR performer have resulted in having lower performance of voice especially at MH and H traffic intensities. The HTTP results also demonstrated that HCPR was the best performer at L and ML intensities, for higher traffic intensities, HCPR also outperformed other routing protocols at 30- and 40-network sizes.

15.4.5 Infrastructure Wireless Networks: QoS Analysis

We now present a QoS analysis of the results obtained through simulating video applications over an infrastructure wireless network. We will see that, in contrast to the results for ad hoc networks presented in the earlier part of this section (Section 15.4), the networks will be able to support much larger video intensity. The input parameters used in these simulations include three different voice codecs and three different traffic intensity levels for each of video and HTTP application. In addition, there are 12 different variation levels in (latency, packet discard ratio) pair for metropolitan area network (MAN). We have performed well over 100 network simulations by using different input configurations obtained through variations in these input parameters' values.

The network topology and architecture is described in Section 15.4.5.1, followed by a description of multimedia traffic, including voice, video, and data used in the QoS analysis in Section 15.4.5.2. The QoS results are discussed in Section 15.4.5.3. As before, the network and applications have been simulated using the OPNET simulator. We will follow the same performance analysis methodology as described in Section 15.4.1, that is, the network is populated with a mix of video, voice, and data applications.

15.4.5.1 The Network Architecture

We are interested in exploring video application performance in MAN environments where the two end hosts access the network through 802.11 Wireless LANs. Specifically, we are interested in investigating video

TABLE 15.2

Traffic Profiles for Wireless Infrastructure Networks (kbps)

	Low (L)	Medium (M)	High (H)
Video	21,600	2800	33,600
Voice	G723.1 (5.3)	GSM-FR (13)	G.711 (64)
HTTP	167	168	169
Total	21,800	28,300	34,400

performance over 802.11 networks under realistic scenarios where the network bandwidth is being shared by other applications including voice and HTTP traffic. Networks are usually heavily loaded with video, data, and other traffic. Hence, in particular, we are interested in investigating multimedia performance over a range of network loads. The structure of the network which we have considered here is described as follows. The network consists of two WLANs which are connected to a common MAN via 1 Gbps Ethernet. We have used 802.11 g (54 Mbps) for each of the two WLANs. For the MAN, we have used an abstract simulation model (a cloud) which allows us to set different parameters for the network including packet discard ratio and latency. A total of 10 nodes are associated to each of the WLANs. These 10 nodes are randomly located within an area of 350×350 m² around each of the two WLANs. The 20 nodes interact with each other across the network using different applications (see the next section).

15.4.5.2 Applications

We populate the network with varying intensity levels of video, HTTP, and voice applications. We use three different traffic intensity levels for each of the video, HTTP, and voice applications; these are listed in Table 15.2. In a similar approach to the one taken in Section 15.4.1.1, the traffic intensity levels are divided into "Low," "High," and "Medium." Columns 2–4 outline the details for each application specific to the three traffic intensity levels. For example, Column 3 lists the voice codec used for the "Medium" intensity traffic case is GSM-FR with 13 kbps bitrate; the video and HTTP traffic in the "Medium" intensity case are 28 Mbps and 168 kbps, with a net traffic of 28.3 Mbps. Similarly, Column 2 and 4 give the details relevant to the "Low" and "High" traffic intensity levels. In each of these cases, 10 nodes from 1 WLAN are connected to the 10 nodes in the other WLAN. Each node in a WLAN is connected to a node in the other WLAN, and hence there are a total of 30 connections, 10 each for HTTP, video, and voice.

15.4.5.3 QoS Analysis

Having described the network architecture and applications set up, we are now ready to analyze QoS for wireless infrastructure networks connected

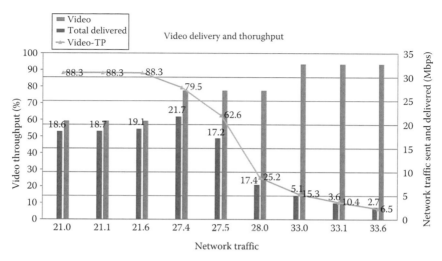

FIGURE 15.22
(See color insert.) Video QoS—traffic delivered and throughput.

through a MAN. First, we discuss the network results while keeping constant the settings for the MAN connecting the WLANs. Secondly, we analyze network QoS against variations in the MAN.

Figure 15.22 plots data about video traffic in the network. The x-axis represents intensity of the total traffic that populates the network. There are two y-axes: the one on the right side represents the video traffic sent and delivered (received) through bar graphs and the other on the left side represents the video throughput in percentage through the line plot. Note that the figure plots the video statistics where the network is also populated with voice and data applications as detailed earlier. We note in the figure that the network is able to deliver 18.6 MB of video per second out of the 21 MB of the transmitted video content, resulting in 88.3% throughput. This value of throughput is maintained for up to 21.6 MB of transmitted video, and then the throughput starts dropping until it reaches to the lowest value of 6.5% for 33.6 MB of video transmitted on the network. The total theoretical capacity of the WLAN is 54 Mbps and therefore it is expected that the network will get saturated at some threshold point and will start dropping the packets in great numbers.

Figure 15.23 plots the network traffic intensity in Mbps for all the network applications as well as the total throughput for the entire network. Note that Figure 15.22 illustrates the video performance alone. There are a total of five plots in the figure; three of these plots represent the traffic produced by the three applications (voice, video, and HTTP), the fourth gives the total traffic intensity, and the fifth gives the throughput obtained against varying network traffic intensity. The total network traffic intensity and total network throughput are also visible in the figure. We note

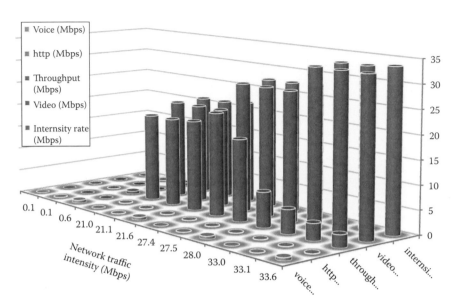

FIGURE 15.23
(See color insert.) Network traffic components—intensities and throughput.

in the graph that initially (i.e., from the left, for lower values of network intensity), the traffic intensity and the throughput bar graphs are of equal heights showing that all the traffic sent to the network is delivered with 100% throughput. However, as the traffic reaches approximately 21 Mbps, the difference in the heights of the two bar graphs become visible with the throughput reaching its lowest value of 2.4 Mbps against the total input traffic of 33.6 Mbps. As for Figure 15.22, this behavior is expected due to the capacity limits of the wireless network.

Figures 15.24 and 15.25 depict end-to-end delay characteristics for (infrastructure) wireless networks connected through a MAN. We now make some variations in the metropolitan network environment which connects the two WLANs. Specifically, we vary the input parameter values for the metro network latency and packet discard ratio. The OPNET software allows us to configure various parameters of the MAN environment through its "Cloud" model. Figure 15.24 plots the average end-to-end delay of the network against variations in traffic intensity and packet discard ratio. We note that the delay increases significantly with increase in the traffic intensity; however, it remains relatively almost constant against changes in the packet discard ratio. Figure 15.25 plots the end-to-end delay against traffic intensity and network latency. As is the case for Figure 15.24, the delay increases with the increase in traffic intensity but remains relatively almost constant against metro network latency. This is due to the fact that the change in the network latency (in MAN environments) is quite low compared to the voice delay caused by the limitations of the WLAN throughput capacity. A detailed

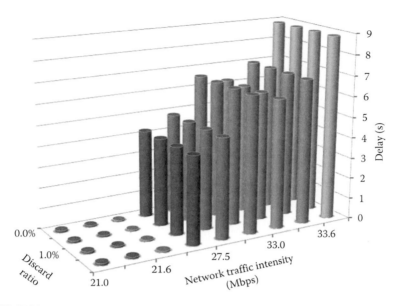

FIGURE 15.24
(See color insert.) Average network delay against MAN packet discard ratio and network traffic.

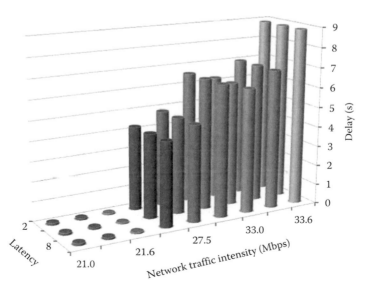

FIGURE 15.25
(See color insert.) The average network delay against MAN latency and network traffic intensity.

analysis of multimedia applications over wireless networks connected within a MAN is presented in Mehmood et al. (2011) with a focus on VoIP applications.

15.5 Conclusions and Future Work

Multimedia applications continue to drive convergence in the telecommunications industry. Video traffic is to become 90% of the overall global network traffic by 2015. People demand anytime, anywhere mobility giving rise to heterogamous networks comprising a range of fixed, wireless, mobile, and ad hoc networks. Multimedia applications pose stringent requirements on networks while ad hoc networks are inherently limited due to their autonomic and multi-hop nature and the lack or scarcity of infrastructure. Supporting multimedia applications over ad hoc networks is a challenge for the industry. The broad aim of our research is to address the challenges associated with the design and analysis of ad hoc networks supporting multimedia applications with scalable QoS.

The aim of this chapter was to present an overview of our work on multimedia wireless networks. We made three contributions in this chapter. Firstly, in Section 15.3, an extensive review of the literature on multimedia ad hoc network design and QoS analysis was presented. Secondly, in Section 15.4, a detailed analysis of multimedia applications over infrastructure and ad hoc networks was presented. Four routing protocols were used in this analysis and over 500 networking and applications scenarios were simulated and it was demonstrated that the HCPR-enabled ad hoc network outperforms the well-known routing schemes, in particular for relatively large networks and high QoS network loads. These results show promising performance because many QoS schemes do work for small networks and low network loads but are unable to sustain performance for large networks and high QoS loads. Thirdly, a detailed end-to-end QoS analysis for multimedia applications was presented over wireless infrastructure networks connected within a MAN environment. All the network analyses presented in this chapter are based on simulations using the OPNET simulator. The HCPR scheme is implemented as an extension (module) to the OPNET simulation software.

We had also contended in this chapter that the vast majority of networks are populated with multimedia traffic, as opposed to video alone, and therefore our analysis in this chapter have intentionally included voice and data. Nevertheless, we have also presented network analysis with focus on video application alone. In this context, a contribution of our work lies in the analysis methodology that we have developed for configuring multimedia applications and networking scenarios. The approach was to simulate and evaluate realistic multimedia applications scenarios for networks with a good number and type of applications—VoIP, video, and HTTP, both elastic and nonelastic—so that the mutual effects of network applications can be taken into account. The work on analysis methodology can be extended and enhanced in the future by using additional types of multimedia traffic models (such as video codecs and models, additional VoIP models, etc.) and network configurations. We would like to formalize our analysis methodology by finding good and bad practices for networks and applications setup found in realistic environments.

Acknowledgment

We are thankful to all the reviewers for their time, very useful recommendations, and their comments for helping us to improve this chapter.

References

Alazawi, Z., Altowaijri, S., Mehmood, R., and Abdljabar, M. B. 2011. Intelligent disaster management system based on cloud-enabled vehicular networks. In: *11th International Conference on Proceedings of the ITS Telecommunications (ITST)*, August 23–25, 2011, St. Petersburg, Russia, pp. 361–368.

Alturki, R. 2011. Multimedia ad hoc networks: Design, QoS, routing and analysis. PhD thesis, Swansea University, Swansea, U.K.

Alturki, R. and Mehmood, R. 2008. Multimedia ad hoc networks: Performance analysis. In: *EMS '08. Second UKSIM European Symposium on Computer Modeling and Simulation*, IEEE Computer Society 2008, September 8–10, 2008, Liverpool, England, UK, pp. 561–566.

Alturki, R. and Mehmood, R. 2012. Cross-layer multimedia QoS and provisioning over ad hoc networks. In: Rashvand, H. F. and Kavian, Y. S. (eds.) *Using Cross-Layer Techniques for Communication Systems: Techniques and Applications*. IGI Global, Hershey, PA, pp. 460–499.

Blake, S., Black, D., Carlson, M., Davies, E., Wang, Z., and Weiss, W. 1998. An Architecture for Differentiated Services. *RFC:2475*. IETF Network Working Group.

Braden, R., Clark, D., and Shenker, S. 1994. Integrated Services in the Internet Architecture: an Overview. RFC 1633. IETF Network Working Group.

Broch, J., Maltz, D. A., Johnson, D. B., Hu, Y.-C., and Jetcheva, J. 1998. A performance comparison of multi-hop wireless ad hoc network routing protocols. In: Osborne, W. P. and Moghe, D. (eds.), *Proceedings of the 4th Annual ACM/IEEE International Conference on Mobile Computing and Networking (MobiCom '98)*, ACM, New York, NY, 1998, pp. 85–97.

Camp, T., Boleng, J. and Davies, V. 2002. A survey of mobility models for ad hoc network research. *Wireless Communications & Mobile Computing (WCMC): Special issue on Mobile Ad Hoc Networking: Research, Trends and Applications*, 2, 483–502.

Chatzistavros, E. and Stamatelos, G. 2009. Comparative performance evaluation of routing algorithms in IEEE 802.11 ad hoc networks. In: *Proceedings of the 16th International Conference on Telecommunications*, May 25–27, 2009, Marrakech, Morocco: IEEE Press, pp. 19–24.

Cisco Systems, I. 2006. Understanding Jitter in Packet Voice Networks (Cisco IOS Platforms). Available: http:/www.cisco.com/en/US/tech/tk652/tk698/technologies_tech_note09186a00800945df.shtml (Accessed October 5, 2010).

Clausen, T. and Jacquet, P. 2003. Optimized link state routing protocol (OLSR). RFC: 3626. IETF Network Working Group.

Gharavi, H. and Ban, K. 2003. Cross-layer feedback control for video communications via mobile ad-hoc networks. In: *IEEE 58th Vehicular Technology Conference*, VTC 2003-Fall, Orlando, FL, Vol. 5, October 6–9, 2003, pp. 2941–2945.

Gottron, C., Konig, A., Hollick, M., Bergstrasser, S., Hildebrandt, T., and Steinmetz, R. 2009. Quality of experience of voice communication in large-scale mobile ad hoc networks. In: *Wireless Days (WD), 2nd IFIP*, December 15–17, 2009, Paris, France, pp. 248–253.

Hofmann, P., An, C., Loyola, L., and Aad, I. 2007. Analysis of UDP, TCP and voice performance in IEEE 802.11b multihop networks. In: *13th European Wireless Conference*, April 1–4, 2007, Paris, France.

Hong, X., Gerla, M., Guangyu, P., and Ching-Chuan, C. 1999. A group mobility model for ad hoc wireless networks. In: *Proceedings of the 2nd ACM International Workshop on Modeling, Analysis and Simulation of Wireless and Mobile Systems*. Seattle, WA. New York: ACM Press, pp. 53–60.

Hongqi, Z., Jiying, Z., and Yang, O. 2008. Adaptive rate control for VoIP in wireless ad hoc networks. In: *IEEE International Conference on Communications, ICC '08*, May 19–23, 2008, Beijing, China, pp. 3166–3170.

IEEE 802.11 Working Group (2007). *IEEE 802.11-2007—Part 11: Wireless LAN Medium Access Control (MAC) and Physical Layer (PHY) Specifications*.

Jeong, Y., Kakumanu, S., Tsao, C.-L., and Sivakumar, R. 2009. VoIP over Wi-Fi networks: performance analysis and acceleration algorithms. *Mobile Networks and Applications*, 14, 523–538.

Kumar, S., Rathy, R. K., and Pandey, D. 2009. Traffic pattern based performance comparison of two reactive routing protocols for Ad hoc networks using NS2. In: *2nd IEEE International Conference on Computer Science and Information Technology, ICCSIT 2009*, August 8–11, 2009, Beijing, China, pp. 369–373.

Layuan, L., Chunlin, L., and Peiyan, Y. 2007. Performance evaluation and simulations of routing protocols in ad hoc networks. *Computer Communications*, 30, 1890–1898.

Lee, G. and Song, H. 2010. Cross layer optimized video streaming based on IEEE 802.11 multi-rate over multi-hop mobile ad hoc networks. *Mobile Networks and Applications*, 15, 652–663.

Li, X. and Cuthbert, L. 2005. Optimal QoS mechanism: integrating multipath routing, DiffServ and distributed traffic control in mobile ad hoc networks. In: Jia, X., Wu, J. and He, Y. (eds.), *Proceedings of the First International Conference on Mobile Ad-hoc and Sensor Networks (MSN'05)*. Berlin, Germany: Springer, pp. 560–569.

Li, J., Jannotti, J., Couto, D. S. J. D., Karger, D. R. and Morris, R. 2000. A scalable location service for geographic ad hoc routing. In: *Proceedings of the 6th Annual International Conference on Mobile Computing and Networking*. Boston, MA. New York: ACM.

Lidong, Z. and Haas, Z. J. 1999. Securing ad hoc networks. *IEEE Network*, 13, 24–30.

Liu, C. and Kaiser, J. 2003. A survey of mobile ad hoc network routing protocols. Department of Computer Structures, University of Ulm, Ulm, Germany.

Mehmood, R. and Alturki, R. 2011. A scalable multimedia QoS architecture for ad hoc networks. *Multimedia Tools and Applications*, 54(3), 551–568.

Mehmood, R., Alturki, R., and Zeadally, S. 2011. Multimedia applications over metropolitan area networks (MANs). *Journal of Network and Computer Applications*, 34, 1518–1529.

Melodia, T. and Akyildiz, I. F. 2010. Cross-layer QoS-aware communication for ultra wide band wireless multimedia sensor networks. *IEEE Journal on Selected Areas in Communications*, 28, 653–663.

Mohapatra, P., Chao, G., and Jian, L. 2004. Group communications in mobile ad hoc networks. *Computer*, 37, 52–59.

de Morais Cordeiro, C., Gossain, H., and Agrawal, D. P. 2003. Multicast over wireless mobile ad hoc networks: Present and future directions. *IEEE Network*, 17, 52–59.

Morgan, Y. L. and Kunz, T. 2005. A proposal for an ad-hoc network QoS gateway. In: *WiMob'2005*, August 2005, Montreal, Canada, Vol. 3, pp. 221–228.

Mundarath, J. C., Ramanathan, P., and Veen, B. D. V. 2009. A quality of service aware cross-layer approach for wireless ad hoc networks with smart antennas. *Ad Hoc Networking*, 7, 891–903.

Murthy, C. S. R. and Manoj, B. S. 2004. *Ad Hoc Wireless Networks: Architectures and Protocols*. Upper Saddle River, NJ: Prentice Hall, PTR.

Ng, P. C. and Liew, S. C. 2007. Throughput analysis of IEEE802.11 multi-hop ad hoc networks. *IEEE/ACM Transactions on Networking*, 15, 309–322.

O'Hara, B. and Petrick, A. 2005. *The IEEE 802.11 Handbook: A Designer's Companion*, 2nd edn. New York: Standards Information Network, IEEE Press.

OPNET Technologies Inc. 2008. OPNET Modeler-Educational Version, 14.5 ed.

Perkins, C., Belding-Royer, E., and Das, S. 2003. Ad hoc on-demand distance vector (AODV) routing. RFC: 3561. IETF Network Working Group.

Pompili, D. and Akyildiz, I. F. 2010. A multimedia cross-layer protocol for underwater acoustic sensor networks. *IEEE Transactions on Wireless Communications*, 9, 2924–2933.

Pucha, H., Das, S. M., and Hu, Y. C. 2007. The performance impact of traffic patterns on routing protocols in mobile ad hoc networks. *Computer Networks*, 51(12), 3595–3616.

Qadri, N. N. and Llotta, A. 2010. Analysis of pervasive mobile ad hoc routing protocols. In: Aboul-Ella, H., Jemal, A., Ajith, A., Hani, H. (eds.), *Pervasive Computing, Computer Communications and Networks*, Part 4, pp. 433–453.

Ramanathan, R. and Redi, J. 2002. A brief overview of ad hoc networks: challenges and directions. *IEEE Communications Magazine*, 40, 20–22.

Rosen, E., Viswanathan, A. and Callon, R. 2001. Multiprotocol label switching architecture. RFC 3031. IETF Network Working Group.

Santos, N. P. 2009. Voice traffic over mobile ad hoc networks: A performance analysis of the optimized link state routing protocol. Master's thesis, Air Force Institute of Technology, Wright-Patterson AFB, OII.

Satyanarayanan, M. 2010. Mobile computing: The next decade. In: *Proceedings of the 1st ACM Workshop on Mobile Cloud Computing and Services: Social Networks and Beyond, (MCS '10), ACM,* New York, NY, Article 5, p.6.

Setton, E., Taesang, Y., Xiaoqing, Z., Goldsmith, A., and Girod, B. 2005. Cross-layer design of ad hoc networks for real-time video streaming. *IEEE Wireless Communications Magazine,* 12, 59–65.

Sondi, P., Gantsou, D. and Lecomte, S. 2010. Performance evaluation of multimedia applications over an OLSR-based mobile ad hoc network using OPNET. In: *12th International Conference on Computer Modelling and Simulation (UKSim),* March 24–26, 2010, Cambridge, U.K., pp. 567–572.

Takagi, H. and Kleinrock, L. 1984. Optimal transmission ranges for randomly distributed packet radio terminals. *IEEE Transactions on Communications,* 32, 246–257.

Utsu, K., Chow, C. and Ishii, H. 2010. A study on video performance of multipoint-to-point video streaming with multiple description coding over ad hoc networks. *Electrical Engineering in Japan,* 170, 43–50.

Xue, Q. and Ganz, A. 2003. Ad hoc QoS on-demand routing (AQOR) in mobile ad hoc networks. *Journal of Parallel and Distributed Computing,* 63, 154–165.

Index

Printed and bound by CPI Group (UK) Ltd, Croydon, CR0 4YY

18/10/2024

01776236-0013